BC

Environmental Impact of Water Resources Projects

by Larry Canter

Environmental and Ground Water Institute
University of Oklahoma
Norman, Oklahoma

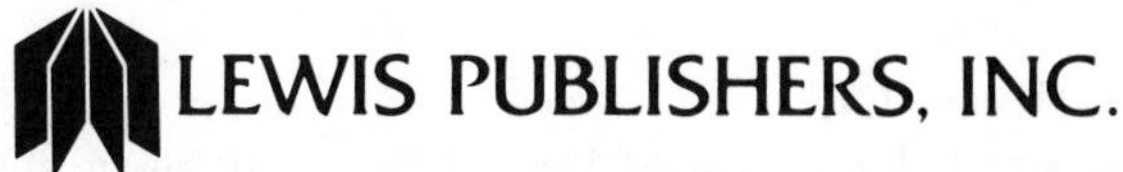

Library of Congress Cataloging-in-Publication Data

Canter, Larry W.
Environmental impacts of water resources projects.

Bibliography: p.
Includes index.
1. Hydraulic engineering--Environmental aspects.
2. Hydraulic structures--Design and construction--
Environmental aspects. 3. Environmental impact
analysis. I. Title. II. Title: Water resources projects.
TD195.H93C35 1985 627 85-23130
ISBN 0-87371-015-0

Second Printing 1986

LEWIS PUBLISHERS, INC.
121 South Main Street, P.O. Drawer 519, Chelsea, Michigan 48118

PRINTED IN THE UNITED STATES OF AMERICA

LARRY W. CANTER, P.E., is the Sun Company Professor of Ground Water Hydrology, and Director, Environmental and Ground Water Institute, at the University of Oklahoma, Norman, Oklahoma, in the USA. Dr. Canter received his Ph.D. in Environmental Health Engineering from the University of Texas in 1967, MS in Sanitary Engineering from the University of Illinois in 1962, and BE in Civil Engineering from Vanderbilt University in 1961. Before joining the faculty of the University of Oklahoma in 1969, he was on the faculty at Tulane University and was a sanitary engineer in the U.S. Public Health Service. He served as Director of the School of Civil Engineering and Environmental Science at the University of Oklahoma from 1971 to 1979.

Dr. Canter has published several books and has written chapters in other books; he is also the author or co-author of numerous papers and research reports. His research interests include environmental impact assessment and ground water pollution control. In 1982 he received the Outstanding Faculty Achievement in Research Award from the College of Engineering, and in 1983 the Regent's Award for Superior Accomplishment in Research.

Dr. Canter currently serves on the U.S. Army Corps of Engineers Environmental Advisory Board. He has conducted research, presented short courses, or served as advisor to institutions in Mexico, Panama, Colombia, Venezuela, Peru, Scotland, The Netherlands, France, Germany, Italy, Greece, Turkey, Kuwait, Thailand, and the People's Republic of China.

PREFACE

Water resources projects such as dams and reservoirs, channelization, and dredging can represent large-scale engineering works or activities which can cause significant impacts on physical-chemical, biological, cultural, and socio-economic components of the environment. Environmental impact studies for such projects should be planned and conducted in a scientifically defensible manner. This book summarizes information from key technical literature related to the implementation of such studies. These studies should be conducted in a manner so as to yield projects which are more compatible with the environment. All too often, environmental impact studies have been used to delay or stop projects.

This book is organized into five chapters and 16 appendices. Following an introductory chapter which focuses on summary information from 434 included references, Chapters 2 through 4 are related to environmental impact studies for dam and reservoir projects, channelization projects, and dredging projects, respectively. Chapter 5 addresses some other project types such as irrigation and shoreline structures. Abstracts of the 434 included references are divided into 16 appendices as follows: impacts of impoundment projects; impacts of channelization projects; impacts of dredging projects; impacts of other water resources projects; impacts of nonpoint sources of pollutants on water environment; transport and fate of pollutants in the water environment; baseline studies of the water environment; environmental indices and indicators; water quantity/quality impact prediction and assessment; biological impact prediction and assessment; estuarine impact prediction and assessment; ground water, noise, cultural, visual, and socio-economic impact prediction and assessment; methodologies for trade-off analyses and decision-making; public participation in water resources planning; impact mitigation measures; and related issues and information.

The author wishes to express his appreciation to several individuals instrumental in the development of this book. First, Debby Fairchild of the Environmental and Ground Water Institute at the University of Oklahoma conducted the computer-based literature searches upon which much of the included information was derived. John Stans of the Delft Hydraulics Laboratory in The Netherlands provided encouragement even though he was not aware that he was doing so. This type of unspoken encouragement was also provided by numerous colleagues within the U.S. Army Corps of Engineers. Finally, the author is extremely grateful to Mrs. Leslie Rard of the Environmental and Ground Water Institute for her typing skills, patience, and dedication to the completion of this book.

The author also gratefully acknowledges the support and encouragement of the College of Engineering at the University of

Oklahoma relative to faculty writing endeavors. Most importantly, the author thanks his family for their patience and understanding.

Larry W. Canter
Sun Company Professor
of Ground Water Hydrology
University of Oklahoma
Norman, Oklahoma

July, 1985

To Donna, Doug, Steve, and Greg

CONTENTS

Chapter

Chapter

Chapter

LIST OF TABLES

Table

Table

LIST OF FIGURES

Figure

CHAPTER 1

INTRODUCTION

Water resources projects include dam and reservoir projects, channelization projects, dredging activities, irrigation schemes, and others such as the creation of "new" lands via filling operations. These projects may be very large from a physical perspective, and they are often located in rural areas. As such, it has become necessary to assess the environmental impacts of water resources projects, typically during their planning stages. It can also be useful to consider the environmental impacts of extant projects.

The basic law in the United States requiring environmental impact studies is the National Environmental Policy Act (Public Law 91-190) which became effective on January 1, 1970. Since the passage of this Act with its requirement for considering the environment effects of projects in planning and decision-making, numerous methodologies and approaches for conducting the various components of environmental impact studies have been developed. Key components of environmental impact studies for water resources projects include:

(1) impact identification;
(2) conduction of baseline studies;
(3) prediction of impacts on various environmental factors;
(4) assessment (interpretation) of the predicted impacts;
(5) conduction of trade-off analyses; and
(6) identification and evaluation of mitigation measures.

The planning and conduction of environmental impact studies for water resources projects can best be achieved through using technically sound approaches. In order to appropriately plan and conduct such studies, knowledge of, and usage of, pertinent published information is necessary.

OBJECTIVE AND SCOPE OF THIS BOOK

The objective of this book is to summarize information on key technical literature relating to environmental impact studies for water resources projects. It is recognized that the publication of new information is a continuing process; therefore, this book represents a summary of information available in the mid-1980's. In providing this current summary, it should be recognized that detailed aspects of addressing all potential impacts for every type of water resources project are not included. Annotated bibliographies in 16 appendices provide key information on 434 references. These bibliographies can point the user to more detailed information.

The primary method used to identify the selected references included the conduction of computer-based searches of the literature and systematic review of the procured abstracts. Descriptor words utilized in the computer-based searches included:

Channels
Channel improvement
Channelization
Dams
Dredging
Environmental assessment
Environmental assimilative capacity
Environmental control
Environmental effects
Environmental impacts
Environmental impact analysis
Environmental impact assessment
Environmental impact evaluation
Environmental impact statement
Impact area
Impact studies
Impoundments
Reservoirs
Water pollution effects
Water quality indices
Water resources development
Water resources planning

ORGANIZATION OF THIS BOOK

This book is organized into five chapters and 16 appendices. Following this introductory chapter which focuses on summary information from the 434 references, Chapters 2 through 5 are related to four types of water resources projects: (1) dams and reservoirs; (2) channelization; (3) dredging; and (4) other, including irrigation projects and shoreline projects. Each of these project-oriented chapters address impact identification, baseline studies, impact prediction, impact assessment and mitigation, and methodologies for trade-off analyses and decision-making.

The 434 references were organized into 16 appendices as listed in Table 1. This organization was based on the following aspects of the six components of environmental impact studies for water resources projects:

(1) Impact identification

- Impacts of water resources projects and operations (Appendices A through D)

- Impacts of types of pollutants (Appendices E and F)

(2) Conduction of baseline studies

- Describing the affected environment (Appendices G and H)

(3) Prediction of impacts on various environmental factors, and

(4) Assessment of the predicted impacts

- Impact prediction and assessment (Appendices I through L)

(5) Conduction of trade-off analyses

- Methodologies for impact assessment and decision-making (Appendices M and N)

(6) Identification and evaluation of mitigation measures

- Additional issues and information (Appendices O and P)

Each of the 434 references were assigned to one appendix even though it might include information pertinent to several appendices. This was done to provide the simplest presentation of information, with the responsibility being upon the reader to determine where references might fit into more than one appendix.

Referring to Table 1, a total of 146 references are included which address the impact of water resources projects and operations. This category is used because it is important to have general knowledge about the impacts which can occur from water resources projects, and to utilize this information in identifying potential impacts for new projects, as well as predicting and assessing the importance of the impacts. A total of 44 references addressing the impacts of various types of pollutants are also included. The key division is into nonpoint sources of pollutants, and the transport and fate of specific pollutants.

A total of 46 references are included on describing the affected environment. A considerable amount of effort, both in terms of personnel and monetary expenditures, can go into describing the affected environment, and with the emphasis given through the scoping process in the United States, it is becoming increasingly important to carefully plan baseline environmental studies. This is also true for countries throughout the world in that the major attention in environmental impact studies should be directed toward impact prediction and assessment and decision-making, as opposed to simply describing the environment to be potentially affected.

The category entitled impact prediction and assessment contains 105 selected references. This is an extremely important aspect of environmental impact studies in that it forms the technical basis for project evaluation and decision-making. It has also represented the area of greatest technical deficiency in most environmental impact studies. A total of 54 references are in the category of methodologies for impact assessment and decision-making. These methodologies include approaches for systematically comparing alternatives and evaluating trade-offs as a part of the decision-making process. Public participation is one part of decision-making and it is also included in this category. Finally, 39 references on related issues and information are included. The primary focus of this category is on impact mitigation measures.

IMPACTS OF WATER RESOURCES PROJECTS AND OPERATIONS

Water resources projects and operations can be divided into four groups: (1) impoundments; (2) channelization; (3) dredging; and (4) other, including irrigation projects. Table 2 contains summary comments on each of the 69 references primarily dealing with the impacts of impoundment projects. Appendix A contains abstracts of the 69 references organized by alphabetical order as shown in Table 2. A key point to note from Table 2 is that there are numerous references which describe the impacts of impoundment projects on the physical-chemical, biological, cultural and socio-economic categories of the environment. While Chapter 2 contains detailed information on environmental impact studies for dams and reservoirs, a few key references will be highlighted herein.

Table 1: Summary of Selected Reference Materials

Category	Component	Number of Selected References
Impacts of water resources projects and operations (Appendices A through D)	Impacts of impoundment projects (Table 2)	69
	Impacts of channelization projects (Table 3)	19
	Impacts of dredging projects (Table 4)	30
	Impacts of other water resources projects (Table 5)	28
	Subtotal	146
Impacts of types of pollutants (Appendices E and F)	Nonpoint sources of pollutants (Table 6)	34
	Transport and fate of pollutants (Table 7)	10
	Subtotal	44
Describing the affected environment (Appendices G and H)	Baseline studies of the water environment (Table 8)	30
	Environmental indices and indicators (Table 9)	16
	Subtotal	46
Impact prediction and assessment (Appendices I through L)	Water quantity/quality impact prediction and assessment (Table 10)	40

	Biological impact prediction and assessment (Table 11)		25
	Estuarine impact prediction and assessment (Table 12)		16
	Ground water, noise, cultural, visual and socio-economic impact prediction and assessment (Table 13)		24
		Subtotal	105
Methodologies for impact assessment and decision-making (Appendices M and N)	Methodologies for trade-off analysis and decision-making (Table 14)		35
	Public participation in water resources planning (Table 15)		19
		Subtotal	54
Additional issues and information (Appendices O and P)	Impact mitigation measures (Table 16)		12
	Related issues and information (Table 17)		27
		Subtotal	39
		TOTAL	434

Table 2: Summary of References on the Impacts of Impoundment Projects

Authors (Year)	Comments
Abu-Zeid (1979)	Study of Aswan Dam in Egypt.
Armaly and Lepper (1975)	Diurnal stratification in deep impoundments.
Berkes (1981)	Environmental and social impacts of the James Bay hydroelectric project in Canada.
Biswas (1978)	General discussion of environmental implications of water development projects.
Biswas (1980)	Examples of social and environmental effects of water development projects.
Bombowna, Bucka and Huk (1978)	Study in Poland.
Budweg (1982)	Environmental effects of Brazilian dams.
Buikema and Loeffelman (1980)	Effects of pumpback storage on zooplankton populations.
Burton (1982)	Literature review of microbiological water quality of impoundments.
Byrd and Perona (1979a)	Effect of motorboat usage on lead concentration in California reservoir.
Byrd and Perona (1979b)	Study of lead concentration from recreational boating.
Byrd and Perona (1980)	Temporal and spatial variations of lead levels in a lake as a function of boat usage.
Deudney (1981)	Environmental effects of hydropower projects.
El-Hinnawi (1980)	Review of main environmental changes resulting from development of the Nile River Basin.
Entz (1980)	Succession of aquatic animal and plant communities during first 10 years after completion of Lake Nasser and Lake Nubia.
Environmental Control Technology Corporation (1975)	Prediction of water quality impacts from outboard engines

Table 2: (continued)

Authors (Year)	Comments
Fast and Hulquist (1982)	Discussion of supersaturation of nitrogen gas caused by artificial aeration in reservoirs.
Freeman (1974)	Study in Ghana.
Gallopin, Lee and Nelson (1980)	Environmental and social impacts of construction and operation of dam at Salto Grande on Uruguay River.
Garzon (1974)	Prediction of water quality changes in large tropical reservoirs with long detention times.
Godden, Nicol and Venn (1980)	Environmental impacts of dam on Keiskamma River in South Africa.
Gould (1981)	Case histories and mathematical modeling of impacts of water resources development in the Senegal River Basin.
Grizzle (1981)	Effects of hypolimnetic discharges on fish health below a reservoir.
Grover and Primus (1981)	Environmental effects of hydropower project in Alberta, Canada.
Hafez and Shenouda (1978)	Beneficial and detrimental impacts of Aswan High Dam on Nile River Basin in Egypt and Sudan.
Hagan and Roberts (1980)	Method for calculating energy requirements for construction and operation of a dam in California.
Hefny (1982)	Impacts of Aswan High Dam on land use and management.
Hussong et al. (1979)	Study of microbial impact of Canada geese and whistling swans on aquatic ecosystems.
Interim Committee for Coordination of Investigations of the Lower Mekong Basin (1982)	Guidelines for conducting EIA's on development programs for tropical river basins.
Johnson, Krinitzsky and Dixon (1977)	Study of induced seismicity of reservoirs.

Table 2: (continued)

Authors (Year)	Comments
Kadlec (1962)	Effects of drawdown on a water fowl impoundment.
Kay and McDonald (1980)	Reduction of coliform bacterial densities in two British upland reservoirs.
Keeney (1978)	Impact prediction for proposed impoundment in southwestern Wisconsin.
Kelly, Underwood and Thirumurthi (1980)	Impact of construction of a hydroelectric project on the water quality of 5 lakes in Nova Scotia, Canada.
Kenyon (1981)	General discussion of impacts of small and large hydroelectric projects.
King (1978)	Effects of hydraulic structures.
LaBounty and Roline (1980)	Operational impacts at a pumped storage facility in Colorado.
Lewke and Buss (1977)	Effects on vertebrate animals from project in Washington.
Lewke (1978)	Impacts of Lower Snake River Dam in Washington on aquatic and terrestrial biology.
Manning (1979)	Soil and vegetative impacts from recreational usage of land resources.
Matter et al. (1983)	Movement, transport, and scour of particulate organic matter and aquatic invertebrates downstream from a peaking hydropower project.
McClellan and Frazer (1980)	Origin, distribution, and bioaccumulation of 10 trace metals in Kentucky and Barkley Lakes in Kentucky and Tennessee.
Miracle and Gardner (1980)	Literature review on effects of pumped storage operations on ichthyofauna.
Nelson et al. (1976)	Summary of impacts on fisheries
Nix (1980)	Distribution of trace elements in DeGray Reservoir in Arkansas.
Pastorok, Lorenzen and Ginn (1982)	Environmental aspects of artificial aeration and oxygenation of reservoirs.

Table 2: (continued)

Authors (Year)	Comments
Petts (1980a)	Discussion of first-, second-, and third-order impacts of construction and operation of a dam and reservoir.
Petts (1980b)	Morphological changes in river channels downstream of 14 impoundments in Britain.
Pickering and Andrews (1979)	Evaluation of alternative land developments around New Hampshire lakes.
Ploskey (1982)	367 annotations describing the effects of fluctuating reservoir water levels on fish. Changes on reservoir fisheries and improved management.
Raymond (1979)	Effects on fish migration.
Reynolds and Ujjainwalla (1981)	Beneficial and detrimental impacts of hydro-electric projects and needs for improving associated environmental assessments.
Roseboom et al. (1979)	Study of eutrophy of two impoundments in Illinois.
Sargent and Berke (1979)	Procedure for classifying undeveloped lakeshore areas according to their suitability for public and private uses.
Schreiber and Rausch (1979)	Study of suspended sediment and phosphorus in Missouri flood detention reservoir.
Swanson and Meyer (1977)	Effects on feeding patterns of blue-winged teal.
Teskey and Hinckley (1977a)	Literature review of plant physiological responses.
Teskey and Hinckley (1977b)	Effects on southern forest plant communities.
Teskey and Hinckley (1977c)	Effects on central forest plant communities.
United Nations Environment Program (1978)	Environmental effects of dam construction in river basins and methods of minimizing such effects.
Weiner et al. (1979)	Microbial impact of migratory geese and swans on Chesapeake Bay.

Table 2: (continued)

Authors (Year)	Comments
Williams (1977)	Prediction of hydraulic, hydrologic, and sediment transport effects.
World Health Organization (1979)	Report on a 1978 seminar on environmental health impact assessment held in Greece.
Yousef (1974)	Effects on water quality from boating activities.
Yousef et al. (1978)	Prediction of mixing effects from boating activities.
Yousef, McLellon and Zebuth (1980)	Changes in phosphorus concentrations due to mixing by motor boats in shallow lakes.
Zimmerman, Anderson and Calhoun (1980)	Impacts of Possum Kingdom Reservoir in Texas on forage fishes.

Petts (1980a) describes the environmental consequences of dam and reservoir construction on a river system organized according to first-, second-, and third-order impacts. The immediate and simultaneous effects of the activity (dam building) are first-order impacts, for example, reduction in peak flow, entrapment of sediment load, reduction in sediment and suspended sediment load, induced erosion immediately below the dam, and channel changes. These induce second-order impacts such as changes in channels and invertebrate populations taking place over a longer period after construction--perhaps as long as 50 years. Channel cross-sectional reduction dominates morphological changes in impounded rivers--depth increases from erosion, depth decreases from sedimentation, reductions from redistribution of the flood plain and channel boundary materials, and width reductions from sediment deposition. Petts (1980b) also discussed the channel size and shape changes associated with 14 impounded rivers located throughout Britain. He determined that these changes have halved the water conveyance capability of 11 of the 14 rivers studied. After initial changes in aquatic life during the second-order stage, further adjustments occur as a part of third-order impacts. For example, accumulation of fine sediments in pools discourages growth of invertebrates and encourages establishment of rooted aquatic plants. These plants in turn can further effect channel morphology. Changes in fish habitats, intimately associated with changes in channel width, depth, and sediment composition and in flora and fauna, may continue to occur for many years after initial construction. Environmental impact studies related to the construction and operation of reservoir projects should consider the complex interactions taking place over many years before a morphological and ecological equilibrium is reached.

One of the major concerns associated with reservoir projects is related to the impacts from fluctuating water levels. Ploskey (1982) authored a report containing 367 annotations describing the effects of fluctuating reservoir water levels on fish. Citations on phytoplankton, zooplankton, and water quality effects that pertain to reservoir fisheries are also included.

In a very interesting study focused on energy, Hagan and Roberts (1980) described the conduction of an energy impact analysis for a reservoir project at New Melones, California. They indicated that the total energy impact should be considered, and should include items such as energy consumed in producing the electricity, and energy consumed in producing steel for construction of the dam. The New Melones project yielded estimates of power production from a maximum of 430 million kilowatt hours per year to a net loss of 3.9 million kilowatt hours per year. The proponents of the higher figure did not include energy costs, line losses incurred in delivery of power, and the flooding of the existing power plant in the area. The opponents, who claim the project would produce a loss in available energy, calculated costs in BTU of primary energy, and benefits in BTU of delivered electricity with a benefit-cost ratio of 0.65. When Hagan and Roberts (1980) recalculated these figures and considered the savings produced by inactivation of the old power plant, the benefit-cost ratio increased to 2.7. The key point is that a complete analysis of the energy requirements associated with a hydroelectric project should be made in conjunction with environmental impact studies.

Table 3 contains summary comments on 19 selected references related to the impacts of channelization projects. Appendix B contains abstracts of the 19 references listed in Table 3, and Chapter 3 provides a detailed summary of environmental impact studies for channelization projects. A key reference in Table 3 is Thackston and Sneed (1982), who reviewed the environmental consequences of waterway design and construction practices used by the U.S. Army Corps of Engineers. Consideration was given to channelization as well as the construction of dikes, revetments, levees, and other channel modifications for flood control and navigation purposes. Possible adverse effects included wetlands drainage, loss of native vegetation, cut-off of oxbows and meanders, water table drawdown, increased erosion and sedimentation, and changes in aesthetics. Other possible effects on the aquatic system may include the loss of aquatic habitat, productivity, and species diversity, and the degradation of water quality. Alternatives to traditional channel modification projects were identified and discussed in terms of their relative features and environmental consequences. Current efforts to minimize adverse environmental effects of projects were investigated and summarized (Thackston and Sneed, 1982).

Table 3: Summary of References on the Impacts of Channelization Projects

Authors (Year)	Comments
Bastian (1980)	Salinity effects of deepening dredged channels in Chesapeake Bay.
Benke, Gillespie and Parrish (1979)	Study of impacts on invertebrates in Georgia.
Duvel et al. (1976)	Study of ecological changes in six Pennsylvania streams.

Table 3: (continued)

Authors (Year)	Comments
Erickson, Linder and Harmon (1979)	Study of wetland losses in North and South Dakota.
Frederickson (1979)	Floral and faunal changes in low land hardwood forests in Missouri.
Headrick (1976)	Effects on fish populations in Wisconsin.
Huang and Gaynor (1977)	Model presented for calculating flooding decrease.
Klimas (1982)	Effects of permanently raised water tables on forest overstory vegetation.
Maki, Hazel and Weber (1975)	Effects on bottomland and swamp forest ecosystems in North Carolina.
Parrish et al. (1978)	Effects on fish populations in Hawaii.
Pennington and Baker (1982)	Environmental effects of cutoff bendways.
Possardt and Dodge (1978)	Study of impacts on songbirds and small mammals in Vermont.
Prellwitz (1976)	Effects on terrestrial wildlife in Wisconsin.
Stone and McHugh (1977)	Hydrological effects in Louisiana.
Stone, Bahr and Day (1978)	Effects of projects in Louisiana.
Thackston and Sneed (1982)	Review of physical-chemical and biological impacts from U.S. Army Corps of Engineers waterway design and construction practices.
White and Fox (1980)	Effects on aquatic insects in South Carolina.
Wright (1982)	Biological impacts from physical and chemical changes associated with navigation traffic.
Zimmer and Bachmann (1978)	Study of impacts on invertebrates in Iowa.

Table 4 lists comments on the 30 selected references dealing with the impacts of dredging projects. Appendix C contains abstracts of the 30 references, and Chapter 4 is focused on environmental impact studies for dredging projects. One interesting reference contained a discussion of the effects and impacts of dredge and fill activities on estuaries (Johnston, 1981). The biological effects of turbidity may cause reduced visibility and reductions in the availability of food for fish. High levels of suspended solids can reduce oyster growth and may have toxic effects on various larvae. Dissolved oxygen concentrations are typically lower near dredging and filling sites, and the pH can be reduced. Siltation resulting from these activities can have drastic effects, including the immediate removal of organisms through suffocation and the long-term elimination of many desirable species of flora and fauna. Resuspension of bottom materials can result in the release of nutrients as well as the possible release of toxicants. Various measures for mitigating the deleterious effects of estuarine dredge and fill activities are summarized (Johnston, 1981).

Table 4: Summary of References on the Impacts of Dredging Projects

Authors (Year)	Comments
Allen and Hardy (1980)	Impacts to fish, other aquatic organisms, and wildlife, as well as habitat enhancement opportunities, resulting from new and maintenance dredging.
Bohlen, Cundy and Tramontano (1979)	Field sampling for suspended material distributions in the wake of estuarine channel dredging operations.
Brannon (1978)	Investigation of pollution properties of dredged material.
Chen et al. (1978)	Evaluation of impacts of dredged material disposal on ground water and surface water.
Conner and Simon (1979)	Effects of oyster shell dredging on an estuarine benthic community.
Conrad and Pack (1978)	Methodology for determining land values and associated benefits from dredged material contaminant sites.
Eichenberger and Chen (1980)	Methodology for prediction of effluent water quality from the disposal of dredged material in confined areas.
Engler (1978)	Study of physical effects of aquatic disposal.
Flint (1979)	Study of impacts on fresh water benthos.

Table 4: (continued)

Authors (Year)	Comments
Grimwood and McGhee (1979)	Evaluation of standard elutriate test as a predictive tool.
Gunnison (1978)	Study of mineral cycling in salt marsh-estuaries ecosystems.
Gushue and Kreutziger (1977)	Productive land use at 12 dredged material disposal sites.
Hoeppel (1980)	Monitoring at 9 diked containing areas to determine contaminant mobility.
Holliday (1978)	Prediction of fate of dredged material placed in oceans, estuaries, lakes, and rivers.
Holliday, Johnson and Thomas (1978)	Prediction and monitoring of dredged material movement.
JBF Scientific Corporation (1975)	Summary of dredging technology.
Johanson, Bowen and Henry (1976)	Summary of open water dredged material placement.
Johnston (1981)	Biological and water quality impacts from dredge and fill activities in estuaries.
Landin (1978)	Bibliography on nesting waterbirds and their relationship to dredged material islands.
Laskowski-Hoke and Prater (1981)	Comparison of bulk sediment-chemistry evaluation procedure with a bioassay technique.
Lehmann (1979)	174 abstracts of literature on the environmental and biological effects of dredging.
Maurer et al. (1981)	Study of vertical migration and mortality of benthos in dredged material.
Morrison and Yu (1981)	Impact of dredged material disposal on ground water quality.
National Marine Fisheries Service (1977)	Chemical and biological effects of dredging and spoil disposal.

Table 4: (continued)

Authors (Year)	Comments
Ocean Data Systems, Inc. (1978)	Handbook for terrestrial wildlife habitat development on dredged material disposal sites.
Pavlou et al. (1980)	Impacts of PCB's at a deep water disposal site.
Peterson (1979)	Summary of dredging technology applied to lakes.
Raster et al. (1978)	Methodology for dredged material site selection and design.
Slotta et al. (1974)	Study of physical and biological impacts of dredging.

Table 5 contains summary comments on 28 selected references on water resources projects other than the typically identified projects involving reservoirs, channelization and dredging. Included in Table 5 are references related to irrigation projects and various coastal projects. Appendix D contains abstracts for the 28 selected references, and Chapter 5 summarizes some of the key environmental concerns related to both irrigation and coastal projects.

Table 5: Summary of References on the Impacts of Other Water Resources Projects

Authors (Year)	Comments
Ahmad (1982)	Addresses environmental consequences of irrigation projects in arid and semi-arid areas.
Darnell (1977)	Summarizes mitigation of construction project impacts.
deGroot (1979a)	Effects of marine gravel extraction on herring populations.
deGroot (1979b)	Effects of sand dredging for island construction material.
Diamant (1980)	Environmental impacts of irrigation projects in hot climates.

Table 5: (continued)

Authors (Year)	Comments
Elgershuizen (1981)	Environmental impacts of storm surge barrier across the eastern Scheldt in the Netherlands.
Elkington (1977)	Effects on wetland ecosystems.
Environmental Resources Limited (1983)	Environmental health impact assessmen' of irrigated agricultural development projects.
Gysi (1980)	Energy, environmental, and economic implications of four water resources projects in Alberta, Canada.
Huber and Brezonik (1981)	Water quality impacts from man-made lakes near estuaries in Florida.
Livingstone and Hazlewood (1979)	Analysis of risk in irrigation projects.
Micklin (1977)	International environmental implications of development of Volga River in Russia.
Mulvihill et al (1980)	Composite review of biological impacts of minor shoreline structures such as break waters, bulkheads, and others.
National Oceanic and Atmospheric Administration (1976)	Guidelines for facility development.
Pollard (1981)	Health impacts of irrigation in the Gezira region of Sudan.
Ryner (1978)	Effects of lakefront development in Chicago.
Shabman and Bertelson (1979)	Methodology for estimating the development values of wetlands.
Smies and Huiskes (1981)	Ecological considerations related to storm surge barrier across the eastern Scheldt in the Netherlands.
Smil (1981)	Potential environmental impacts of large-scale irrigation and water transfer plans in China.
Takahasi (1982)	Environmental impacts of water resources projects in Japan.

Table 5: (continued)

Authors (Year)	Comments
Tucker (1983)	Schistosomiasis problems from irrigation projects.
Vendrov (1980)	Impacts of large water resources projects in the USSR.
Watling (1975)	Information needs for impact prediction.
Watling, Pembroke and Lind (1975)	Impact of artificial island on marine ecosystem.
Witten and Bulkley (1975)	Effects on game fish habitat from revetments, retards, and jetties.
Yiqui (1981)	Environmental impact of water transfer project in China.

To provide an illustration of the impacts of coastal projects, Mulvihill et al. (1980) presented information from 555 reference sources on the biological impacts of breakwaters, jetties, groins, bulkheads, revetments, ramps, piers and other support structures, buoys and floating platforms, small craft harbors, bridges, and causeways. The information typically showed that the impact of the structure on the environment is site specific. Fourteen case studies were included, and small boat harbors, bridges, causeways, bulkheads, breakwaters, and jetties were found to have the most potential for causing coastal environment impacts. Revetments, groins, and ramps have moderate impact potential, while buoys and floating platforms, piers, and other support structures have low impact potential. The majority of the information included provides qualitative descriptions of anticipated or realized impacts. Minimal information on the quantitative impacts of specific structures was located as a part of the state-of-the-art review (Mulvihill et al., 1980).

IMPACTS OF TYPES OF POLLUTANTS

Water pollutants can be considered in terms of both nonpoint sources (urban or rural runoff) and point sources. The primary reason for this category is to summarize information on the anticipated environmental effects resulting from specific types of pollutants that can enter the aquatic environment. Table 6 contains summary comments on the 34 selected references dealing with the impacts of nonpoint pollution on the water environment, and Appendix E contains the 34 abstracts. Nonpoint sources of water pollution have been recognized as potential major contributors to the total waste load within the aquatic environment, and it is vitally important in environmental impact studies to consider nonpoint sources along with point sources of water pollution. Nonpoint sources are of particular concern relative to impoundment projects due to their contribution to the total nutrient loading within impoundments.

Table 6: Summary of References on the Impacts of Nonpoint Sources of Pollutants on Water Environment

Authors (Year)	Comments
Bailey and Nicholson (1978)	Methodology for evaluation of models for pesticide transport from agricultural land.
Brookman et al. (1979)	Nonpoint pollution from an industrial site.
Burns (1979)	Sediment model for Piedmont forest area in Georgia.
Cluis, Couillard and Potvin (1979)	Transport model relating land use to mass discharge of total nitrogen and total phosphorus.
Davis and Donigian (1979)	Model for nitrogen and phosphorus from agricultural lands.
Feller (1981)	Effects of clear-cutting and slash-burning on stream water temperatures in British Columbia.
Fowler and Heady (1981)	Estimation of suspended sediment production rates on undisturbed forest land.
Fusillo (1981)	Study of impact of suburban residential development on water resources in New Jersey.
Gaynor (1979)	Phosphorus loading associated with housing in a rural watershed.
Gurtz, Webster and Wallace (1980)	Effects of clear-cutting on suspended particulate matter in two streams in the southern Appalachian Mountains.
Haith (1980)	Mathematical model for estimation of losses of dissolved and solid-phase pesticides in cropland runoff.
Hopkinson and Day (1980)	Use of EPA Storm Water Management Model (SWMM) to examine the relationship between development and storm water and nutrient runoff in Louisiana.
Interstate Commission on the Potomac River Basin (1981)	32 papers in symposium proceedings on tools and techniques for nonpoint pollution control.

Table 6: (continued)

Authors (Year)	Comments
Jewell, Adrian and DiGiano (1980)	Use of linear multiple regression analyses to model urban storm water pollutant loadings.
Larson (1978)	Effect on water quality in a reservoir.
Lusby (1979)	Effects of grazing practices on runoff and sediment yield.
Lynch, Corbett and Sopper (1980)	Effects of forest management practices on the biological and chemical characteristics of streamflow.
Martin, Noel and Federer (1981)	Impacts of clear-cutting on the chemistry of 56 streams in New England.
Mather (1979)	Hydrologic consequences of two case studies involving urbanization.
McCuen et al. (1978)	Computer simulation model for estimating pollutant loading.
Olivieri, Kruse and Kawata (1977)	Study of microorganisms in urban stormwater.
Ongley and Broekhoven (1979)	Regional assessment of agricultural impacts on water quality in southern Ontario, Canada.
Robbins (1978)	Nonpoint pollution from unconfined animal production areas.
Ross, Shanholtz and Contractor (1980)	Finite element model for prediction of erosion and sediment transport.
Schillinger and Stuart (1978)	Quantification of nonpoint pollution from logging, cattle grazing, mining, and subdivision activities.
Schreiber, Duffy and McClurkin (1980)	Aqueous and sediment-phase nitrogen yields from 5 southern pine watersheds.
Smith and Eilers (1978)	Models for effects on receiving stream dissolved oxygen and hydraulic characteristics.
Tubbs and Haith (1981)	Simulation model for agricultural nonpoint source pollution.

Table 6: (continued)

Authors (Year)	Comments
Turner, Brown and Deuel (1980)	Nutrients and associated ion concentrations in irrigation return flow from flooded rice fields.
Unger (1978)	Ecological effects from agriculture and silviculture activities.
U.S. Bureau of Reclamation (1977)	Conjunctive use model for predicting the mineral quality of irrigation return flow.
U.S. Environmental Protection Agency (1976a, b and c)	Three-volume general summary of procedures for predicting impacts of urban stormwater.
Walker (1976)	Literature review of irrigation return flow models.
Watson et al. (1979)	Study of effects of development on lake nutrient budgets.

Two references listed in Table 6 include information on the use of models for predicting the quantity and potential effects of urban runoff (Hopkinson and Day, 1980; and Jewell, Adrian and DiGiano, 1980). Two other references deal with rural runoff and are focused on the effects of clear-cutting activities on nonpoint source pollution (Gurtz, Weber and Wallace, 1980; and Martin, Noel and Federer, 1981). Control of nonpoint pollution is addressed by 32 papers in a symposium proceedings on tools and techniques for nonpoint pollution control (Interstate Commission on the Potomac River Basin, 1981). The 32 papers are classified into five categories: (1) perspectives on nonpoint pollution control; (2) case studies on nonpoint sources of pollution; (3) modeling tools for evaluation of nonpoint pollutants; (4) control measures; and (5) planning an implementation strategy. The majority of the papers are associated with the Potomac River Basin.

Table 7 lists key comments on the ten selected references dealing with the transport and fate of selected pollutants in the water environment. References are included on the fate and effects of metals (Damman, 1979; Drill et al., 1979; Frenet-Robin and Ottman, 1978; and Leland, Luoma and Fielden, 1979); petroleum products (Anderson, 1979; Buikema, McGinniss and Cairns, 1979; and Malins, 1977); and pesticides (Hansen, 1978). Appendix F contains the abstracts of the ten selected references.

Table 7: Summary of References on the Transport and Fate of Pollutants in the Water Environment

Authors (Year)	Comments
Anderson (1979)	Fate and effects of petroleum hydrocarbons in the marine environment.
Buikema, McGinniss and Cairns (1979)	Literature review on effects of phenolics in aquatic ecosystems.
Damman (1979)	Study of heavy metals in freshwater wetlands.
Drill et al. (1979)	Environmental pathways of human exposure to lead.
Frenet-Robin and Ottmann (1978)	Mercury sorption by various clays.
Hansen (1978)	Laboratory studies of impacts of pesticides on estuarine animals.
Hoover (1978)	Literature review of health effects and analytical needs.
Iwamoto et al. (1978)	Literature review on the physical and biological effects of sedimentation in streams.
Leland, Luoma and Fielden (1979)	Literature review on bioaccumulation and toxicity of heavy metals and related trace elements in aquatic ecosystems.
Malins (1977)	Two-volume book on the effects of petroleum on marine environments and organisms.

DESCRIBING THE AFFECTED ENVIRONMENT

Table 8 provides summary comments on 30 selected references dealing with the planning and conduction of baseline studies of the water environment. Abstracts for each of the 30 selected references are contained in Appendix G. A review of Table 8 indicates that some references are primarily focused on water quality considerations (for example, Adrian et al., 1980; Brown, 1977; Liebetrau, 1979; Loftis and Ward, 1980a; and Loftis and Ward, 1980b); while others deal with comprehensive planning for aquatic biology surveys (for example, Hellawell, 1978; and Ward, 1978).

Table 8: Summary of References on Baseline Studies of the Water Environment

Authors (Year)	Comments
Adrian et al. (1980)	Procedures for locating stream sampling stations.
Bingham et al. (1982)	Grab samplers for benthic macroinvertebrates.
Bogucki and Gruendling (1978)	Study of Lake Champlain wetlands and effects due to fluctuating water levels.
Brown (1977)	Study of lake water quality in California.
Burke (1978)	Bibliography of information on wildlife used by 17 western states and various Federal agencies.
Burns (1978)	Proceedings of symposium on planning for environmental assessment studies.
Cairns and Gruber (1980)	Biological monitoring systems based on measuring the ventilatory behavior of fish.
Cermak, Feldman and Webb (1979)	Identifies land use classifications for use in hydrologic models.
Collotzi and Dunham (1978)	Systematic approach for inventorying aquatic habitat.
Colwell et al. (1978)	Study of waterfowl habitat quality.
Fry and Pflieger (1978)	Assessment of aquatic habitat in Missouri.
Gonor and Kemp (1978)	Quantitative studies of marine intertidal benthic environments.
Groves and Coltharp (1977)	Study for monitoring land use and land use changes.
Haugen, McKim and Marlar (1976)	Study to assess effects of land use on sediment loading of streams.
Hellawell (1978)	Handbook for the biological monitoring of rivers.
Hundemann (1978)	Annotated bibliography with 156 abstracts on remote sensing applied to environmental pollution problems.

Table 8: (continued)

Authors (Year)	Comments
Hyman, Lorda and Saila (1977)	Study to produce a standard program for ichthyoplankton sampling.
Jacobs and Grant (1978)	Methods of zooplankton sampling and analysis for quantitative surveys.
James, Woods and Blanz (1976)	Methodology for using Landsat for the environmental evaluation of impoundments and channelization projects.
Liebetrau (1979)	Statistical considerations in water quality sampling.
Loftis and Ward (1980a)	Selection of sampling frequencies for regulatory water quality monitoring.
Loftis and Ward (1980b)	Practical sampling frequency considerations for water quality monitoring.
McNeely, Neimanis and Dwyer (1979)	General discussion of 70 water quality parameters.
Persoone and DePauw (1978)	Systems of biological indicators for water quality assessment.
States et al. (1978)	Planning for ecological baseline studies related to energy development projects in the western United States.
Stofan and Grant (1978)	Methods of phytoplankton sampling and analysis for quantitative surveys.
Stout et al. (1978)	Planning for integrated baseline studies of the environment.
Villeneuve et al. (1979)	Use of kriging in the design of streamflow sampling networks.
Ward (1978)	Book on the planning, conduction and interpretation of biological impact studies.
Weiderholm (1980)	Use of profundal benthic communities as integral measures of autotrophic and heterotrophic lake processes.

Table 9 summarizes 16 selected references on the use of environmental indices and indicators in environmental impact studies. Appendix H contains the 16 abstracts. The development and usage of water quality indices are described in several references in Table 9, with the usage of indices being of value in summarizing water quality information and communicating this information in an understandable fashion to reviewers of environmental impact reports (Inhaber, 1976; and Ott, 1978). In addition, the use of water quality indices can aid in determining the magnitude of anticipated impacts. Polivannaya and Sergeyeva (1978) described the use of zooplankters as indicators of water quality.

Table 9: Summary of References on Environmental Indices and Indicators

Authors (Year)	Comments
Ball and Church (1980)	Usage and limitations of water quality indices.
Booth, Carubia and Lutz (1976)	Methodology for comparative evaluation of indices.
Chiaudani and Pagnotta (1978)	Use of the ATP/chlorophyll ratio as an index.
Dunnette (1979)	Index based on geographical characteristics of river basins.
House and Ellis (1981)	Advantages of water quality indices.
Ibbotson and Adams (1977)	Formulation and testing of a water quality index using a matrix format.
Inhaber (1976)	Book describing economic, air quality, water, land, biological, aesthetic, and other environmental indices.
Keilani, Peters and Reynolds (1974)	Water quality index for economic decision-making regarding preventive measures and treatment programs.
Landwehr (1979)	Statistical analyses of several indices.
Lee, Wang and Kuo (1978)	Use of community diversity index of benthic macroinvertebrates and fish.
Ott (1978)	Book describing the structure of environmental indices, including air pollution and water pollution indices.
Polivannaya and Sergeyeva (1978)	Use of zooplankters as bioindicators of water quality.

Table 9: (continued)

Authors (Year)	Comments
Provencher and Lamontagne (1979)	Method for establishing a water quality index for different uses.
Reynolds (1975)	Index based on water uses and water quality objectives.
Thomas (1976)	Survey of professionals relative to use of water quality indices.
Yu and Fogel (1978)	Index based on use-oriented benefits and treatment costs analysis.

IMPACT PREDICTION AND ASSESSMENT

The most important technical activity in environmental impact studies involves the scientific prediction of the effects of various project actions, and the assessment or interpretation of the significance or importance of those effects. Numerous mathematical models and other scientific approaches are available for predicting and assessing the impacts of water resources projects on various features of the environment. Table 10 contains summary comments on 40 selected references on models and methodologies for water quantity and quality impact prediction for river systems. The cited references are primarily associated with developed models for predicting impacts on flow and quality, and Appendix I contains the 40 abstracts. To serve as an example reference from Table 10, French and Krenkel (1981) discussed several factors affecting the effectiveness of water quality models. Some processes are not easy to model, for example, erosion, eutrophication, and toxicity relationships; however, dissolved oxygen, temperature and dissolved solids are relatively easy to model. Indicator bacteria, sediment transport, algal growth, metal transport, nutrient transport, and pesticide transport are intermediate in complexity. Model calibration and verification are necessary to insure that the results obtained are appropriate to the situation being analyzed. An important usage of river models is associated with sensitivity analysis, that is, showing the effect of variations of a given parameter on the output if all other factors are assumed to be constant.

Table 10: Summary of References on Water Quantity/Quality Impact Prediction and Assessment

Authors (Year)	Comments
Abbott (1977)	Calibration and use of the Storage, Treatment, Overflow Runoff Model (STORM)

Table 10: (continued)

Authors (Year)	Comments
Ahlgren (1980)	Hydraulic dilution model for nitrogen and phosphorus in four eutrophied lakes near Stockholm, Sweden.
Ahmed and Schiller (1981)	Use of model for computing loading estimates from nonpoint sources in 16 lakes in Connecticut and Massachusetts.
Austin, Landers and Dougal (1978)	Models for simulation of effects of fluctuating water levels in Iowa.
Austin, Riddle and Landers (1979)	Model for shoreline vegetative succession.
Baca et al. (1974)	Multisegment deep reservoir water quality simulation model for prediction of algal and DO-BOD dynamics.
Baca et al. (1977a and b)	Parts 2 and 3 describe eutrophication model and limnological model for water supply reservoirs in Australia.
Booth (1975)	Model for calculating radionuclide transport between receiving waters and bottom sediments.
Bourne, Day and Debo (1978)	Use of Hydrocomp Simulation Program (HSP) for continuously simulating the hydrologic and water quality responses of a watershed.
Brandstetter et al. (1977)	Summary of Baca et al. (1977a and b), Parts 2 and 3.
Brown et al. (1981)	Catchment, lake and channel models for hydrologic regime of Upper Nile River Basin.
Carrigan (1979)	200 abstracts on water quality modeling for hydrological and limnological systems.
Charlton (1980)	Productivity and morphometry effects on hypoliminion oxygen consumption in lakes.
Freedman, Canale and Pendergast (1980)	Model for prediction of the transient impact of storm loads on phosphorus, fecal coliform, and dissolved oxygen concentrations in an eutrophic lake.
French and Krenkel (1981)	Discussion of 8 factors affecting the effectiveness of river models.

Table 10: (continued)

Authors (Year)	Comments
Ford and Stefan (1980)	Mixed-layer model for seasonal temperature predictions in lakes.
Hoopes et al. (1979)	Vertical mixing in stratified impoundment from submerged discharge of heated water.
Horst (1980)	Mathematical model for assessment of effects on zooplankton populations.
Huang (1979)	Changes in channel geometry and channel capacity of alluvial streams below large impoundment structures.
Johanson and Leytham (1977)	Sediment transport model for natural channels.
Jorgensen (1980)	Development and implementation of a series of water quality models for the Nile River Basin.
Karim, Croley and Kennedy (1979)	Computer-based model for calculation of amounts, rates, and spatial distributions of sediment in lakes and reservoirs.
Lehmann (1978a)	192 abstracts on citizen perceptions of water resources projects and programs.
Lehmann (1978b)	185 abstracts on water quality modeling for hydrological and limnological systems.
McCuen, Cook and Powell (1980)	Water quality/quantity model for proposed reservoir in Maryland.
Meinholz et al. (1979)	Use of a modified Harper's water quality model for verifying observed impacts in the Milwaukee River.
Noble (1979)	Model for prediction of natural temperatures in rivers.
Onishi (1981)	Modeling of sediment transport of kepone in the James River Estuary in Virginia.
Orlob (1977)	Two-volume literature review of mathematical modeling of surface water impoundments.
Rahman (1979)	Two-layer model for water temperature prediction in stratified reservoirs.

Table 10: (continued)

Authors (Year)	Comments
Rosendahl and Waite (1978)	Phosphorus movement in channelized and meandering streams.
Snodgrass and Holloran (1977)	One-dimensional temperature-oxygen model for reservoirs.
Thomann (1979)	Mass balance model for PCB distribution in a lake.
U.S. Environmental Protection Agency (1980)	Input-output phosphorus lake model to quantify the relationship between land use and lake trophic quality.
Uzzell and Ozisik (1978)	Model for prediction of temperature distribution in lakes resulting from thermal discharges.
Vick et al. (1977)	Pre-impoundment study in Georgia.
Ward (1981)	Conceptual design model for predicting the sediment trapping performance of small impoundments.
Webster, Benfield and Cairns (1978)	Model for effects of impoundment on particulate organic matter dynamics in a river-reservoir ecosystem.
Woodward, Fitch and Fontaine (1981)	Model for prediction of fate of heavy metals downstream from wastewater discharges.
Wycoff and Singh (1980)	Application of the Continuous Storm-Water Pollution Simulation System (CSPSS) in a Philadelphia urban area discharging into the Delaware Estuary.

Biological impact prediction and assessment is addressed in Table 11 via the inclusion of summary comments on 25 selected references. Appendix J contains the 25 abstracts. A valuable reference is the book on environmental biology written for nonbiologists working on environmental impact studies (Camougis, 1981). In addition, the U.S. Fish and Wildlife Service (1979) has published a procedural manual for quantitatively estimating and comparing development project impacts on fish and wildlife resources.

Table 11: Summary of References on Biological Impact Prediction and Assessment

Authors (Year)	Comments
Belyakova (1980)	Model for seasonal dynamics of biomass and production of the main trophic groups of aquatic organisms, and concentrations of biogenic elements.
Bovee (1974)	Methodology for recommendation of minimum discharges for a warm water fishery.
Bovee and Cochnauer (1977)	Methodology using weighted criteria to assess impacts on stream habitats.
Brungs (1977)	Methodology for assignment of relative ecological values to areas using maps.
Camougis (1981)	Book on environmental biology for non-biologists working in EIA.
Casti et al. (1979)	Polyhedral dynamics for modeling the aquatic ecosystem.
Cowardin et al. (1979)	Classification scheme for wetlands and deepwater habitats of the United States.
Elwood and Eyman (1976)	Model for predicting body burdens of trace contaminants in aquatic consumers.
Fieterse and Toerien (1978)	Eutrophication model based on relationship between phosphate phosphorus and chlorophyll-a for Roodeplatt Dam in South Africa.
Hazel et al. (1976)	47 case studies of California water projects that altered streamflows and causally affected fish and wildlife.
O'Connor, DiToro and Thomann (1975)	Seasonal distribution of phytoplankton relative to eutrophication problems.
Oglesby and Schaffner (1978)	Phosphorus loadings to 15 New York lakes and ecological responses.
O'Neill (1978)	Evaluation of compartmental analysis in ecosystem modeling.
Ostrofsky and Duthie (1978)	Methodology for modeling productivity in reservoirs.

Table 11: (continued)

Authors (Year)	Comments
Scavia and Robertson (1979)	Series of papers on lake ecological model usage, possible model improvements, and new directions for development.
Schnoor and O'Conner (1980)	Steady state eutrophication model for predicting partitioning of nutrients among organic, inorganic, and phytoplankton fractions.
Stalnaker and Arnette (1976)	Methodologies for determining instream flow requirements for fish, terrestrial wildlife, and water quality.
Taylor (1979)	Model for predicting chemical pollutant toxicity in fish.
Thomas et al. (1978)	Statistical procedures for assessing biological impact studies at 3 plants.
U.S. Fish and Wildlife Service (1979)	Procedural manual for estimating and comparing development project impacts on fish and wildlife resources.
Vieth, DeFoe and Bergstedt (1979)	Model for estimating bioconcentration of organic chemicals in fish.
Walters (1980)	Model for chemical, physical, and biological processes in deep stratified lakes in the temperature zone.
Watanabe (1978)	Closed, energy-driven, matter-flow loop of nitrogen and phosphorus cycles.
Williams (1978)	Post-impoundment study in Pennsylvania.
Yahnke (1981)	Nutrient loading models and temperature simulation model for predicting trophic state, productivity, and duration of stratification.

Table 12 contains summary comments on 16 selected references on models and methodologies for addressing project impacts on estuaries, and Appendix K contains the associated abstracts. Several references are associated with analyzing the impacts of decreased fresh water inflows into estuarine systems (Armstrong, 1980; Armstrong and Wart, 1981; Browder and Moore, 1981; Klein et al., 1981; Neu, 1982; and Shea et al., 1981). Onishi and Wise (1978) described a finite element model to simulate sediment and pollutant transport in conjunction with studying the transport and fate of the pesticide kepone deposited in the James River Estuary, Virginia, in the early 1970's.

Table 12: Summary of References on Estuarine Impact Prediction and Assessment

Authors (Year)	Comments
Armstrong (1980)	Overview of relevant factors for determining needed fresh water inflows to estuarine systems.
Armstrong and Wart (1981)	Physical and chemical effects of reduced inflows to Matagorda Bay, Texas.
Bella and Williamson (1980)	Simulation model for sulfur cycle in estuaries.
Browder and Moore (1981)	Quantitative relationship between fishery production and the flow of fresh water to estuaries.
Chu and Yeh (1980)	Vertically averaged two-dimensional numerical models for estuarine hydrodynamics and salinity.
Green (1978)	Model for the Chesapeake Bay ecosystem with submodels on wetlands, plankton, seagrasses, other benthos, and fish trophic levels.
Jennings (1981)	Use of statistical studies and river basin modeling for determining modifications from river basin development.
Klein et al. (1981)	Field studies of effects of fresh water inflows to Chesapeake Bay.
Lauria and O'Melia (1980)	Two steady-state, one-dimensional nutrient models for Pamlico Estuary in North Carolina.
Linton and Appan (1981)	Dynamic methodology for characterizing and monitoring estuarine ecosystems.
Najarian and Harleman (1977)	Model of nitrogen-cycle dynamics in an estuarine system.
Neu (1982)	Estuarine impacts of water storage projects.
Onishi and Wise (1978)	Application of finite element sediment and contaminant transport model to the James River Estuary in Virginia.

Table 12: (continued)

Authors (Year)	Comments
Ozturk (1979)	Modeling of dissolved oxygen in estuaries.
Radford and Joint (1980)	Application of an ecosystem model to the Bristol channel and Severn Estuary in England.
Shea et al. (1981)	Assessment of impacts of low fresh water inflows to Chesapeake Bay.

The impacts of water resources projects on ground water, noise, cultural, visual, and socio-economic features of the environment are addressed in Table 13. Twenty-four references are summarized in Table 13, and Appendix L contains the pertinent abstracts. The cultural environment consists of both historic and archaeological resources as well as aesthetic features. Altshul (1980) suggested that the approach for conducting cultural resource analyses should include seven steps: (1) contact with the State Historic Preservation Office; (2) contact with state historical societies, museums, universities, and other recognized institutions; (3) contact with a qualified agency or consulting firm; (4) preparation of a list of historic and prehistoric sites and finds in the area; (5) incorporation of a report from the consultants into the initial environmental assessment; (6) modification of plans incorporating the assessment into the cultural resources section of the final environmental impact report; and (7) assessment of the need for recovery of significant data. Three case studies are used to illustrate the application of these seven steps (Altshul, 1980).

Table 13: Summary of References on Ground Water, Noise, Cultural, Visual and Socio-economic Impact Prediction and Assessment

Authors (Year)	Comments
Altshul (1980)	Seven steps for conducting cultural resource analyses for water resources projects.
Carlson and Sargent (1979)	Follow-up study of the social impacts of Boise Project in Idaho.
Chang and Beard (1979)	Social impacts from two reservoirs in Texas.
Coughlin (1982)	Evaluating the effects of water resources projects on aesthetic resources.
Daneke and Priscoli (1979)	Discussion of quality of life accounting methodologies.

Table 13: (continued)

Authors (Year)	Comments
Dickens and Hill (1978)	16 papers on cultural resources planning and management.
Eckhardt (1979)	Description of cultural resource inventory on about 15,000 acres in California.
Felleman (1975)	Review of numerical, geometric and geomorphic landform description approaches for evaluating scenic quality.
Fletcher and Busnel (1978)	Book summarizing effects of noise on aquatic and terrestrial wildlife.
Harper (1975)	Use-oriented method for visual quality evaluation of the coastal zone.
Harvey and Emmett (1980)	Model for predicting water table rises from proposed Prosperity Reservoir in Missouri.
Hitchcock (1977)	Review of research reports on social impacts of water resources projects.
Hoffman (1977)	Socio-economic impacts of the proposed Rochester Dam in Kentucky.
Kessler et al. (1978)	Evaluation of construction site noise.
King (1978)	Describes methods for conducting archaeological surveys.
Leatherberry (1979)	Assessment of features, or conditions, in riparian environments that may provide recreational, or preservational and aesthetic values.
Michalson (1977)	Quantification of the aesthetics of wild and scenic rivers based on recreation demand.
Munter and Anderson (1981)	Use of ground water flow models for estimating lake seepage rates.
Nelson, Warnick and Potratz (1979)	Follow-up study of the economic impacts of the "without" project conditions for the Boise Project in Idaho.
Nieman (1975)	Difficulties in assessing the visual quality of the coastal zone.

Table 13: (continued)

Authors (Year)	Comments
Pickering and Andrews (1979)	Economic and environmental evaluation of commercial and residential land development patterns around lakes.
Ricci, Laessig and Glaser (1978)	Pre-operational property price changes around Pennsylvania reservoir.
Shapiro, Luecks and Kuhner (1978)	Evaluation of the infrastructure requirements resulting from secondary development.
Sloane and Dickinson (1979)	Computer simulation modeling of the socio-economic impacts of land use policies in the Lake Tahoe Basin.

METHODOLOGIES FOR IMPACT ASSESSMENT AND DECISION-MAKING

The 1979 Council on Environmental Quality Regulations in the United States gave major emphasis to the evaluation of alternatives in project planning and decision-making, with particular reference to the systematic comparison of the environmental effects of the alternatives. Table 14 summarizes the 35 selected references dealing with methodologies for impact assessment and decision-making. Included are matrix and checklist methods, modeling, and multiattribute decision-making. Abstracts for each of the identified references are contained in Appendix M.

Table 14: Summary of References on Methodologies for Trade-off Analyses and Decision-making

Authors (Year)	Comments
Ahmed, Husseiny and Cho (1979)	Checklist for development of index of site acceptability for nuclear power plants.
Anderson (1981)	Cascaded Trade-offs as a method for ranking of alternatives on the basis of public values.
Baram and Webster (1979)	Computerized matrix for identifying impacts to be addressed for U.S. Army military activities.
Bohm and Henry (1979)	Use of cost-benefit analysis along with environmental effects in three case studies.

Table 14: (continued)

Authors (Year)	Comments
Brown, Quinn and Hammond (1980)	Evaluation of four types of measurement scales for alternative plans.
Bryant (1978)	Computer model used to trace flow of resources generated by various processes.
Budge (1981)	Use of matrix and ordinal rankings of alternatives for water resources projects.
Burnham, Nealey and Maynard (1975)	Weighting checklist for combining societal and technical judgments relative to nuclear power plant siting.
Davos (1977)	Priority-tradeoff-scanning using 3 types of matrices.
Duckstein et al. (1977)	Methodology for including uncertainty in environmental impact assessment.
ESSA Environmental and Social Systems Analysts, Ltd. (1982)	Review and evaluation of the adaptive environmental assessment and management method.
French et al. (1980)	Use of interactive computer graphics in water resources planning.
Herzog (1973)	Dynamic matrix for assessing impacts of technological changes.
Hill (1976a)	Linear programming model for economic evaluations of wetlands.
Hill (1976b)	Resource allocation model for evaluation of wastewater management alternatives.
Hodgins, Wisner and McBean (1977)	Simulation model for determining the impacts of a series of reservoirs.
Keeney (1976)	Decision-making using multi-attribute utility techniques.
Kemp and Boynton (1976)	Seven basic steps for the environmental evaluation of projects.
Lincoln and Rubin (1979)	Human preference model developed for use with an environmental emissions model.
Loran (1975)	Matrix displaying interrelated clusters of high-valued ratings.

Table 14: (continued)

Authors (Year)	Comments
Meyers (1977)	Evaluation of water resources planning and decision-making using an energetics approach.
Motayed (1980)	Use of weighting-scaling checklist for power plant site selection.
Okenik (1978)	Heirarchical multiobjective optimization for water resources planning projects.
Peterson, Clinton and Chambers (1979)	Field test of weighting-scaling checklist on channel project in Louisiana.
Rubinstein and Horn (1978)	Methodology for including risk analysis in environmental impact assessment.
Schrender, Rustagi and Bare (1976)	Simulation models for evaluating the impacts of alternative wildland use decisions.
Schwind (1977)	Matrix used to evaluate impacts of alternative land uses in terms of cost-benefit approaches.
Seaver (1979)	Applications and evaluation of decision analysis in water resources planning.
Sellers and North (1979)	Matrix for evaluation of trade-offs between economic and environmental objectives in water resources planning.
Sicherman (1978)	Use of computer analysis for defining preferences of different interest groups.
Sondheim (1978)	Scaling checklist for evaluation of a proposed dam project.
Tamblyn and Cederborg (1975)	Matrix for nuclear power plant site selection.
Whitlatch (1976)	Use of matrix or stepped matrix approaches in conjunction with linear vector or nonlinear evaluation systems.
Yapijakas and Molof (1981)	Decision-making method for evaluating alternatives for a multinational river basin development.

Table 14: (continued)

Authors (Year)	Comments
Yorke (1978)	Matrix for summarizing impacts of water resources projects on stream characteristics.

Many types of methodologies have been developed and used in environmental impact studies. The majority of the methodologies can be divided into either matrix or checklist approaches. Matrix methodologies are characterized by systematic displays of the activities associated with an alternative relative to environmental factors or descriptors. Checklists range from simple listings of environmental factors or effects to comprehensive methodologies which include importance weighting and impact scaling. Budge (1981) describes a methodology which includes the identification of environmental impacts through the use of a matrix, the collection and collation of information on each impact, and the comparison of options using a combination of ordinal rankings in preference to cost-benefit techniques. Brown, Quinn and Hammond (1980) describe four types of measurement scales which can be used in scaling the impacts of alternative plans. The concepts of reliability and validity are discussed in detail along with the concept of measurement standards.

An important element in environmental decision-making is related to public participation. Table 15 summarizes 19 selected references on public participation in water resources planning, and Appendix N contains the associated abstracts. Of particular interest is the book by Sargent (1978) describing the various components of public participation. Environmental mediation as a means of resolving conflicts over projects can also be an important element in decision-making, and information is becoming available on this subject (Lake, 1980; and Ostrom, 1976).

Table 15: Summary of References on Public Participation in Water Resources Planning

Authors (Year)	Comments
Albert (1978)	Seminar on education of water resources planners and managers for effective public participation.
Arnett and Johnson (1976)	Geographical-based differences in perceptions around a potential Kentucky reservoir project.
Brown (1979)	210 abstracts on citizen perceptions of water resources projects and programs.

Table 15: (continued)

Authors (Year)	Comments
Bultena, Rogers and Conner (1975)	Study of public knowledge about a water resource development issue.
Dinius (1981)	Development and use of a Visual Perception Test by the general public in evaluating water quality.
Edgmon (1979)	Study of citizen participation approaches used in the Urban Studies Program of three U.S. Army Corps of Engineers District Offices.
Ertel and Koch (1977)	Evaluation of three public participation programs conducted by the New England River Basins Commission.
Ertel (1979)	Identification of training needs for fulfilling public participation responsibilities in water resources planning and development of programs to meet the needs.
Fusco (1980)	Seven elements of a methodology for public participation, including citizen attitude surveys.
Lake (1980)	Book describing several approaches for environmental mediation.
Lehmann (1978c)	86 abstracts on water quality modeling for hydrological and limnological systems.
Ortolano and Wagner (1977)	Evaluation of public involvement techniques on a flooding study in California.
Ostrom (1976)	Review of conflict resolution models in water resources management.
Potter and Norville (1979)	Citizen perceptions of the effectiveness of public participation.
Potter, Grossman and Taylor (1980)	Descriptive comparisons of environmental perceptions held by community leaders and the general public.
Sargent (1978)	Book describing various facets of public participation in environmental decision-making.

Table 15: (continued)

Authors (Year)	Comments
Schimpeler, Gay and Roark (1977)	Summarizes several techniques for public participation.
Shanley (1976)	Use of citizen advisory committees for a proposed park in Massachusetts.
Silberman (1977)	Public participation program for a flood control project in Minnesota.

ADDITIONAL ISSUES AND INFORMATION

Table 16 summarizes 12 selected references dealing with impact mitigation. Tourbier and Westmacott (1980) prepared a handbook containing descriptions of measures in urban development to prevent, reduce, or ameliorate potential problems that would otherwise adversely effect water resources. These problems consist of runoff increases, decreases in infiltration, and a greater degree of erosion and sedimentation, runoff pollution, and discharge of sewage effluents. Measures are presented in groups and related directly to the problem identified. Each group is preceded by a flow chart that relates measures and can aid in the selection of alternative techniques. Each measure is described and cites characteristics to which its applicability is identified. The application, advantages and disadvantages, design criteria and specifications, cost guidelines and maintenance, and legal implementation of each measure are individually covered.

Table 16: Summary of References on Impact Mitigation Measures

Authors (Year)	Comments
Anton and Bunnell (1976)	Guidelines for minimizing erosion from construction projects.
Darnell (1977)	Impact of construction activities in wetlands.
Deiner (1979)	Mitigation measures for man-induced modifications in estuaries.
Gangstad (1978)	Control of aquatic weeds in rivers and waterways.
Mulla, Majori and Arata (1979)	Impact of mosquito control agents on non-target aquatic biota.

Table 16: (continued)

Authors (Year)	Comments
Ripken, Killen and Gulliver (1977)	Methods for separation of sediment from storm water at construction sites.
Therrien (1982)	Mitigation measures for Canadian water resources projects.
Tourbier and Westmacott (1980)	Handbook of measures to reduce or ameliorate potential problems from urban development that would adversely affect water resources.
U.S. Environmental Protection Agency (1981)	Workshop on the state-of-the-art of chemical, biological, mechanical, and integrated control of aquatic weeds.
Walter, Steenhuis and Haith (1979)	Effects of soil and water conservation practices on minimizing nonpoint source pollution.
Whalen (1977)	Guidance for controlling nonpoint pollution.
Whisler et al. (1979)	Summarizes agricultural management practices in terms of minimizing runoff and sediment production.

Table 17 contains summary comments on 27 selected references dealing with related issues and information. Abstracts for these references are in Appendix P. The majority of the references in Table 17 are related to post-project construction studies (post-EIS audits) to verify predicted impacts. These studies are not only important in verifying predicted impacts, they can also aid in adjusting the prediction approaches used in future studies. A post-development audit of a water resources project in the United Kingdom is described by the PADC Environmental Impact Assessment and Planning Unit (1983). A summary of the adequacy and predictive efficacy of fish and wildlife planning at 20 U.S. Army Corps of Engineers reservoir projects has been prepared (Martin, Prosser and Radonski, 1983). Finally, a growing area of importance in environmental impact studies is the need to identify monitoring activities which should be conducted from the early stages of a project. These monitoring activities can be the basis for a continuing monitoring program during project construction and operation (Marcus, 1979).

Table 17: Summary of References on Related Issues and Information

Authors (Year)	Comments
Ciliberti (1980)	Analysis of the effects of the Libby Hydroelectric power project in Montana.

Table 17: (continued)

Authors (Year)	Comments
Golden et al. (1979)	Data reference book for environmental impact studies.
Hendrey and Barvenik (1979)	Effects of acid rainfall on plant communities in lakes.
Marcus (1979)	Methodology for planning a post-EIS monitoring program.
Martin, Prosser and Radonski (1983)	Summary of 20 case studies to evaluate the adequacy and predictive efficacy of fish and wildlife planning at water reservoir projects.
Ortolano (1984)	Book on environmental planning and decision-making, including chapter on assessing impacts on water resources.
PADC Environmental Impact Assessment and Planning Unit (1983)	Post-development audits of impact prediction methods.
Rau and Wooten (1980)	Comprehensive handbook for conduction of environmental impact studies.
Reuss (1980)	Simulation model for prediction of the most likely effects of rainfall acidity on the leaching of cations from noncalcareous soils.
Sport Fishing Institute (1976)	Effects on fish and wildlife from project in South Dakota.
Sport Fishing Institute (1977)	Fish and wildlife impacts of Ice Harbor Lock and Dam Project.
Sport Fishing Institute (1979a)	Fish and wildlife impacts of Keystone Lake Project.
Sport Fishing Institute (1979b)	Fish and wildlife impacts of Okatibbee Lake Project.
Sport Fishing Institute (1981a)	Fish and wildlife impacts of Dworshak Project.
Sport Fishing Institute (1981b)	Fish and wildlife impacts of Beltzville Reservoir Project.

Table 17: (continued)

Authors (Year)	Comments
Sport Fishing Institute (1981c)	Fish and wildlife impacts of Beaver Reservoir Project.
Sport Fishing Institute (1982a)	Fish and wildlife impacts of Allegheny Reservoir Project.
Sport Fishing Institute (1982b)	Fish and wildlife impacts of Eufaula Reservoir Project.
Sport Fishing Institute (1983a)	Fish and wildlife impacts of Deer Creek Lake Project.
Sport Fishing Institute (1983b)	Fish and wildlife impacts of Pine Flat Lake Reservoir Project.
Sport Fishing Institute (1983c)	Fish and wildlife impacts of Pat Mayse Lake Project.
Sport Fishing Institute (1983d)	Fish and wildlife impacts of J. Percy Priest Reservoir Project.
Sposito, Page and Frink (1980)	Computer model for calculating the effects of acid precipitation on soil leachate quality.
Vick et al. (1976)	Post-impoundment study in Georgia.
Vlachos and Hendricks (1977)	Book on technology assessments for water resources projects.
Watson, Barr and Allenson (1977)	Model for estimating atmospheric contaminant rainout.
Williams et al. (1978)	Summary of data collected on 418 lakes east of the Mississippi River as part of the National Eutrophication Survey.

SUMMARY

The first 15 years following the passing of the National Environmental Policy Act can be characterized as involving the publication of many reference materials for the preparation of environmental impact reports on water resources projects. Based on the fact that 434 pertinent references are included in this book, it can be stated that the technical literature is continuing to grow at a rapid rate relative to information on the conduction of environmental impact studies. The primary need is to begin to more systematically incorporate technical approaches and findings into environmental impact studies for water resources projects. This suggests that there is a continuing need for technology transfer to inform practitioners of the availability and usability of information. This book presents screening-type information on a number of references related to the conduction of environmental impact studies on water resources projects. Detailed information is not included herein; however, summary information on the 434 references is included, and should provide the reader with an opportunity of deciding where he/she might go to obtain additional information.

SELECTED REFERENCES

Abbott, J., "Guidelines for Calibration of STORM", Training Document No. 8, 1977, Hydrologic Engineering Center, U.S. Army Corps of Engineers, Davis, California.

Abu-Zeid, M., "Short and Long-Term Impacts of the River Nile Projects", Water Supply and Management, Vol. 3, No. 4, 1979, pp. 275-283.

Adrian, D.D. et al., "Cost Effective Stream and Effluent Monitoring", Publication No. 118, Sept. 1980, Water Resources Research Center, University of Massachusetts, Amherst, Massachusetts.

Ahlgren, I., "A Dilution Model Applied to a System of Shallow Eutrophic Lakes After Diversion of Sewage Effluents", Archive fur Hydrobiologie, Vol. 89, No. 1/2, June 1980, pp. 17-32.

Ahmad, Y.J., "Irrigation in Arid and Semi-Arid Areas", 1982, United Nations Environment Programme, Nairobi, Kenya.

Ahmed, R. and Schiller, R.W., "A Methodology for Estimating the Loads and Impacts of Non-Point Sources on Lake and Stream Water Quality", Proceedings of a Technical Symposium on Non-Point Pollution Control--Tools and Techniques for the Future, Technical Publication 81-1, Jan. 1981, Interstate Commission on the Potomac River Basin, Rockville, Maryland, pp. 154-162.

Ahmed, S., Husseiny, A.A. and Cho, H.Y., "Formal Methodology for Acceptability Analysis of Alternate Sites for Nuclear Power Stations", Nuclear Engineering Design, Vol. 51, No. 3, Feb. 1979, pp. 361-388.

Albert, H.E., editor, "Education of Water Resources Planners and Managers for Effective Public Participation", Report No. 71, Feb. 1978, Water Resources Research Institute, Clemson University, Clemson, South Carolina.

Allen, K.O. and Hardy, J.W., "Impacts of Navigational Dredging on Fish and Wildlife: A Literature Review", FWS/OBS-80/07, Sept. 1980, U.S. Fish and Wildlife Service, Washington, D.C.

Altshul, D.A., "Guidelines: The Use of Cultural Resource Information in Water Resource Environmental Impact Reports", M.S. Thesis, 1980, Department of Hydrology and Water Resources, University of Arizona, Tucson, Arizona.

Anderson, B.F., "Cascaded Tradeoffs: A Multiple-Objective, Multiple Publics Method for Alternatives Evaluation in Water Resources Planning", Aug. 1981, U.S. Bureau of Reclamation, Denver, Colorado.

Anderson, J.W., "An Assessment of Knowledge Concerning the Fate and Effects of Petroleum Hydrocarbons in the Marine Environment", in: Marine Pollution, Functional Responses, Vernberg, W.D. et al., Editors, 1979, Academic Press, New York, New York, pp. 3-21.

Anton, W.F. and Bunnell, J.L., "Environmental Protection Guidelines for Construction Projects", Journal of American Water Works Association, Vol. 68, No. 12, Dec. 1976, pp. 643-646.

Armaly, B.F. and Lepper, S.P., "Diurnal Stratification of Deep Water Impoundments", Report No. 75-HT-35, 1975, American Society of Mechanical Engineers, New York, New York.

Armstrong, N.E., "Effects of Altered Fresh Water Inflows on Estuarine Systems", Proceedings of the Gulf of Mexico Coastal Ecosystems Workshop, FWS/OBS-80/30, May 1980, U.S. Fish and Wildlife Service, Washington, D.C., pp. 17-31.

Armstrong, N.E. and Wart, Jr., G.H., "Effects of Alternatives of Fresh Water Inflows into Madagorda Bay, Texas", Proceedings of the National Symposium on Fresh Water Inflows to Estuaries, FWS/OBS-81-04, Vol. II, Oct. 1981, U.S. Fish and Wildlife Service, Washington, D.C., pp. 179-196.

Arnett, W.E. and Johnson, S., "Dams and People: Geographic Impact Area Analysis", Research Report No. 97, Sept. 1976, Kentucky Water Resources Research Institute, University of Kentucky, Lexington, Kentucky.

Austin, T.A., Landers, R.Q. and Dougal, M.D., "Environmental Management of Multipurpose Reservoirs Subject to Fluctuating Flood Pools", Technical Completion Report No. ISWRRI-84, June 1978, Water Resources Research Institute, Iowa State University, Ames, Iowa.

Austin, T.A., Riddle, W.F. and Landers, Jr., R.Q., "Mathematical Modeling of Vegetative Impacts from Fluctuating Flood Pools", Water Resources Bulletin, Vol. 15, No. 5, Oct. 1979, pp. 1265-1280.

Baca, R.G. et al., "A Generalized Water Quality Model for Eutrophic Lakes and Reservoirs", Nov. 1974, Battelle Pacific Northwest Laboratory, Richland, Washington.

Baca, R.G. et al., "Water Quality Models for Municipal Water Supply Reservoirs, Part 2. Model Formulation, Calibration and Verification", Jan. 1977a, Battelle Pacific Northwest Laboratory, Richland, Washington.

Baca, R.G. et al., "Water Quality Models for Municipal Water Supply Reservoirs. Part 3. User's Manual", Jan. 1977b, Battelle Pacific Northwest Laboratory, Richland, Washington.

Bailey, G.W. and Nicholson, H.P., "Predicting and Simulating Pesticide Transport from Agricultural Land: Mathematical Model Development and Testing", Symposium on Environmental Transport and Transformation of

Pesticides, Oct. 1976, Tbilis, USSR, EPA/600/9-78-003, Feb. 1978, Environmental Research Laboratory, U.S. Environmental Protection Agency, Athens, Georgia, pp. 30-37.

Ball, R.O. and Church, R.L., "Water Quality Indexing and Scoring", Journal of the Environmental Engineering Division, American Society of Civil Engineers, Vol. 106, No. EE4, Aug. 1980, pp. 757-771.

Baram, R. and Webster, R.D., "Interactive Environmental Impact Computer System (EICS) User Manual", CERL-TR-N-80, Sept. 1979, U.S. Army Construction Engineering Research Laboratory, Champaign, Illinois.

Bastian, D.F., "The Salinity Effects of Deepening the Dredged Channels in the Chesapeake Bay", Report NWS-81-S1, Dec. 1980, U.S. Army Institute for Water Resources, Fort Belvoir, Virginia.

Bella, D.A. and Williamson, K.J., "Simulation of Sulfur Cycle in Estuarine Sediments", Journal of the Environmental Engineering Division, American Society of Civil Engineers, Vol. 106, No. EE1, Feb. 1980, pp. 125-143.

Belyakova, O.V., "Model of the Seasonal Dynamics of an Ecosystem of a Shallow Lake", Water Resources (English Translation), Vol. 7, No. 5, Sept./Oct. 1980, pp. 450-457.

Benke, A.C., Gillespie, D.M. and Parrish, F.K., "Biological Basis for Assessing Impacts of Channel Modification: Invertebrate Production, Drift, and Fish Feeding in a Southeastern Blackwater River", Report No. ERC 06-79, 1979, Environmental Resources Center, Georgia Institute of Technology, Atlanta, Georgia.

Berkes, F., "Some Environmental and Social Impacts of the James Bay Hydroelectric Project, Canada", Journal of Environmental Management, Vol. 12, No. 2, Mar. 1981, pp. 157-172.

Bingham, C.R. et al., "Grab Samplers for Benthic Macroinvertebrates in the Lower Mississippi River", Misc. Paper E-82-3, July 1982, U.S. Army Engineer Waterways Experiment Station, Vicksburg, Mississippi.

Birtles, A.B. and Brown, S.R.A., "Computer Prediction of the Changes in River Quality Regimes Following Large Scale Inter Basin Transfers", Proceedings: Baden Symposium on Modeling the Water Quality of the Hydrological Cycle, IIASA (Laxenburg, Austria) and IAHS (United Kingdom), IAHS-AISH Publication No. 125, Sept. 1978, Reading, England, pp. 288-298.

Biswas, A.K., "Environmental Implications of Water Development for Developing Countries", Water Supply and Management, Vol. 2, No. 4, 1978, pp. 233-297.

Biswas, A.K., "Environment and Water Development in Third World", Journal of the Water Resources Planning and Management Division, American Society of Civil Engineers, Vol. 106, No. WR1, Mar. 1980, pp. 319-332.

Bogucki, D.J. and Gruendling, G.K., "Remote Sensing to Identify, Assess, and Predict Ecological Impact on Lake Champlain Wetlands", Final Report, 1978, State University of New York at Plattsburgh, Plattsburgh, New York.

Bohlen, W.F., Cundy, D.F. and Tramontano, J.M., "Suspended Material Distributions in the Wake of Estuarine Channel Dredging Operations", Estuarine and Coastal Marine Science, Vol. 9, No. 6, Dec. 1979, pp. 699-711.

Bohm, P. and Henry, C., "Cost-Benefit Analysis and Environmental Effects", Ambio, Vol. 8, No. 1, 1979, pp. 18-24.

Bombowna, M., Bucka, H. and Huk, W., "Impoundments and Their Influence on the Rivers Studied by Bioassays", Proceedings: Congress in Denmark 1977, Part 3; Internationale Vereingung fur Theoretische und Angewandte Limnologie, Vol. 20, 1978, Polish Academy of Sciences, Krakow, Poland, pp. 1629-1633.

Booth, R.S., "A Systems Analysis Model for Calculating Radionuclide Transport Between Receiving Waters and Bottom Sediments", Proceedings of the 18th Rochester International Conference on Environmental Toxicity, 1975, Oak Ridge National Laboratory, Oak Ridge, Tennessee.

Booth, W.E., Carubia, P.C. and Lutz, F.C., "A Methodology for Comparative Evaluation of Water Quality Indices", 1976, Worcester Polytechnic Institute, Worcester, Massachusetts.

Bourne, R.G., Day, G.N. and Debo, T.N., "Water Quality Modeling Using Hydrocomp Simulation Programming (HSP)", Proceedings of the 26th Annual Hydraulics Division Specialty Conference on Verification of Mathematical and Physical Models in Hydraulic Engineering, 1978, American Society of Civil Engineers, New York, New York, pp. 358-362.

Bovee, K.D., "The Determination, Assessment, and Design of 'In-Stream Value' Studies for the Northern Great Plains Region", Sept. 1974, Department of Geology, Montana University, Missoula, Montana.

Bovee, K.D. and Cochnauer, T., "Development and Evaluation of Weighted-Criteria, Probability-of-Use Curves for Instream Flow Assessments: Fisheries", Report No. FWS/OBS-77/63, IFIP-3, Dec. 1977, U.S. Fish and Wildlife Service, Fort Collins, Colorado.

Bradt, P.T. and Wieland, III, G.E., "The Impact of Stream Reconstruction and a Gabion Installation on the Biology and Chemistry of a Trout Stream", Completion Report, Jan. 1978, Department of Biology, Lehigh University, Bethlehem, Pennsylvania.

Brandstetter, A. et al., "Water Quality Models for Municipal Water Supply Reservoirs, Part I. Summary", Jan. 1977, Battelle Pacific Northwest Laboratory, Richland, Washington.

Brannon, J.M., "Evaluation of Dredged Material Pollution Potential", Technical Report No. DS-78-6, Aug. 1978, U.S. Army Engineer Waterways Experiment Station, Vicksburg, Mississippi.

Brookman, G.T. et al., "Technical Manual for the Measurement and Modelling of Non-Point Sources at an Industrial Site on a River", EPA/600/7-79/049, Feb. 1979, Industrial Environments Research Laboratory, Research Triangle Park, North Carolina.

Browder, J.A. and Moore, D., "A New Approach to Determining the Quantitative Relationship Between Fishery Production and the Flow of Fresh Water to Estuaries", Proceedings of the National Symposium on Fresh Water Inflow to Estuaries, FWS/OBS-81-04, Vol. I, Oct. 1981, U.S. Fish and Wildlife Service, Washington, D.C., pp. 403-430.

Brown, C.A., Quinn, R.J. and Hammond, K.R., "Scaling Impacts of Alternative Plans", June 1980, Center for Research on Judgment and Policy, University of Colorado, Boulder, Colorado.

Brown, J.A.H. et al., "A Mathematical Model of the Hydrologic Regime of the Upper Nile Basin", Journal of Hydrology, Vol. 51, No. 1-4, May 1981, pp. 97-107.

Brown, R.J., "Public Opinion and Sociology of Water Resource Development (A Bibliography with Abstracts)", U.S. Department of Commerce, NTIS/PS-79/0515/1WP, June 1979, National Technical Information Service, Springfield, Virginia.

Brown, R.L., "Monitoring Water Quality by Remote Sensing", Final Report No. NASA CR 154 259, July 1977, California State Department of Water Resources, Sacramento, California.

Brungs, W.A. and Jones, B.R., "Temperature Criteria for Freshwater Fish: Protocol and Procedures", EPA/600/3-77/061, May 1977, U.S. Environmental Protection Agency, Duluth, Minnesota.

Bryant, J.W., "Modelling for Natural Resource Utilization Analysis", Journal Operations Research Society, Vol. 29, No. 7, July 1978, pp. 667-676.

Budge, A.L., "Environmental Input to Water Resources Selection", Water Science and Technology, Vol. 13, No. 6, 1981, pp. 39-46.

Budweg, F.M., "Reservoir Planning for Brazilian Dams", International Water Power and Dam Construction, Vol. 34, No. 5, May 1982, pp. 48-49.

Buikema, Jr., A.L., McGinniss, M.J. and Cairns, Jr., J., "Phenolics in Aquatic Ecosystems: A Selected Review of Recent Literature", Marine Environmental Research, Vol. 2, No. 2, Apr. 1979, pp. 87-181.

Buikema, Jr., A.L. and Loeffelman, P.H., "Effects of Pumpback Storage on Zooplankton Populations", Proceedings of the Clemson Workshop on Environmental Impacts of Pumped Storage Hydroelectric Operations, FWS/OBS-80/28, Apr. 1980, U.S. Fish and Wildlife Service, Washington, D.C., pp. 109-124.

Bultena, G.L., Rogers, D.L. and Conner, K.A., "Characteristics and Correlates of Public Knowledge About a Water Resource Development Issue", OWRTB-020-IA(9), 1975, Iowa State University, Ames, Iowa.

Burke, H.D., "Bibliography of Manual and Handbooks from Natural Resource Agencies", FWS/OBS-78/22, Mar. 1978, Thorne Ecological Institute, Boulder, Colorado.

Burnham, J.B., Nealey, S.M. and Maynard, W.S., "Method for Integrating Societal and Technical Judgments in Environmental Decision Making", Nuclear Technology, Vol. 25, No. 4, Apr. 1975, pp. 675-681.

Burns, E.A., "Symposium Proceedings of Process Measurements for Environmental Assessment Held at Atlanta, on February 13-15, 1978. Final Task Rept., Apr. 1977-Feb. 1978", EPA/600/7-78/168, Aug. 1978, TRW Systems Group, Redondo Beach, California.

Burns, R.G., "An Improved Sediment Delivery Model for Piedmont Forests", Technical Completion Report No. ERC 03-79, June 1979, Environmental Resources Center, Georgia Institute of Technology, Atlanta, Georgia.

Burton, Jr., G.A., "Microbiological Water Quality of Impoundments: A Literature Review", Misc. Paper E-82-6, Dec. 1982, U.S. Army Engineer Waterways Experiment Station, Vicksburg, Mississipppi.

Byrd, J.E. and Perona, M.J., "The Effect of Recreation on Water Quality", Technical Completion Report, Jan. 1979a, California Water Resources Center, University of California, Davis, California.

Byrd, J.E. and Perona, M.J., "Water Quality Effects of Lead from Recreational Boating", Technical Completion Report, Dec. 1979b, California Water Resources Center, University of California, Davis, California.

Byrd, J.E. and Perona, M.J., "The Temporal Variations of Lead Concentration in a Fresh Water Lake", Water, Air, and Soil Pollution, Vol. 13, No. 2, June 1980, pp. 207-220.

Cairns, Jr., J. and Gruber, D., "A Comparison of Methods and Instrumentation of Biological Early Warning Systems", Water Resources Bulletin, Vol. 16, No. 2, Apr. 1980, pp. 261-266.

Camougis, G., Environmental Biology for Engineers, 1981, McGraw-Hill Book Company, Inc., New York, New York.

Carlson, J.E. and Sargent, M.J., "A Dynamic Regional Impact Analysis of Federal Expenditures of a Water and Related Land Resource Project--The Boise Project of Idaho, Part IV: A Social Impact Analysis of Federal Expenditures on a Water Related Resource Project: Boise Project, Social Subproject", Technical Completion Report, Mar. 1979, Water Resources Research Institute, Idaho University, Kimberly, Idaho.

Carrigan, B., "Water Quality Modelling--Hydrological and Limnological Systems. Volume 3. July 1977-June 1979 (A Bibliography with Abstracts)", 1979, National Technical Information Service, U.S. Department of Commerce, Springfield, Virginia.

Casti, J. et al., "Lake Ecosystems: A Polyhedral Dynamics Representation", Ecological Modeling, Vol. 7, No. 3, Sept. 1979, pp. 223-237.

Cermak, R.J., Feldman, A.D. and Webb, R.P., "Hydrologic Land Use Classification Using Landsat", Technical Paper No. 67, Oct. 1979, U.S. Army Engineers Hydrologic Engineering Center, Davis, California.

Chang, S. and Beard, L.R., "Social Impact Studies: Belton and Stillhouse Hollow Reservoirs", Technical Report No. CRWR-164, June 1979, University of Texas, Austin, Texas.

Charlton, M.N., "Hypolimnion Oxygen Consumption in Lakes: Discussion of Productivity and Morphometry Effects", Canadian Journal of Fisheries and Aquatic Sciences, Vol. 37, No. 10, Oct. 1980, pp. 1531-1539.

Chen, K.Y. et al., "Confined Disposal Area Effluent and Leachate Control Laboratory and Field Investigations", Technical Report No. DS-78-7, Oct. 1978, U.S. Army Engineer Waterways Experiment Station, Vicksburg, Mississippi.

Chiaudani, G. and Pagnotta, R., "Ratio of ATP/Chlorophyll as an Index of Rivers' Water Quality", Proceedings: Congress in Denmark 1977 Part 3; Internationale Vereinigung fur Theoretische und Angewandte Limnologie, Instituto di Ricerca sulle Acque, Rome, Italy, Vol. 20, 1978, pp. 1897-1901.

Chu, W.S. and Yeh, W.W., "Two-Dimensional Tidally Averaged Estuarine Model", Journal of the Hydraulics Division, American Society of Civil Engineers, Vol. 106, No. HY4, Apr. 1980, pp. 501-518.

Ciliberti, Jr., V.A., "Libby Dam Project: Ex-Post Facto Analysis of Selected Environmental Impacts, Mitigation Commitments, Recreation Usage, and Hydroelectric Power Production", Report No. 106, 1980, Water Resources Research Center, Montana State University, Bozeman, Montana.

Cluis, D.A., Couillard, D. and Potvin, L., "A Square Grid Transport Model Relating Land Use Exports to Nutrient Loads in Rivers", Water Resources Research, Vol. 15, No. 3, June 1979, pp. 630-636.

Collotzi, A.W. and Dunham, D.K., "Inventory and Display of Aquatic Habitat", Classification, Inventory, and Analysis of Fish and Wildlife Habitat--The Proceedings of a Natural Symposium, Jan. 24-27, 1977, Phoenix, Arizona, FWS/OBS-78/76, 1978, U.S. Forest Service, Washington, D.C., pp. 533-542.

Colwell, J.E. et al., "Use of Landsat Data to Assess Waterfowl Habitat Quality", Jan. 1978, Environmental Research Institute, University of Michigan, Ann Arbor, Michigan.

Conner, W.G. and Simon, J.L., "The Effects of Oyster Shell Dredging on an Estuarine Benthic Community", Estuarine and Coastal Marine Science, Vol. 9, No. 6, Dec. 1979, pp. 749-758.

Conrad, E.T. and Pack, A.J., "A Methodology for Determining Land Value and Associated Benefits Created from Dredged Material Containment", Technical Report No. D-78-19, June 1978, U.S. Army Engineer Waterways Experiment Station, Vicksburg, Mississippi.

Coughlin, R.E. et al., "Assessing Aesthetic Attributes in Planning Water Resource Projects", Environmental Impact Assessment Review, Vol. 3, No. 4, 1982, pp. 406-416.

Cowardin, L.M. et al., "Classification of Wetlands and Deepwater Habitats of the United States", FWS/OBS-79/31, Dec. 1979, U.S. Fish and Wildlife Service, Washington, D.C.

Damman, W.H., "Mobilization and Accumulation of Heavy Metals in Freshwater Wetlands", Research Project Technical Report, 1979, Institute of Water Resources, Connecticut University, Storrs, Connecticut.

Daneke, G.A. and Priscoli, J.D., "Social Assessment and Resource Policy Lessons from Water Planning", Natural Resources Journal, Vol. 19, No. 2, Apr. 1979, pp. 359-375.

Darnell, R.M., "Minimization of Construction Impacts on Wetlands: Dredge and Fill, Dams, Dikes, and Channelization", Proceedings of the National Wetland Protection Symposium, June 6-8, 1977, Reston, Virginia, 1978a, Texas A and M University, College Station, Texas, pp. 29-36.

Darnell, R.M., "Overview of Major Development Impacts on Wetlands", Proceedings of the National Wetland Protection Symposium, June 6-8, 1977, Reston, Virginia, 1978b, Department of Oceanography, Texas A and M University, College Station, Texas.

Davis, Jr., H.H. and Donigian, Jr., A.S., "Simulating Nutrient Movement and Transformations with the Arm Model", Transactions of the American Society of Agricultural Engineers, Vol. 22, No. 5, Sept.-Oct. 1979, pp. 1081-1086.

Davos, C.A., "A Priority-Tradeoff-Scanning Approach to Evaluation in Environmental Management", Journal of Environmental Management, Vol. 5, No. 3, 1977, pp. 259-273.

deGroot, S.J., "The Potential Environmental Impact of Marine Gravel Extraction in the North Sea", Ocean Management, Vol. 5, 1979a, pp. 233-249.

deGroot, S.J., "An Assessment of the Potential Environmental Impact of Large-Scale Sand-Dredging for the Building of Artificial Islands in the North Sea", Ocean Management, Vol. 5, No. 3, Oct. 1979b, pp. 211-232.

Deudney, D., "Hydropower--An Old Technology for a New Era", Environment, Vol. 23, No. 7, Sept. 1981, pp. 16-20, 37-45.

Diamant, B.Z., "Environmental Repercussions of Irrigation Development in Hot Climates", Environmental Conservation, Vol. 7, No. 1, Spring 1980, pp. 53-58.

Dickens, Jr., R.S. and Hill, C.E., editors, Cultural Resources--Planning and Management, 1978, Westview Press, Boulder, Colorado.

Diener, R.A., "Man-induced Modifications in Estuaries of the Northern Gulf of Mexico: Their Impacts on Fishery Resources and Measures of Mitigation", Proceedings of the Mitigation Symposium: A National Workshop on Mitigating Losses of Fish and Wildlife Habitats, Technical Report RM-65, 1979, U.S. Fish and Wildlife Service, Washington, D.C., pp. 115-120.

Dinius, S.H., "Public Perceptions in Water Quality Evaluation", Water Resources Bulletin, Vol. 17, No. 1, Feb. 1981, pp. 116-121.

Drill, S. et al., "The Environmental Lead Problem: An Assessment of Lead in Drinking Water from a Multi-Media Perspective", EPA/570/9-79/003, May 1979, Mitre Corporation, McLean, Virginia.

Duckstein, L. et al., "Practical Use of Decision Theory to Assess Uncertainties about Actions Affecting the Environment", Completion Report, Feb. 1977, Department of Systems and Industrial Engineering, Arizona University, Tucson, Arizona.

Dunnette, D.A., "A Geographically Variable Water Quality Index Used in Oregon", Journal of the Water Pollution Control Federation, Vol. 51, No. 1, Jan. 1979, pp. 53-61.

Duvel, W.A. et al., "Environmental Impact of Stream Channelization", Water Resources Bulletin, Vol. 12, No. 4, Aug. 1976, pp. 799-812.

Eckhardt, W.T., "Cultural Resource Inventory of Areas Affected by Reject Stream Replacement Projects", July 1979, Westec Services, Inc., San Diego, California.

Edgmon, T.D., "A Systems Resource Approach to Citizen Participation: The Case of the Corps of Engineers", Water Resources Bulletin, Vol. 15, No. 5, Oct. 1979, pp. 1341-1352.

Eichenberger, B.A. and Chen, K.Y., "Methodology for Effluent Water Quality Prediction", Journal of the Environmental Engineering Division, American Society of Civil Engineers, Vol. 106, No. EE1, Feb. 1980, pp. 197-209.

Elgershuizen, J.H., "Some Environmental Impacts of a Storm Surge Barrier", Marine Pollution Bulletin, Vol. 12, No. 8, Aug. 1981, pp. 265-271.

El-Hinnawi, E.E., "The State of the Nile Environment: An Overview", Water Supply and Management, Vol. 4, No. 1-2, 1980, pp. 1-11.

Elkington, J.B., "The Impact of Development Projects on Estuarine and Other Wetland Ecosystems", Environmental Conservation, Vol. 4, No. 2, Summer 1977, pp. 135-144.

Elwood, J.W. and Eyman, L.D., "Test of a Model for Predicting the Body Burden of Trace Contaminants in Aquatic Consumers", Journal of the Fisheries Research Board of Canada, Vol. 33, 1976, pp. 1162-1166.

Engler, R.M., "Impacts Associated with the Discharge of Dredged Material Into Open Water", Proceedings of the Third U.S.-Japan Expert's Meeting on Management of Bottom Sediments Containing Toxic Substances, Report No. EPA-600/3-78-084, 1978, U.S. Environmental Protection Agency, Washington, D.C., pp. 213-223.

Entz, B., "Ecological Aspects of Lake Nasser-Nubia", Water Supply and Management, Vol. 4, No. 1-2, 1980, pp. 67-72.

Environmental Control Technology Corporation, "Analysis of Pollution from Marine Engines and Effects on the Environment", Environmental Protection Technology Series No. EPA-670/2-75-062, June 1975, Ann Arbor, Michigan.

Environmental Resources Limited, "Environmental Health Impact Assessment of Irrigated Agricultural Development Projects", Dec. 1983, World Health Organization Regional Office for Europe, Copenhagen, Denmark.

Erickson, R.E., Linder, R.L. and Harmon, K.W., "Stream Channelization (PL. 83-566) Increased Wetland Losses in the Dakotas", Wildlife Society Bulletin, Vol. 7, No. 2, Summer 1979, pp. 71-78.

Ertel, M.O. and Koch, S.G., "Public Participation in Water Resources Planning: A Case Study and Literature Review", Publication No. 89, July 1977, Water Resources Research Center, University of Massachusetts, Amherst, Massachusetts.

Ertel, M.O., "Identifying and Meeting Training Needs for Public Participation Responsibilities in Water Resources Planning", Publication No. 107, 1979, Water Resources Research Center, University of Massachusetts, Amherst, Massachusetts.

ESSA Environmental and Social Systems Analysts, Ltd., "Review and Evaluation of Adaptive Environmental Assessment and Management", Oct. 1982, Environment Canada, Vancouver, British Columbia.

Fast, A.W. and Hulquist, R.G., "Supersaturation of Nitrogen Gas Caused by Artificial Aeration in Reservoirs", Tech. Report E-82-9, Sept. 1982, U.S. Army Engineer Waterways Experiment Station, Vicksburg, Mississippi.

Felleman, J.P., "Coastal Landforms and Scenic Analysis: A Review", Proceedings, The First Annual Conference of the Coastal Society, Nov. 1975, Arlington, Virginia, State University of New York, College of Environmental Science and Forestry, Syracuse, New York, pp. 203-217.

Feller, M.C., "Effects of Clearcutting and Slash-Burning on Stream Temperature in Southwestern British Columbia", Water Resources Bulletin, Vol. 17, No. 5, Oct. 1981, pp. 863-867.

Fieterse, A.J.H. and Toerien, D.F., "The Phosphorus-Chlorophyll Relationship in Roodeplaat Dam", Water SA (Pretoria), Vol. 4, No. 3, 1978, pp. 105-112.

Fletcher, J.L. and Busnel, R.G., Effects of Noise on Wildlife, 1978, Academic Press, New York, New York.

Flint, R.W., "Responses of Freshwater Benthos to Open-Lake Dredged Spoils Disposal in Lake Erie", Journal of Great Lakes Research, Vol. 5, No. 3-4, 1979, pp. 264-275.

Ford, D.E. and Stefan, H.G., "Thermal Predictions Using Integral Energy Model", Journal of the Hydraulics Division, American Society of Civil Engineers, Vol. 106, No. HY1, Jan. 1980, pp. 39-55.

Fowler, J.M. and Heady, E.O., "Suspended Sediment Production Potential on Undisturbed Forest Land", Journal of Soil and Water Conservation, Vol. 36, No. 1, Jan.-Feb. 1981, pp. 47-50.

Frederickson, L.H., "Floral and Faunal Changes in Low Land Hardwood Forests in Missouri Resulting from Channelization, Drainage, and Impoundment", FWS/OBS-78-91, Jan. 1979, U.S. Fish and Wildlife Service, Washington, D.C.

Freedman, P.L., Canale, R.P. and Pendergast, J.F., "Modeling Storm Overflow Impacts on a Eutrophic Lake", Journal of Environmental Engineering Division, American Society of Civil Engineers, Vol. 106, No. EE2, Apr. 1980, pp. 335-349.

Freeman, P.H., "The Environmental Impact of a Large Tropical Reservoir: Guidelines for Policy and Planning, Based Upon a Case Study of Lake Volta, Ghana, in 1973 and 1974", 1974, Smithsonian Institution, Washington, D.C.

French, P.N. et al., "Water Resources Planning Using Computer Graphics", Journal of the Water Resources Planning and Management Division, American Society of Civil Engineers, Vol. 106, No. WR1, Mar. 1980, pp. 21-42.

French, R.H. and Krenkel, P.A., "Effectiveness of River Models", Water Science and Technology, Vol. 13, No. 3, 1981, pp. 99-113.

Frenet-Robin, M. and Ottmann, F., "Comparative Study of the Fixation of Inorganic Mercury on the Principal Clay Minerals and the Sediments of the Loire Estuary", Estuarine and Coastal Marine Science, Vol. 7, No. 5, Nov. 1978, pp. 425-436.

Fry, J.P. and Pflieger, W.L., "Habitat Scarcity, A Basis for Assigning Unit Values for Assessment of Aquatic Wildlife Habitat", Classification, Inventory, and Analysis of Fish and Wildlife Habitat--The Proceedings of a National Symposium, Jan. 24, 1977, Phoenix, Arizona, FWS/OBS-78/76, 1978, U.S. Forest Service, Washington, D.C., pp. 491-494.

Fusco, S.M., "Public Participation in Environmental Statements", Journal of the Water Resources Planning and Management Division, American Society of Civil Engineers, Vol. 106, No. 1, Mar. 1980, pp. 123-125.

Fusillo, T.V., "Impact of Suburban Residential Development on Water Resources in the Area of Winslow Township, Camden, County, New Jersey", Water Resources Investigation 81-27, 1981, U.S. Geological Survey, Trenton, New Jersey.

Gallopin, G., Lee, T.R. and Nelson, M., "The Environmental Dimension in Water Management: The Case of the Dam at Salto Grande", Water Supply and Management, Vol. 4, No. 4, 1980, pp. 221-241.

Gangstad, E.O., Weed Control Methods for River Basin Management, 1978, CRC Press, West Palm Beach, Florida.

Garzon, C.E., "Water Quality in Hydroelectric Projects--Considerations for Planning in Tropical Forest Regions", Tech. Paper No. 20, 1984, The World Bank, Washington, D.C.

Gaynor, J.D., "Phosphorus Loading Associated with Housing in a Rural Watershed", Journal of Great Lakes Research, Vol. 5, No. 2, 1979, pp. 124-130.

Godden, G.F., Nicol, S.M. and Venn, A.C., "Environmental Aspects of Rural Development with Particular Reference to the Keiskamma River Basin Study", Civil Engineer in South Africa, Vol. 22, No. 5, May 1980, pp. 111-116.

Golden, J. et al., Environmental Impact Data Book, 1979, Ann Arbor Science Publishers, Inc., Ann Arbor, Michigan.

Gonor, J.J. and Kemp, P.F., "Procedures for Quantitative Ecological Assessments in Intertidal Environments", EPA/600/3-78/087, Sept. 1978, School of Oceanography, Oregon State University, Corvallis, Oregon.

Gould, M.S., "A Water Quality Assessment of Development in the Senegal River Basin", Water Resources Bulletin, Vol. 17, No. 3, June 1981, pp. 466-473.

Green, K.A., "A Conceptual Ecological Model for Chesapeake Bay", FWS/OBS-78/69, Sept. 1978, U.S. Fish and Wildlife Service, Washington, D.C.

Grimwood, C. and McGhee, T.J., "Prediction of Pollutant Release Resulting from Dredging", Journal of the Water Pollution Control Federation, Vol. 51, No. 7, July 1979, pp. 1811-1815.

Grizzle, J.M., "Effects of Hypolimnetic Discharge on Fish Health Below a Reservoir", Transactions of the American Fisheries Society, Vol. 110, No. 1, Jan. 1981, pp. 29-43.

Grover, B. and Primus, C., "Investigating Whether a Large Hydro Development Can Be Environmentally Compatible: The Slave River Hydro Feasibility Study", Canadian Water Resources Journal, Vol. 6, No. 3, 1981, pp. 47-62.

Groves, D.H. and Coltharp, G.B., "Remote Sensing of Effects of Land Use Practices on Water Quality", Final Report, May 1977, Department of Forestry, University of Kentucky, Lexington, Kentucky.

Gunnison, D., "Mineral Cycling in Salt Marsh-Estuarine Ecosystems; Ecosystem Structure, Function, and General Compartmental Model Describing Mineral Cycles", Technical Report No. D-78-3, Jan. 1978, U.S. Army Engineer Waterways Experiment Station, Vicksburg, Mississippi.

Gurtz, M.E., Webster, J.R. and Wallace, J.B., "Seston Dynamics in Southern Appalachian Streams: Effects of Clear-cutting", Canadian Journal of Fisheries and Aquatic Sciences, Vol. 37, No. 4, Apr. 1980, pp. 624-631.

Gushue, J.J. and Kreutziger, K.M., "Case Studies and Comparative Analyses of Issues Associated with Productive Land Use at Dredged Material Disposal Sites", Technical Report No. D-77-43, Dec. 1977, Two Volumes, U.S. Army Engineer Waterways Experiment Station, Vicksburg, Mississippi.

Gysi, M., "Energy, Environmental, and Economic Implications of Some Recent Alberta Water Resources Projects", Water Resources Bulletin, Vol. 16, No. 4, Aug. 1980, pp. 676-680.

Hafez, M. and Shenouda, W.K., "The Environmental Impacts of the Aswan High Dam", Proceedings of the United Nations Water Conference on Water Management and Development, 1978, Vol. 1, Part 4, Pergamon Press, New York, New York, pp. 1777-1786.

Hagan, R.M. and Roberts, E.B., "Energy Impact Analysis in Water Project Planning", Journal of the Water Resources Planning and Management Division, American Society of Civil Engineers, Vol. 106, No. WR1, Mar. 1980, pp. 289-302.

Haith, D.A., "A Mathematical Model for Estimating Pesticide Losses in Runoff", Journal of Environmental Quality, Vol. 9, No. 3, July-Sept. 1980, pp. 428-433.

Hansen, D.J., "Impact of Pesticides on the Marine Environment", First American-Soviet Symposium on the Biological Effects of Pollution on Marine Organisms, 20-24 September 1976, Gulf Breeze, Florida, EPA-600/9-78-007, May 1978, Environmental Research Laboratory, U.S. Environmental Protection Agency, Gulf Breeze, Florida, pp. 126-137.

Harper, D.B., "Focusing on Visual Quality of the Coastal Zone", Proceedings, The First Annual Conference of the Coastal Society, Nov. 1975, Arlington, Virginia, State University of New York, College of Environmental Science and Forestry, Syracuse, New York, pp. 218-224.

Harvey, E.J. and Emmett, L.F., "Hydrology and Model Study of the Proposed Prosperity Reservoir, Center Creek Basin, Southwestern Missouri", Geological Survey Water Resources Investigation 80-7, June 1980, U.S. Geological Survey, Rolla, Missouri.

Haugen, R.K., McKim, H.L. and Marlar, T.L., "Remote Sensing of Land Use and Water Quality Relationships--Wisconsin Shore, Lake Michigan", Report No. 76-30, Aug. 1976, Cold Regions Research and Engineering Laboratory, U.S. Department of the Army, Hanover, New Hampshire.

Hazel, C. et al., "Assessment of Effects of Altered Stream Flow Characteristics on Fish and Wildlife, Part B: California, Case Studies", FWS/OBS-76/34, Dec. 1976, U.S. Fish and Wildlife Service, Washington, D.C.

Headrick, M.R., "Effects of Stream Channelization on Fish Populations in the Buena Vista Marsh, Portage County, Wisconsin", Sept. 1976, U.S. Fish and Wildlife Service, Stevens Point, Wisconsin.

Hefny, K., "Land Use and Management Problems in the Nile Delta", Nature and Resources, Vol. 18, No. 2, Apr./June 1982, pp. 22-27.

Hellawell, J.M., Biological Surveillance of Rivers: A Biological Monitoring Handbook, 1978, Water Research Center, Stevenage, England.

Hendrey, G.R. and Barvenik, F.W., "Impacts of Acid Precipitation on Decomposition and Plant Communities in Lakes", CONF-7805164-1, 1978, Brookhaven National Laboratory, Upton, New York.

Herzog, Jr., H.W., "Environmental Assessment of Future Production-Related Technological Change: 1970-2000 (An Input-Output Approach)", Technological Forecasting, Vol. 5, No. 1, 1973, pp. 75-90.

Hill, D., "A Modeling Approach to Evaluate Tidal Wetlands", Transactions 41st North American Wildlife and Natural Resource Conference, Mar. 21-25, 1976, Washington, D.C., 1976a, Wildlife Management Institute, Washington, D.C., pp. 105-117.

Hill, D., "A Resource Allocation Model for the Evaluation of Alternatives in Section 208 Planning Considering Environmental, Social and Economic Effects", Proceedings of the Conference on Environmental Modeling and Simulation, Apr. 19-22, 1976, Cincinnati, Ohio, EPA 600/9-76-016, July 1976b, U.S. Environmental Protection Agency, Washington, D.C., pp. 401-406.

Hirschberg, R.I., Goodling, J.S. and Maples, G., "The Effects of Diurnal Mixing on Thermal Stratification of Static Impoundments", Water Resources Bulletin, Vol. 12, No. 6, Dec. 1976, pp. 1151-1159.

Hitchcock, H., "Analytical Review of Research Reports on Social Impacts of Water Resources Development Projects", IWR Contract Report 77-3, Mar. 1977, Program of Policy Studies in Science and Technology, George Washington University, Washington, D.C.

Hodgins, D.B., Wisner, P.E. and McBean, E.A., "A Simulation Model for Screening a System of Reservoirs for Environmental Impact", Canadian Journal of Civil Engineering, Vol. 4, No. 1, Mar. 1977, pp. 1-9.

Hoeppel, R.E., "Contaminant Mobility in Diked Containment Areas", Proceedings of the 5th United States-Japan Experts Meeting on Management of Bottom Sediments Containing Toxic Substances, EPA-600/9-80-044, Sept. 1980, U.S. Environmental Protection Agency, Washington, D.C., pp. 175-207.

Hoffman, W.L., "A Socio-Economic Feasibility Study of the Proposed Rochester Dam", Technical Assistance Report, Nov. 1977, U.S. Department of Commerce, Washington, D.C.

Holliday, B.W., "Processes Affecting the Fate of Dredged Material", Technical Report No. DS-78-2, Aug. 1978, U.S. Army Engineer Waterways Experiment Station, Vicksburg, Mississippi.

Holliday, B.W., Johnson, B.H. and Thomas, W.A., "Predicting and Monitoring Dredged Material Movement", Technical Report No. DS-78-3, Dec. 1978, U.S. Army Engineer Waterways Experiment Station, Vicksburg, Mississippi.

Hoopes, J.A. et al., "Selective Withdrawal and Heated Water Discharge: Influence on the Water Quality of Lakes and Reservoirs, Part II--Induced Mixing with Submerged, Heated Water Discharge", Technical Report No. WIS SRC 79-04, 1979, Water Resources Center, University of Wisconsin, Madison, Wisconsin.

Hoover, T.B., "Inorganic Species in Water: Ecological Significance and Analytical Needs, A Literature Review", EPA/600/3-78/064, July 1978, U.S. Environmental Protection Agency, Environmental Research Laboratory, Athens, Georgia.

Hopkinson, Jr., C.S. and Day, Jr., J.W., "Modeling the Relationship Between Development and Storm Water and Nutrient Runoff", Environmental Management, Vol. 4, No. 4, July 1980, pp. 315-324.

Horst, T.J., "A Mathematical Model to Assess the Effects of Passage of Zooplankton on Their Respective Populations", Proceedings of the Clemson Workshop on Environmental Impacts of Pumped Storage Hydroelectric Operations, FWS/OBS-80/28, Apr. 1980, U.S. Fish and Wildlife Service, Washington, D.C., pp. 177-189.

House, M. and Ellis, J.B., "Water Quality Indices: An Additional Management Tool", Water Science and Technology, Vol. 13, No. 7, 1981, pp. 413-423.

Huang, T., "Changes in Channel Geometry and Channel Capacity of Alluvial Streams Below Large Impoundment Structures", M.S. Thesis, 1979, Department of Civil Engineering, University of Kansas, Lawrence, Kansas.

Huang, Y.H. and Gaynor, R.K., "Effects of Stream Channel Improvements on Downstream Floods", Research Report No. 102, Jan. 1977, Kentucky Water Resources Research Institute, University of Kentucky, Lexington, Kentucky.

Huber, W.C. and Brezonik, P.L., "Water Budget and Projected Water Quality and Proposed Man-Made Lakes Near Estuaries in the Marco Island Area, Florida", Proceedings of the National Symposium of Fresh Water Inflow to Estuaries, FWS/OBS-81-04, Vol. I, Oct. 1981, U.S. Fish and Wildlife Service, Washington, D.C., pp. 241-251.

Hundemann, A.S., "Remote Sensing Applied to Environmental Pollution Detection and Management (A Bibliography with Abstracts)", NTIS/PS-78/9789/4WP, Aug. 1978, National Technical Information Service, U.S. Department of Commerce, Springfield, Virginia.

Hussong, D. et al., "Microbial Impact of Canada Geese (Branta Canadensis) and Whistling Swans (Cygnus Columbianum Columbianus) on Aquatic Ecosystems", Applied and Environmental Microbiology, Vol. 37, No.1, Jan. 1979, pp. 14-20.

Hyman, M.A.M., Lorda, E. and Saila, S.B., "A Standard Program for Environmental Impact Assessment: Phase I--Ichthyoplankton Sampling", Proceedings of Program Review of Environmental Effects of Energy Related Activities on Marine/Estuarine Ecosystems, Report No. EPA-600/7-77-111, Oct. 1977, U.S. Environmental Protection Agency, Washington, D.C., pp. 153-159.

Ibbotson, B. and Adams, B.J., "Formulation and Testing of a New Water Quality Index", Water Pollution Research in Canada 1977, Proceedings of Twelfth Canadian Symposium on Water Pollution Research, 1977, Toronto University, Department of Civil Engineering, Toronto, Ontario, Canada, pp. 101-119.

Inhaber, H., Environmental Indices, John Wiley and Sons, Inc., 1976, New York, New York.

Interim Committee for Coordination of Investigations of the Lower Mekong Basin, Environmental Impact Assessment--Guidelines for Application to Tropical River Basin Development, 1982, ESCAP, Bangkok, Thailand.

Interstate Commission on the Potomac River Basin, Proceedings of a Technical Symposium on Non-Point Pollution Control--Tools and Techniques for the Future, Technical Publication 81-1, Jan. 1981, Rockville, Maryland.

Iwamoto, R. et al., "Sediment and Water Quality: A Review of the Literature Including a Suggested Approach for Water Quality Criteria with Summary of Workshop and Conclusions and Recommendations", EPA/910/9-78/048, Feb. 1978, U.S. Environmental Protection Agency, Washington, D.C.

Jacobs, F. and Grant, G.C., "Guidelines for Zooplankton Sampling in Quantitative Baseline and Monitoring Programs", EPA/600-3-78/026, Feb. 1978, Virginia Institute of Marine Science, Gloucester Point, Virginia.

James, W.P., Woods, C.E. and Blanz, R.E., "Environmental Evaluation of Water Resources Development", Completion Report TR-76, July 1976, Texas Water Resources Institute, Texas A and M University, College Station, Texas.

JBF Scientific Corporation, "Dredge Disposal Study, San Francisco Bay and Estuary. Appendix M--Dredging Technology", Sept. 1975, U.S. Army Corps of Engineers, San Francisco, California.

Jennings, M.E., "Characterization of Fresh Water Inflow Modification to Estuaries Resulting from River Basin Development", Proceedings of the National Symposium on Fresh Water Inflow to Estuaries, FWS/OBS-81-04, Vol. II, Oct. 1981, U.S. Fish and Wildlife Service, Washington, D.C., pp. 375-384.

Jewell, T.K., Adrian, D.D. and DiGiano, F.A., "Urban Storm Water Pollutant Loadings", Publication No. 113, 1980, Water Resources Research Center, University of Massachusetts, Amherst, Massachusetts.

Johanson, E.E., Bowen, S.P. and Henry, G., "State-of-the-Art Survey and Evaluation of Open-Water Dredged Material Placement Methodology", Contract Report No. D-76-3, Apr. 1976, U.S. Army Engineer Waterways Experiment Station, Vicksburg, Mississippi.

Johanson, R.C. and Leytham, K.M., "Modeling Sediment Transport in Natural Channels", Watershed Research in Eastern North America, A Workshop to Compare Results, Volume II, Feb. 28-Mar. 3, 1977, Report No. NSF/RA-770255, 1977, Chesapeake Bay Center for Environmental Studies, Edgewater, Maryland, pp. 861-885.

Johnson, S.J., Krinitzsky, E.L. and Dixon, N.A., "Reservoirs and Induced Seismicity at Corps of Engineers Projects", Miscellaneous Paper S-77-3, Jan. 1977, U.S. Army Engineer Waterways Experiment Station, Vicksburg, Mississippi.

Johnston, Jr., S.A., "Estuarine Dredge and Fill Activities: A Review of Impacts", Environmental Management, Vol. 5, No. 5, Sept. 1981, pp. 427-440.

Jorgensen, S.E., "Water Quality and Environmental Impact Model of the Upper Nile Basin", Water Supply and Management, Vol. 4, No. 3, 1980, pp. 147-153.

Kadlec, J.A., "Effects of a Drawdown on a Water Fowl Impoundment", Ecology, Vol. 43, No. 2, Spring 1962, pp. 267-281.

Karim, F., Croley, II, T.E. and Kennedy, J.F., "A Numerical Model for Computation of Sedimentation in Lakes and Reservoirs", Completion Report No. 105, 1979, Water Resources Research Institute, Iowa State University, Ames, Iowa.

Kay, D. and McDonald, A., "Reduction of Coliform Bacteria in Two Upland Reservoirs: The Significance of Distance Decay Relationships", Water Research, Vol. 14, No. 4, 1980, pp. 305-318.

Keeney, D.R., "A Prediction of the Quality of Water in a Proposed Impoundment in Southwestern Wisconsin, USA", Environmental Geology, Vol. 2, No. 6, 1978, pp. 341-349.

Keeney, R.L., "Preference Models of Environmental Impact", IIASA-RM-76-4, Jan. 1976, International Institute for Applied Systems Analysis, Laxenburg, Austria.

Keilani, W.M., Peters, R.H. and Reynolds, P.J., "A Water Quality Economic Index", Proceedings of the 9th Canadian Symposium on Water Pollution Research, 1974, Department of the Environment, Ottawa, Ontario, Canada, pp. 1-24.

Kelly, D.M., Underwood, J.F. and Thirumurthi, D., "Impact of Construction of a Hydroelectric Project on the Water Quality of Five Lakes in Nova Scotia", Canadian Journal of Civil Engineering, Vol. 7, No. 1, 1980, pp. 173-184.

Kemp, W.H. and Boynton, W.R., "Integrating Scientific Data into Environmental Planning and Impact Analysis, General Methodology and a Case Study", The Environmental Impact of Freshwater Wetland Alterations on Coastal Estuaries, Conference held at Savannah, Georgia on June 23, 1976, Florida University, Gainesville, Florida, pp. 61-86.

Kenyon, G.F., "The Environmental Effects of Hydroelectric Projects", Canadian Water Resources Journal, Vol. 6, No. 3, 1981, pp. 309-314.

Kessler, F.M. et al., "Construction-Site Noise Control Cost-Benefit Estimating Procedures", CERL-IR-N-36, Jan. 1978, U.S. Army Construction Engineering Research Laboratory, Champaign, Illinois.

King, D.L., "Environmental Effects of Hydraulic Structures", Journal of Hydraulics Division, American Society of Civil Engineers, Vol. 104, No. 2, Feb. 1978, pp. 203-221.

King, T.F., "The Archaeological Survey: Methods and Uses", 1978, Heritage Conservation and Recreation Service, U.S. Department of the Interior, Washington, D.C.

Klein, C.J. et al., "Assessment Methodologies for Fresh Water Inflows to Chesapeake Bay", Proceedings of the National Symposium on Fresh Water Inflow to Estuaries, FWS/OBS-81-004, Vol. I, Oct. 1981, U.S. Fish and Wildlife Service, Washington, D.C., pp. 185-199.

Klimas, C.V., "Effects of Permanently Raised Water Tables on Forest Overstory Vegetation in the Vicinity of the Tennessee-Tombigbee Waterway", Misc. Paper E-82-5, Aug. 1982, U.S. Army Engineer Waterways Experiment Station, Vicksburg, Mississippi.

LaBounty, J.F. and Roline, R.A., "Studies of the Effects of Operating the Mt. Elbert Pumped Storage Powerplant", Proceedings of the Clemson Workshop on Environmental Impacts of Pumped Storage Hydroelectric Operations, FWS/OBS-80/28, Apr. 1980, U.S. Fish and Wildlife Service, Washington, D.C., pp. 54-66.

Lake, L.M., editor, Environmental Mediation, 1980, Westview Press, Boulder, Colorado.

Landin, M.C., "A Selected Bibliography of the Life Requirements of Colonial Nesting Waterbirds and Their Relationship to Dredged Material Islands", Misc.

Paper No. D-78-5, 1978, U.S. Army Engineer Waterways Experiment Station, Vicksburg, Mississippi.

Landwehr, J.M., "A Statistical View of a Class of Water Quality Indices", Water Resources Research, Vol. 15, No. 2, Apr. 1979, pp. 460-468.

Larson, F.C., "The Impact of Urban Stormwater on the Water Quality Standards of a Regulated Reservoir", Research Report No. 62, Mar. 1978, Water Resources Research Center, University of Tennessee, Knoxville, Tennessee.

Laskowski-Hoke, R.A. and Prater, B.L., "Dredged Material Evaluations: Correlations Between Chemical and Biological Evaluation Procedures", Journal of the Water Pollution Control Federation, Vol. 53, No. 7, July 1981, pp. 1260-1262.

Lauria, D.T. and O'Melia, C.R., "Nutrient Models for Engineering Management of Pamlico Estuary, North Carolina", Report No. 146, July 1980, Water Resources Research Institute, University of North Carolina, Raleigh, North Carolina.

Leatherberry, E.C., "River Amenity Evaluation: A Review and Commentary", Water Resources Bulletin, Vol. 15, No. 5, Oct. 1979, pp. 1281-1292.

Lee, C.D., Wang, S.B. and Kuo, C.L., "Benthic Macroinvertebrate and Fish as Biological Indicators of Water Quality, With Reference to Community Diversity Index", Water Pollution Control in Developing Countries. Proceedings of the International Conference, Held at Bangkok, Thailand, Feb. 1978, 1978, Pergamon Press, Inc., New York, New York, pp. 233-238.

Lehmann, E.J., "Public Opinion and Sociology of Water Resource Development (A Bibliography with Abstracts)", NTIS/PS-78/0437/OWP, May 1978a, National Technical Information Service, U.S. Department of Commerce, Springfield, Virginia.

Lehmann, E.J., "Water Quality Modelling--Hydrological and Limnological Systems, Volume 2, 1975-June 1977 (A Bibliography with Abstracts)", June 1978b, National Technical Information Service, U.S. Department of Commerce, Springfield, Virginia.

Lehmann, E.J., "Water Quality Modelling--Hydrological and Limnological Systems. Volume 3, July 1977-June 1978 (A Bibliography with Abstracts)", 1978c, National Technical Information Service, U.S. Department of Commerce, Springfield, Virginia.

Lehmann, E.J., "Dredging: Environmental and Biological Effects (Citations from the Engineering Index Data Base)", Dec. 1979, National Technical Information Service, U.S. Department of Commerce, Springfield, Virginia.

Leland, H.V., Luoma, S.N. and Fielden, J.M., "Bioaccumulation and Toxicity of Heavy Metals and Related Trace Elements", Journal of the Water Pollution Control Federation, Vol. 51, No. 6, June 1979, pp. 1592-1616.

Lewke, R.E. and Buss, I.O., "Impacts of Impoundment to Vertebrate Animals and Their Habitats in the Snake River Canyon, Washington", Northwest Science, Vol. 51, No. 4, 1977, pp. 219-270.

Lewke, R.E., "Dams and Wildlife", Passenger Pigeon, Vol. 40, No. 3, Fall 1978, pp. 429-442.

Liebetrau, A.M., "Water Quality Sampling: Some Statistical Considerations", Water Resources Research, Vol. 15, No. 6, Dec. 1979, pp. 1717-1725.

Lincoln, D.R. and Rubin, E.S., "Cross-Media Environmental Impacts of Coal-Fired Power Plants: An Approach Using Multi-Attribute Utility Theory", IEEE Transactions for Systematic Management of Cybernetics, Vol. SMC-9, No. 5, May 1979, pp. 285-289.

Linton, T.L. and Appan, S.G., "A Dynamic Methodology for Characterizing and Monitoring Estuarine Ecosystems", Proceedings of the National Symposium on Fresh Water Inflow to Estuaries, FWS/OBS-81-04, Vol. II, Oct. 1981, U.S. Fish and Wildlife Service, Washington, D.C., pp. 448-462.

Livingstone, I. and Hazlewood, A., "The Analysis of Risk in Irrigation Projects in Developing Countries", Oxford Bulletin of Economics and Statistics, Vol. 41, No. 1, Feb. 1979, pp. 21-35.

Loftis, J.C. and Ward, R.C., "Sampling Frequency Selection for Regulatory Water Quality Monitoring", Water Resources Bulletin, Vol. 16, No. 3, June 1980a, pp. 501-507.

Loftis, J.C. and Ward, R.C., "Water Quality Monitoring--Some Practical Sampling Frequency Considerations", Environmental Management, Vol. 4, No. 6, Nov. 1980b, pp. 521-526.

Loran, B., "Quantitative Assessment of Environmental Impact", Journal of Environmental Systems, Vol. 5, No. 4, 1975, pp. 247-256.

Lusby, G.C., "Effects of Grazing on Runoff and Sediment Yield from Desert Rangeland at Badger Wash in Western Colorado, 1953-73", Water Supply Paper 1532-I, 1979, U.S. Geological Survey, Washington, D.C.

Lynch, J.A., Corbett, E.S. and Sopper, W.E., "Evaluation of Management Practices of the Biological and Chemical Characteristics of Streamflow and Forested Water Sheds", 1980, Institute for Research on Land and Water Resources, Pennsylvania State University, University Park, Pennsylvania.

Maki, T.E., Hazel, W. and Weber, A.J., "Effects of Stream Channelization on Bottomland and Swamp Forest Ecosystems", Completion Report, Oct. 1975, School of Forest Resources, North Carolina State University, Raleigh, North Carolina.

Malins, D.C., "Effects of Petroleum on Arctic and Subarctic Environments and Organisms, Volume 1: Nature and Fate of Petroleum, Volume 2: Biological Effects", 1977, Academic Press, New York, New York.

Manning, R.E., "Impacts of Recreation on Riparian Soils and Vegetation", Water Resources Bulletin, Vol. 15, No. 1, Feb. 1979, pp. 30-43.

Marcus, L.G., "Methodology for Post-EIS (Environmental Impact Statement) Monitoring", Circ. No. 782, 1979, U.S. Geological Survey, Washington, D.C.

Martin, C.W., Noel, D.S. and Federer, C.A., "The Effect of Forest Clear-Cutting in New England on Stream Water Chemistry and Biology", Research Report 34, July 1981, Water Resources Research Center, University of New Hampshire, Durham, New Hampshire.

Martin, R.G., Prosser, N.S. and Radonski, G.C., "Adequacy and Predictive Value of Fish and Wildlife Planning Recommendations at Corps of Engineers Reservoir Projects", Dec. 1983, Sport Fishing Institute, Washington, D.C.

Massoglia, M.F., "Dredging in Estuaries--A Guide for Review of Environmental Impact Statements", Report No. NSF/RA-770284, 1977, Oregon State University, Corvallis, Oregon.

Mather, J.R., "The Influence of Land Use Change on Water Resources", June 1979, Water Resources Center, University of Delaware, Newark, Delaware.

Matter, W. et al., "Movement, Transport, and Scour of Particulate Organic Matter and Aquatic Invertebrates Downstream from a Peaking Hydropower Project", Tech. Report E-83-12, May 1983, U.S. Army Engineer Waterways Experiment Station, Vicksburg, Mississippi.

Maurer, D. et al., "Vertical Migration and Mortality of Benthos in Dredged Material, Part I: Mollusca", Marine Environmental Research, Vol. 4, No. 4, 1981, pp. 299-319.

McClellan, B.E. and Frazer, K.J., "An Environmental Study of the Origin, Distribution, and Bioaccumulation of Selenium in Kentucky and Barkley Lakes", Research Report No. 122, 1980, Water Resources Research Institute, University of Kentucky, Lexington, Kentucky.

McCuen, R.H. et al., "Estimates of Nonpoint Source Pollution by Mathematical Modeling", Technical Report No. 43, Mar. 1978, Maryland Water Resources Research Center, University of Maryland, College Park, Maryland.

McCuen, R.H., Cook, D.E. and Powell, R.L., "Water Quality Projections: Preimpoundment Case Study", Water Resources Bulletin, Vol. 16, No. 1, Feb. 1980, pp. 79-85.

McNeeley, R.N., Neimanis, V.P. and Dwyer, L., "Water Quality Sourcebook: A Guide to Water Quality Parameters", 1979, Water Quality Branch, Department of the Environment, Ottawa, Ontario, Canada.

Meinholz, T.L. et al., "Verification of the Water Quality Impacts of Combined Sewer Overflow", EPA-600/2-79-155, Dec. 1979, U.S. Environmental Protection Agency, Washington, D.C.

Meyers, C.D., "Energetics: Systems Analysis with Application to Water Resources Planning and Decision Making", IWR Contract Report 77-6, Dec. 1977, U.S. Army Engineer Institute for Water Resources, Fort Belvoir, Virginia.

Michalson, E.L., "An Attempt to Quantify the Esthetics of Wild and Scenic Rivers in Idaho", Proceedings: River Recreation Management and Research Symposium, Jan. 24-27, 1977, Minneapolis, Minnesota, U.S. Forest Service General Technical Report No. NC-28, U.S. Forest Service, Department of Agriculture, St. Paul, Minnesota, pp. 320-328.

Micklin, P.P., "International Environmental Implications of Soviet Development of the Volga River", Human Ecology, Vol. 5, No. 2, June 1977, pp. 113-135.

Miracle, R.D. and Gardner, Jr., J.A., "Review of the Literature on the Effects of Pumped Storage Operations on Ichthyofauna", Proceedings of the Clemson Workshop on Environmental Impacts of Pumped Storage Hydroelectric Operations, FWS/OBS-80/28, Apr. 1980, U.S. Fish and Wildlife Service, Washington, D.C., pp. 40-53.

Morrison, R.D. and Yu, K.Y., "Impact of Dredged Material Disposal Upon Ground Water Quality", Ground Water, Vol. 19, No. 3, May/June 1981, pp. 265-270.

Motayed, A.K., "Alternative Evaluation of Power Plant Sites", Journal of the Energy Division, American Society of Civil Engineers, Vol. 106, No. EY2, Oct. 1980, pp. 229-234.

Mulla, M.S., Majori, G. and Arata, A.A., "Impact of Biological and Chemical Mosquito Control Agents on Non-Target Biota in Aquatic Ecosystems", Residue Reviews, Vol. 71, 1979, pp. 121-173.

Mulvihill, E.L. et al., "Biological Impacts of Minor Shoreline Structures on the Coastal Environment: State-of-the-Art Review, Volume I", FWS/OBS-77-51, Mar. 1980, U.S. Fish and Wildlife Service, Washington, D.C.

Munter, J.A. and Anderson, M.P., "The Use of Ground Water Flow Models for Estimating Lake Seepage Rates", Ground Water, Vol. 19, No. 6, Nov./Dec. 1981, pp. 608-616.

Najarian, T.O. and Harleman, D.R.F., "A Real Time Model of Nitrogen-Cycle Dynamics in an Estuarine System", Progress in Water Technology, Vol. 8, No. 4-5, 1977, pp. 323-345.

National Marine Fisheries Service, "Physical, Chemical and Biological Effects of Dredging in the Thames River (CT) and Spoil Disposal at the New London (CT) Dumping Ground", Final Report, Apr. 1977, Division of Environmental Assessment, Highlands, New Jersey.

National Oceanic and Atmospheric Administration, "Coastal Facility Guidelines: A Methodology for Development with Environmental Case Studies on Marinas and Power Plants", Working Paper, Aug. 1976, Rockville, Maryland.

Nelson, T.L., Warnick, C.C. and Potratz, C.J., "A Dynamic Regional Impact Analysis of Federal Expenditures of a Water and Related Land Resource Project --The Boise Project of Idaho, Part III: Economic Scenario of the Boise Region 'Without' a Federal Irrigation Project, Economics Subproject", Technical Completion Report, Mar. 1979, Water Resources Research Institute, University of Idaho, Moscow, Idaho.

Nelson, W. et al., "Assessment of Effects of Stored Stream Flow Characteristics on Fish and Wildlife, Part A: Rocky Mountains and Pacific Northwest (Executive Summary)", Publication No. FWS/OBS-76/28, Aug. 1976, Environmental Control, Inc., Rockville, Maryland.

Neu, H.J., "Man-Made Storage of Water Resources--A Liability to the Ocean Environment, Part II", Marine Pollution Bulletin, Vol. 13, No. 2, Feb. 1982, pp. 44-47.

Nieman, T.J., "Assessing the Visual Quality of the Coastal Zone", Proceedings: The First Annual Conference of the Coastal Society, Nov. 1975, Arlington, Virginia, State University of New York, College of Environmental Science and Forestry, Syracuse, New York, pp. 247-251.

Nix, J., "Distribution of Trace Elements in a Warm Water Release Impoundment", Oct. 1980, Water Resources Research Center, University of Arkansas, Fayetteville, Arkansas.

Noble, R.D., "Analytical Prediction of Natural Temperatures in Rivers", Journal of the Environmental Engineering Division, American Society of Civil Engineers, Vol. 105, No. EE5, Oct. 1979, pp. 1014-1018.

Ocean Data Systems, Inc., "Handbook for Terrestrial Wildlife Habitat Development of Dredged Material", Technical Report No. D-78-37, July 1978, U.S. Army Engineer Waterways Experiment Station, Vicksburg, Mississippi.

O'Connor, D.J., Di Toro, D.M. and Thomann, R.V., "Phytoplankton Models and Eutrophication Problems", Ecological Modeling in a Resource Management Framework, 1975, Resources for the Future, Inc., Washington, D.C., pp. 149-209.

Oglesby, R.T. and Schaffner, W.R., "Phosphorus Loadings to Lakes and Some of Their Responses. Part 2. Regression Models of Summer Phytoplankton Standing Crops, Winter Total P, and Transparency of New York Lakes with Known Phosphorus Loadings", Limnology and Oceanography, Vol. 23, No. 1, Jan. 1978, pp. 135-145.

Olenik, S.C., "A Hierarchical Multiobjective Method for Water Resources Planning", M.S. Thesis, 1978, School of Engineering, Case Western Reserve University, Cleveland, Ohio.

Olivieri, V.P., Kruse, C.W. and Kawata, K., "Micro-organisms in Urban Stormwater", EPA/600/2-77/087, July 1977, U.S. Environmental Protection Agency, Cincinnati, Ohio.

O'Neill, R.V., "Review of Compartmental Analysis in Ecosystem Science", CONF-780839-1, 1978, Oak Ridge National Laboratory, Oak Ridge, Tennessee.

Ongley, E.D. and Broekhoven, L.H., "Data Filtering Techniques and Regional Assessment of Agricultural Impacts Upon Water Quality Southern Onterio", Progress in Water Technology, Vol. 11, No. 6, 1979, pp. 551-577.

Onishi, Y. and Wise, S.E., "Mathematical Modeling of Sediment and Contaminant Transport in the James River Estuary", Proceedings of the 26th Annual Hydraulics Division Specialty Conference on Verification of Mathematical and Physical Models in Hydraulic Engineering, 1978, American Society of Civil Engineers, New York, New York, pp. 303-310.

Onishi, Y., "Sediment-Contaminant Transport Model", Journal of the Hydraulics Division, American Society of Civil Engineers, Vol. 107, No. HY9, Sept. 1981, pp. 1089-1107.

Orlob, G.T., "Mathematical Modeling of Surface Water Impoundments, Volume I, and II", 1977, Resource Management Associates, Lafayette, California.

Ortolano, L. and Wagner, T.P., "Field Evaluation of Some Public Involvement Techniques", Water Resources Bulletin, Vol. 13, No. 6, Dec. 1977, pp. 1131-1139.

Ortolano, L., Environmental Planning and Decision Making, 1984, John Wiley and Sons, New York, New York.

Ostrofsky, M.L. and Duthie, H.C., "An Approach to Modelling Productivity in Reservoirs", Proceedings: Congress in Denmark 1977, Part 3: Internationale Vereingung fur Theoretische und Angewandte Limnologie, Vol. 20, 1978, pp. 1562-1567.

Ostrom, A.R., "A Review of Conflict Resolution Models in Water Resources Management", Workshop on the Vistula and Tisza River Basins, Feb. 11-13, 1975, International Institute for Applied Systems Analysis, Apr. 1976, Laxenburg, Austria, pp. 95-105.

Ott, W.R., Environmental Indices--Theory and Practice, 1978, Ann Arbor Science Publishers, Inc., Ann Arbor, Michigan.

Ozturk, Y.F., "Mathematical Modeling of Dissolved Oxygen in Mixed Estuaries", Journal of the Environmental Engineering Division, American Society of Civil Engineers, Vol. 105, No. EE5, Oct. 1979, pp. 883-904.

PADC Environmental Impact Assessment and Planning Unit, "Post-development Audits to Test the Effectiveness of Environmental Impact Prediction Methods and Techniques", 1983, University of Aberdeen, Aberdeen, Scotland.

Parrish, J.D. et al., "Stream Channelization Modification in Hawaii, Part D: Summary Report", Report No. FWS/OBS-78/19, Oct. 1978, U.S. Fish and Wildlife Service, Hawaii Cooperative Fishery Research Unit, Honolulu, Hawaii.

Pastorok, R.A., Lorenzen, M.W. and Ginn, T.C., "Environmental Aspects of Artificial Aeration and Oxygenation of Reservoirs: A Review of Theory, Techniques, and Experiences", Tech. Report E-82-3, May 1982, U.S. Army Engineer Waterways Experiment Station, Vicksburg, Mississippi.

Pavlou, S.P. et al., "Release, Distribution, and Impacts of Polychlorinated Biphenyls (PCB) Induced by Dredged Material Disposal Activities at a Deep Water Estuarine Site", Proceedings of the 5th United States-Japan Experts Meeting on Management of Bottom Sediments Containing Toxic Substances, EPA-600/9-80-044, Sept. 1980, U.S. Environmental Protection Agency, Washington, D.C., pp. 129-174.

Pennington, C.H. and Baker, J.A., "Environmental Effects of Tennessee-Tombigbee Project Cutoff Bendways", Misc. Paper E-82-4, Aug. 1982, U.S. Army Engineer Waterways Experiment Station, Vicksburg, Mississippi.

Persoone, G. and DePauw, N., "Systems of Biological Indicators for Water Quality Assessment", Biological Aspects of Fresh Water Pollution, O. Ravera, Editor, Pergamon Press, New York, New York, 1978, pp. 39-75.

Peterson, J.H., Clinton, C.A. and Chambers, E., "A Field Test of Environmental Impact Assessment in the Tensas Basin", Proceedings of the 14th Annual Mississippi Water Resources Conference, Sept. 1979, Mississippi State University, Mississippi State, Mississippi, pp. 27-32.

Peterson, S.A., "Dredging and Lake Restoration", Report No. EPA 440/5-79-001, 1979, U.S. Environmental Protection Agency, Corvallis Environmental Research Laboratory, Corvallis, Oregon, pp. 105-144.

Petts, G.E., "Long-Term Consequences of Upstream Impoundment", Environmental Conservation, Vol. 7, No. 4, Winter 1980a, pp. 325-332.

Petts, G.E., "Morphological Changes of River Channels Consequent Upon Headwater Impoundment", Journal of the Institution of Water Engineers and Scientists, Vol. 34, No. 4, July 1980b, pp. 374-382.

Pickering, J.A. and Andrews, R.A., "An Economic and Environmental Evaluation of Alternative Land Development Around Lakes", Water Resources Bulletin, Vol. 15, No. 4, Aug. 1979, pp. 1039-1049.

Ploskey, G.R., "Fluctuating Water Levels in Reservoirs: An Annotated Bibliography on Environmental Effects and Management for Fisheries", Tech. Report E-82-5, May 1982, U.S. Army Engineer Waterways Experiment Station, Vicksburg, Mississippi.

Ploskey, G.R., "A Review of the Effects of Water-level Changes on Reservoir Fisheries and Recommendations for Improved Management", Tech. Report E-83-3, Feb. 1983, U.S. Army Engineer Waterways Experiment Station, Vicksburg, Mississippi.

Polivannaya, M.F. and Sergeyeva, O.A., "Zooplankters as Bioindicators of Water Quality", Hydrobiological Journal, Vol. 14, No. 3, 1978, pp. 39-43.

Pollard, N., "The Gezira Scheme--A Study in Failure", Ecologist, Vol. 11, No. 1, Jan.-Feb. 1981, pp. 21-31.

Possardt, E.E. and Dodge, W.E., "Stream Channelization Impacts on Songbirds and Small Mammals in Vermont", Wildlife Society Bulletin, Vol. 6, No. 1, Spring 1978, pp. 18-24.

Potter, H.R. and Norville, H.J., "Perceptions of Effective Public Participation in Water Resources Decision Making and Their Relationship to Levels of Participation", OWRT-A-043-IND(1), Jan. 1979, Water Resources Research Center, Purdue University, West Lafayette, Indiana.

Potter, H.R., Grossman, G.M. and Taylor, A.K., "Participation in Water Resources Planning: Leader and Non-leader Comparisons", Technical Report No. 107, 1980, Water Resources Research Center, Purdue University, West Lafayette, Indiana.

Prellwitz, D.M., "Effects of Stream Channelization on Terrestrial Wildlife and Their Habitats in the Buena Vista Marsh, Wisconsin", Report FWS/OBS-76-25, Dec. 1976, Wisconsin Cooperative Fishery Research Unit, Stevens Point, Wisconsin.

Provencher, M. and Lamontagne, M.P., "A Method for Establishing a Water Quality Index for Different Uses (IQE)", July 1979, Environmental Protection Services, Montreal, Quebec, Canada.

Radford, P.J. and Joint, I.R., "The Application of an Ecosystem Model to the Bristol Channel and Severn Estuary", Water Pollution Control, Vol. 79, No. 2, 1980, pp. 244-254.

Rahman, M., "Temperature Structure in Large Bodies of Water: Analytical Investigation of Temperature Structure in Large Bodies of Stratified Water", Journal of Hydraulic Research, Vol. 17, No. 3, 1979, pp. 207-215.

Raster, T.E. et al., "Development of Procedures for Selecting and Designing Reusable Dredged Material Disposal Sites", Technical Report No. D-78-22, June 1978, U.S. Army Engineer Waterways Experiment Station, Vicksburg, Mississippi.

Rau, J.G. and Wooten, D.C., Environmental Impact Analysis Handbook, 1980, McGraw-Hill Book Company, New York, New York.

Raymond, H.L., "Effects of Dams and Impoundments on Migrations of Juvenile Chinook Salmon and Steelhead from the Snake River, 1966 to 1975", Transactions of the American Fisheries Society, Vol. 108, 1979, pp. 505-529.

Reuss, J.O., "Simulation of Soil Nutrient Losses Resulting from Rainfall Acidity", Ecological Modeling, Vol. 11, No. 1, Oct. 1980, pp. 15-38.

Reynolds, P.J., "Environmental Indicators in River Basin Management", Proceedings of International Symposium on Hydrological Characteristics of River Basins and the Effects of These Characteristics on Better Waste Management, Tokyo, Japan, Dec. 1-8, 1975, IAHS-AISH Pub. No. 117, 1975, International Association of Hydrological Scientists, Paris, France, pp. 557-569.

Reynolds, P.J. and Ujjainwalla, S.H., "Environmental Implications and Assessments of Hydroelectric Projects", Canadian Water Resources Journal, Vol. 6, No. 3, 1981, pp. 5-19.

Ricci, P.F., Laessig, R.E. and Glaser, E.R., "The Preoperational Effects of a Water-Resources Project on Property Prices", Water Resources Bulletin, Vol. 14, No. 3, June 1978, pp. 524-531.

Ripken, J.F., Killen, J.M. and Gulliver, J.S., "Methods for Separation of Sediment from Storm Water at Construction Sites", EPA-600/2-77-033, 1977, U.S. Environmental Protection Agency, Washington, D.C.

Robbins, J.W.D., "Environmental Impact Resulting from Unconfined Animal Production", EPA-600/2-78-046, Feb. 1978, U.S. Environmental Protection Agency, Washington, D.C.

Roseboom, D.P. et al., "Effect of Bottom Conditions on Eutrophy of Impoundments", Illinois State Water Survey Circular 139, 1979, Illinois State Water Survey, Urbana, Illinois.

Rosendahl, P.C. and Waite, T.D., "Transport Characteristics of Phosphorus in Channelized and Meandering Streams", Water Resources Bulletin, Vol. 14, No. 5, Oct. 1978, pp. 1227-1238.

Ross, B.B., Shanholtz, V.O. and Contractor, D.N., "A Spatially Responsive Hydrologic Model to Predict Erosion and Sediment Transport", Water Resources Bulletin, Vol. 16, No. 3, June 1980, pp. 538-545.

Rubinstein, S. and Horn, R.L., "Risk Analysis in Environmental Studies. I. Risk Analysis Methodology: A Statistical Approach; II. Data Management for Environmental Studies", CONF-780316-8, Mar. 1978, Atomics International Division, Rockwell Hanford Operations, U.S. Department of Energy, Richland, Washington.

Ryner, P.C., "Chicago Lakefront Demonstration Project. Environmental Impact Handbook", 1978, Illinois Coastal Zone Management Program, Chicago, Illinois.

Sargent, F.O. and Berke, P.R., "Planning Undeveloped Lakeshore: A Case Study on Lake Champlain, Ferrisburg, Vermont", Water Resources Bulletin, Vol. 15, No. 3, June 1979, pp. 826-837.

Sargent, H., Fishbowl Management: A Participative Approach to Systemic Management, 1978, John Wiley and Sons, Inc., New York, New York.

Scavia, D. and Robertson, A., Editors, Perspectives on Lake Ecosystem Modeling, 1979, Ann Arbor Science, Ann Arbor, Michigan.

Schillinger, J.E. and Stuart, D.G., "Quantification of Non-Point Water Pollutants from Logging, Cattle Grazing, Mining, and Subdivision Activities", Report No. 93, 1978, Water Resources Research Center, Montana State University, Bozeman, Montana.

Schimpeler, C.C., Gay, M. and Roark, A.L., "Public Participation in Water Quality Management Planning", Handbook of Water Quality Management Planning, 1977, Van Nostrand Reinhold Company, New York, New York, pp. 336-372.

Schnoor, J.L. and O'Connor, D.J., "A Steady State Eutrophication Model for Lakes", Water Research, Vol. 14, No. 11, Nov. 1980, pp. 1651-1665.

Schreiber, J.D. and Rausch, D.L., "Suspended Sediment-Phosphorus Relationships for the Inflow and Outflow of a Flood Detention Reservoir", Journal of Environmental Quality, Vol. 8, No. 4, Oct.-Dec. 1979, pp. 510-514.

Schreiber, J.D., Duffy, P.D. and McClurkin, D.C., "Aqueous and Sediment-Phase Nitrogen Yields from Five Southern Pine Watersheds", Soil Science Society of America, Vol. 44, No. 2, Mar.-Apr. 1980, pp. 401-407.

Schrender, G.F., Rustagi, K.P. and Bare, B.B., "A Computerized System for Wild Land Use Planning and Environmental Impact Assessment", Computers and Operations Research, Vol. 3, No. 2/3, Aug. 1976, pp. 217-228.

Schwind, P.J., "Environmental Impacts of Land Use Change", Journal of Environmental Systems, Vol. 6, No. 2, 1977, pp. 125-145.

Seaver, D.A., "Applications and Evaluation of Decision Analysis in Water Resources Planning", Dec. 1979, Office of Water Research and Technology, U.S. Department of Interior, Washington, D.C.

Sellers, J. and North, R.M., "A Viable Methodology to Implement the Principles and Standards", Water Resources Bulletin, Vol. 15, No. 1, Feb. 1979, pp. 167-181.

Shabman, L. and Bertelson, M.K., "The Use of Development Value Estimates for Coastal Wetland Permit Decisions", Land Economics, Vol. 55, No. 2, May 1979, pp. 213-222.

Shanley, R.A., "Attitudes and Interactions of Citizen Advisory Groups and Governmental Officials in the Water Resources Planning Process", Publication No. 78, Aug. 1976, Massachusetts Water Resources Research Center, University of Massachusetts, Amherst, Massachusetts.

Shapiro, M., Luecks, D.F. and Kuhner, J., "Assessment of the Environmental Infrastructure Required by Large Public and Private Investments", Journal of Environmental Management, Vol. 7, No. 2, Sept. 1978, pp. 157-176.

Shea, G.B. et al., "Aspects of Impact Assessment of Low Fresh Water Inflows to Chesapeake Bay", Proceedings of the National Symposium on Fresh Water Inflows to Estuaries, FWS/OBS-81-04, Vol. I, Oct. 1981, U.S. Fish and Wildlife Service, Washington, D.C., pp. 128-148.

Sicherman, A., "General Methodology and Computer Tool for Environmental Impact Assessment with Two Case Study Examples", Proceedings of the International Conference of the Cybernetics Society, Tokyo and Kyoto, Japan, Nov. 1978, IEEE, Vol. 1, 1978, New York, New York, pp. 638-642.

Silberman, E., "Public Participation in Water Resource Development", Journal of the Water Resources Planning and Management Division, American Society of Civil Engineers, Vol. 103, No. WR1, May 1977, pp. 111-123.

Sloane, B.A. and Dickinson, T.E., "Computer Modeling for the Lake Tahoe Basin: Impacts of Extreme Land Use Policies on Key Environmental Variables", Journal of Environmental Systems, Vol. 9, No. 1, 1979, pp. 39-56.

Slotta, L.S. et al., "An Examination of Some Physical and Biological Impacts of Dredging in Estuaries", Interim Progress Report to the National Science Foundation, Dec. 1974, Oregon State University, School of Oceanography, Corvallis, Oregon.

Smies, M. and Huiskes, A.H., "Holland's Eastern Scheldt Estuary Barrier Scheme: Some Ecological Considerations", Ambio, Vol. 10, No. 4, 1981, pp. 158-165.

Smil, V., "China's Agro-Ecosystem", Agro-Ecosystems, Vol. 7, No. 1, 1981, pp. 27-46.

Smith, R. and Eilers, R.G., "Stream Models for Calculating Pollutional Effects of Stormwater Runoff", EPA-600/2-78-148, Aug. 1978, U.S. Environmental Protection Agency, Cincinnati, Ohio.

Snodgrass, W.J. and Holloran, M.F., "Utilization of Oxygen Models in Environmental Impact Analysis", Proceedings of 12th Canadian Symposium on Water Pollution Research, 1977, McMaster University, Hamilton, Ontario, Canada, pp. 135-156.

Sondheim, M.W., "A Comprehensive Methodology for Assessing Environmental Impact", Journal of Environmental Management, Vol. 6, No. 1, Jan. 1978, pp. 27-42.

Sport Fishing Institute, "Evaluation of Planning for Fish and Wildlife: Lake Sharpe Reservoir Project", Oct. 1976, U.S. Department of the Army, Office of the Chief of Engineers, Washington, D.C.

Sport Fishing Institute, "Evaluation of Planning for Fish and Wildlife at Corp of Engineers Reservoirs--Ice Harbor Lock and Dam Project, Washington", Nov. 1977, Washington, D.C.

Sport Fishing Institute, "Evaluation of Planning for Fish and Wildlife at Corps of Engineers Reservoirs--Keystone Lake Project, Oklahoma", Feb. 1979a, Washington, D.C.

Sport Fishing Institute, "Evaluation of Planning for Fish and Wildlife at Corps of Engineers Reservoirs--Okatibbee Lake Project, Mississippi", Feb. 1979b, Washington, D.C.

Sport Fishing Institute, "Evaluation of Planning for Fish and Wildlife at Corps of Engineers Reservoirs--Dworshak Reservoir Project, Idaho", Feb. 1981a, Washington, D.C.

Sport Fishing Institute, "Evaluation of Planning for Fish and Wildlife at Corps of Engineers Reservoirs--Beltzville Reservoir Project, Pennsylvania", Apr. 1981b, Washington, D.C.

Sport Fishing Institute, "Evaluation of Planning for Fish and Wildlife at Corps of Engineers Reservoirs--Beaver Reservoir Project, Arkansas", Sept. 1981c, Washington, D.C.

Sport Fishing Institute, "Evaluation of Planning for Fish and Wildlife at Corps of Engineers Reservoirs--Allegheny Reservoir Project, Pennsylvania", July 1982a, Washington, D.C.

Sport Fishing Institute, "Evaluation of Planning for Fish and Wildlife at Corps of Engineers Reservoirs--Eufaula Reservoir Project, Oklahoma", Aug. 1982b, Washington, D.C.

Sport Fishing Institute, "Evaluation of Planning for Fish and Wildlife at Corps of Engineers Reservoirs--Deer Creek Lake Project, Ohio", Jan. 1983a, Washington, D.C.

Sport Fishing Institute, "Evaluation of Planning for Fish and Wildlife at Corps of Engineers Reservoirs--Pine Flat Lake Reservoir Project, California", Jan. 1983b, Washington, D.C.

Sport Fishing Institute, "Evaluation of Planning for Fish and Wildlife at Corps of Engineers Reservoirs--Pat Mayse Lake Project, Texas", July 1983c, Washington, D.C.

Sport Fishing Institute, "Evaluation of Planning for Fish and Wildlife at Corps of Engineers Reservoirs--J. Percy Priest Reservoir Project, Tennessee", Sept. 1983d, Washington, D.C.

Sposito, G., Page, A.L. and Frink, M.E., "Effects of Acid Precipitation on Soil Leachate Quality, Computer Calculations", EPA 600/3-80-015, Jan. 1980, U.S. Environmental Protection Agency, Washington, D.C.

Stalnaker, C.B. and Arnette, J.L., "Methodologies for the Determination of Stream Resource Flow Requirements: An Assessment", FWS/OBS-76/03, Apr. 1976, U.S. Fish and Wildlife Service, Washington, D.C.

States, J.B. et al., "A Systems Approach to Ecological Baseline Studies", FWS/OBS-78/21, Mar. 1978, U.S. Fish and Wildlife Service, Fort Collins, Colorado.

Stofan, P.E. and Grant, G.C., "Phytoplankton Sampling in Quantitative Baseline and Monitoring Programs", EPA/600/3-78-025, Feb. 1978, Virginia Institute of Marine Science, Gloucester Point, Virginia.

Stone, J.H. and McHugh, G.F., "Simulated Hydrologic Effects of Canals in Barataria Basin: A Preliminary Study of Cumulative Impacts", Final Report, June 1977, Louisiana State University Center for Wetland Resources, Baton Rouge, Louisiana.

Stone, J.H., Bahr, Jr., L.M. and Way, Jr., J.W., "Effects of Canals on Freshwater Marshes in Coastal Louisiana and Implications for Management", Freshwater Wetlands: Ecological Processes and Management Potential, 1978, Academic Press, New York, New York, pp. 299-320.

Stout, G.E. et al., "Baseline Data Requirements for Assessing Environmental Impact", IIEQ-78-05, May 1978, Institute for Environmental Studies, University of Illinois, Urbana-Champaign, Illinois.

Swanson, G.A. and Meyer, M.I., "Impact of Fluctuating Water Levels on Feeding Ecology of Breeding Blue-Winged Teal", Journal of Wildlife Management, Vol. 41, No. 3, July 1977, pp. 426-433.

Takahasi, Y., "Changes and Processes of Water Resource Development and Flood Control in Post-Second World War Japan", Water Supply and Management, Vol. 6, No. 5, 1982, pp. 375-386.

Tamblyn, T.A. and Cederborg, E.A., "Environmental Assessment Matrix as a Site-Selection Tool--A Case Study", Nuclear Technology, Vol. 25, No. 4, Apr. 1975, pp. 598-606.

Taylor, M.H., "An Indicator-Prediction Model for Ecosystem Parameters of Water Quality", Technical Completion Report, Nov. 1979, Water Resources Center, University of Delaware, Newark, Delaware.

Teskey, R.O. and Hinckley, T.M., "Impact of Water Level Changes on Woody Riparian and Wetland Communities. Volume I: Plant and Soil Responses to Flooding", Office of Biological Services Report 77/58, Dec. 1977a, U.S. Fish and Wildlife Service, Washington, D.C.

Teskey, R.O. and Hinckley, T.M., "Impact of Water Level Changes on Woody Riparian and Wetland Communities. Volume II: The Southern Forest Region", Report No. 77/59, Dec. 1977b, Office of Biological Services, U.S. Fish and Wildlife Service, Washington, D.C.

Teskey, R.O. and Hinckley, T.M., "Impact of Water Level Changes on Woody Riparian and Wetland Communities. Volume III: The Central Forest Region", Report No. 77/60, Dec. 1977c, Office of Biological Services, U.S. Fish and Wildlife Service, Washington, D.C.

Thackston, E.L. and Sneed, R.B., "Review of Environmental Consequences of Waterway Design and Construction Practices as of 1979", Tech. Report E-82-4, Apr. 1982, U.S. Army Engineer Waterways Experiment Station, Vicksburg, Mississippi.

Therrien, D., "Environmental Corrective Programs at the La Grande Complex", Canadian Water Resources Journal, Vol. 7, No. 2, 1982, pp. 147-162.

Thomann, R.V., "An Analysis of PCB in Lake Ontario Using a Size-Dependent Food Chain Model", Perspectives in Lake Ecosystem Modeling, 1979, Manhattan College, Bronx, New York, pp. 293-320.

Thomas, J.M. et al., "Statistical Methods Used to Assess Biological Impact at Nuclear Power Plants", Journal of Environmental Management, Vol. 7, No. 3, Nov. 1978, pp. 269-290.

Thomas, W.A., "Attitudes of Professionals in Water Management Toward the Use of Water Quality Indices", Journal of Environmental Management, Vol. 4, 1976, pp. 325-338.

Tourbier, J.T. and Westmacott, R., "Water Resources Protection Measures in Land Development--A Handbook-Revised Addition", OWRT TT/81-5, Aug. 1980, Office of Water Research and Technology, U.S. Department of Interior, Washington, D.C.

Tubbs, L.J. and Haith, D.A., "Simulation Model for Agricultural Non-Point Source Pollution", Journal of the Water Pollution Control Federation, Vol. 53, No. 9, Sept. 1981, pp. 1425-1433.

Tucker, J.B., "Schistosomiasis and Water Projects: Breaking the Link", Environment, Vol. 25, No. 7, Sept. 1983, pp. 17-20.

Turner, F.T., Brown, K.W. and Deuel, L.E., "Nutrients and Associated Ion Concentrations in Irrigation Return Flow from Flooded Rice Fields", Journal of Environmental Quality, Vol. 9, No. 2, Apr.-June 1980, pp. 256-260.

Unger, S.G., "Environmental Implications of Trends in Agriculture and Silviculture. Volume II: Environmental Effects of Trends", EPA/600/3-78/102, Dec. 1978, Development Planning and Research Associates, Inc., Manhattan, Kansas.

United Nations Environment Program, "Environmental Issues in River Basin Development", Proceedings of the United Nations Water Conference on Water Management and Development, 1978, Vol. 1, Part 3, Pergamon Press, New York, New York, pp. 1163-1172.

U.S. Bureau of Reclamation, "Prediction of Mineral Quality of Irrigation Return Flow: Volume I. Summary Report and Verification", Report No. EPA-600/2-77-179a, Aug. 1977, Denver, Colorado.

U.S. Environmental Protection Agency, "Areawide Assessment Procedures Manual, Volume I", EPA/600/9-76/014-1, July 1976a, Municipal Environmental Research Laboratory, Cincinnati, Ohio.

U.S. Environmental Protection Agency, "Areawide Assessment Procedures Manual, Volume II", EPA/600/9-76/014-2, July 1976b, Municipal Environmental Research Laboratory, Cincinnati, Ohio.

U.S. Environmental Protection Agency, "Areawide Assessment Procedures Manual, Volume III", EPA/600/9-76/014-3, July 1976c, Municipal Environmental Research Laboratory, Cincinnati, Ohio.

U.S. Environmental Protection Agency, "Modeling Phosphorus Loading and Lake Response Under Uncertainty: A Manual and Compilation of Export Coefficients", EPA-4405-80-011, June 1980, Washington, D.C.

U.S. Environmental Protection Agency, Proceedings of the Workshop on Aquatic Weeds, Control and Its Environmental Consequences, EPA-600/9-81-010, Feb. 1981, Washington, D.C.

U.S. Fish and Wildlife Service, "Habitat Evaluation Procedures", Mar. 1979, Division of Ecological Services, Fort Collins, Colorado.

Uzzell, Jr., J.C. and Ozisik, M.N., "Three-Dimensional Temperature Model for Shallow Lakes", Journal of the Hydraulics Division, American Society of Civil Engineers, Vol. 104, No. HY12, Dec. 1978, pp. 1635-1645.

Veith, G.D., DeFoe, D.L. and Bergstedt, B.V., "Measuring and Estimating the Bio-Concentration Factor of Chemicals in Fish", Journal of the Fisheries Research Board of Canada, Vol. 36, 1979, pp. 1040-1048.

Vendrov, S.L., "Interaction of Large Hydraulic Engineering Systems with the Environment", Hydrotechnical Construction, No. 2, Feb. 1980, pp. 175-181.

Vick, H.C. et al., "West Point Lake Impoundment Study", EPA Report No. 904/9-77-004, Nov. 1976, National Technical Information Service, U.S. Department of Commerce, Springfield, Virginia.

Vick, H.C. et al., "Preimpoundment Study: Cedar Creek Drainage Basin: Evans County Watershed: Evans, Tattnall, and Candler Counties, Georgia", EPA

Report No. 904/9-77-006, Mar. 1977, U.S. Environmental Protection Agency, Athens, Georgia.

Villeneuve, J.P. et al., "Kriging in the Design of Streamflow Sampling Networks", Water Resources Research, Vol. 15, No. 6, Dec. 1979, pp. 1833-1840.

Vlachos, E. and Hendricks, D.W., Technology Assessment for Water Supplies, 1977, Water Resource Publications, Fort Collins, Colorado.

Walker, W.R., "Assessment of Irrigation Return Flow Models", EPA Report No. 600/2/76-219, Oct. 1976, Department of Agricultural and Chemical Engineering, Colorado State University, Fort Collins, Colorado.

Walter, M.F., Steenhuis, T.S. and Haith, D.A., "Nonpoint Source Pollution Control by Soil and Water Conservation Practices", Transactions of the American Society of Agricultural Engineers, Vol. 22, No. 4, July-Aug. 1979, pp. 834-840.

Walters, R.A., "A Time- and Depth-Dependent Model for Physical, Chemical and Biological Cycles in Temperate Lakes", Ecological Modeling, Vol. 8, Jan. 1980, pp. 79-96.

Ward, A.D., "A Verification Study on a Reservoir Sediment Deposition Model", Transactions of the American Society of Agricultural Engineers, Vol. 24, No. 2, Mar./Apr. 1981, pp. 340-352.

Ward, D.V., Biological Environmental Impact Studies: Theory and Methods, 1978, Academic Press, New York, New York.

Watanabe, M., "Modeling of the Eutrophication Process in Lakes and Reservoirs", Proceedings of the Baden Symposium on Modeling the Water Quality of the Hydrological Cycle, Publication No. 125, 1978, International Association of Hydrological Sciences, pp. 200-210.

Watling, L., "Artificial Islands: Information Needs and Impact Criteria", Marine Pollution Bulletin, Vol. 6, No. 9, Sept. 1975, pp. 139-141.

Watling, L., Pembroke, A. and Lind, H., "Environmental Assessment", Final Technical Report No. NSF-RE-E-054A, 1975, College of Marine Studies, University of Delaware, Lewes, Delaware, pp. 294-431.

Watson, C.W., Barr, S. and Allenson, R.E., "Rainout Assessment: The ACRA System and Summaries of Simulation Results", LA-6763, Sept. 1977, Los Alamos Scientific Laboratory, Los Alamos, New Mexico.

Watson, V.J. et al., "Impact of Development on Watershed Hydrologic and Nutrient Budgets", Journal of the Water Pollution Control Federation, Vol. 51, No. 12, Dec. 1979, pp. 2875-2885.

Webster, J.R., Benfield, E.F. and Cairns, J., "Model Predictions of Effects of Impoundment of Particulate Organic Matter Transport in a River System", The Ecology of Regulated Streams, 1978, Plenum Publishing Corporation, New York, New York, pp. 339-364.

Weiner, R.M. et al., "Microbial Impact of Canada Geese (Branta canadensis) and Whistling Swans (Cygnus columbianus columbianus) on Aquatic Ecosystems", Applied and Environmental Microbiology, Vol. 37, No. 1, Jan. 1979, pp. 14-20.

Whalen, N.A., "Nonpoint Source Control Guidance, Hydrologic Modifications", Feb. 1977, U.S. Environmental Protection Agency, Washington, D.C.

Whisler, F.D. et al., "Agricultural Management Practices to Effect Reductions in Runoff and Sediment Production", Oct. 1979, Water Resources Research Institute, Mississippi State University, Starkville, Mississippi.

White, T.R. and Fox, R.C., "Recolonization of Streams by Aquatic Insects Following Channelization", Technical Report 87, Vol. I, May 1980, Water Resources Research Institute, Clemson University, Clemson, South Carolina.

Whitlatch, Jr., E.E., "Systematic Approaches to Environmental Impact Assessment: An Evaluation", Water Resources Bulletin, Vol. 12, No. 1, Feb. 1976, pp. 123-137.

Wiederholm, T., "Use of Benthos in Lake Monitoring", Journal of the Water Pollution Control Federation, Vol. 52, No. 3, Mar. 1980, pp. 537-547.

Williams, D.F., "Postimpoundment Survey of Water Quality Characteristics of Raystown Lake, Huntington and Bedford Counties, Pennsylvania", Geological Survey Water-Resources Investigation 78-42, July 1978, National Technical Information Service, U.S. Department of Commerce, Springfield, Virginia.

Williams, D.T., "Effects of Dam Removal: An Approach to Sedimentation", Technical Paper No. 50, Oct. 1977, U.S. Army Corps of Engineers, Hydrologic Engineering Center, Davis, California.

Williams, L.R. et al., "Relationships of Productivity and Problem Conditions to Ambient Nutrients: National Eutrophication Survey Findings for 418 Lakes", EPA 600/3-78-002, Jan. 1978, Environmental Monitoring and Support Laboratory, U.S. Environmental Protection Agency, Las Vegas, Nevada.

Witten, A.L. and Bulkley, R.V., "A Study of the Effects of Stream Channelization and Bank Stabilization on Warmwater Sport Fish in Iowa. Subproject No. 2. A Study of the Impact of Selected Bank Stabilization Structures on Game Fish and Associated Organisms", Report No. 76/12, May 1975, Iowa Cooperative Fishery Research Unit, Ames, Iowa.

Woodward, F.E., Fitch, Jr., J.J. and Fontaine, R.A., "Modeling Heavy Metal Transport in River Systems", Apr. 1981, Land and Water Resources Center, University of Maine, Orono, Maine.

World Health Organization, "Environmental Health Impact Assessment", EURO Reports and Studies No. 7, 1979, 31 pp., Copenhagen, Denmark.

Wright, T.D., "Potential Biological Impacts of Navigation Traffic", Misc. Paper E-82-2, June 1982, U.S. Army Engineer Waterways Experiment Station, Vicksburg, Mississippi.

Wycoff, R.L. and Singh, U.P., "Application of the Continuous Storm Water Pollution Simulation System (CSPSS): Philadelphia Case Study", Water Resources Bulletin, Vol. 16, No. 3, June 1980, pp. 463-470.

Yahnke, J.W., "Water Quality of the Proposed Norden Reservoir, Nebraska, and Its Implications for Fishery Management", REC-ERC-81-8, May 1981, U.S. Bureau of Reclamation, Denver, Colorado.

Yapijakas, C. and Molof, A.H., "A Comprehensive Methodology for Project Appraisal and Environmental Protection in Multinational River Basin Development", Water Science and Technology, Vol. 13, No. 7, 1981, pp. 425-436.

Yiqui, C., "Environmental Impact Assessment of China's Water Transfer Project", Water Supply and Management, Vol. 5, No. 3, 1981, pp. 253-260.

Yorke, T.H., "Impact Assessment of Water Resource Development Activities: A Dual Matrix Approach", FWS/OBS-78/82, Sept. 1978, U.S. Fish and Wildlife Service, Kearneysville, West Virginia.

Yousef, Y.A., "Assessing Effects on Water Quality by Boating Activity", EPA/670/2-74-072, Oct. 1974, U.S. Environmental Protection Agency, Cincinnati, Ohio.

Yousef, Y.A. et al., "Mixing Effects Due to Boating Activities in Shallow Lakes", Technical Report No. 78-10, June 1978, Environmental Systems Engineering Institute, Florida Technological University, Orlando, Florida.

Yousef, Y.A., McLellon, W.M. and Zebuth, H.H., "Changes in Phosphorus Concentrations Due to Mixing by Motor Boats in Shallow Lakes", Water Research, Vol. 14, No. 7, July 1980, pp. 841-852.

Yu, J.K. and Fogel, M.M., "The Development of a Combined Water Quality Index", Water Resources Bulletin, Vol. 14, No. 5, Oct. 1978, pp. 1239-1250.

Zimmer, D.W. and Bachmann, R.W., "Channelization and Invertebrates in Some Low Streams", Water Resources Bulletin, Vol. 14, No. 4, Aug. 1978, pp. 868-883.

Zimmerman, E.G., Anderson, K.A. and Calhoun, S.W., "Impact of Discharge from Possum Kingdom Reservoir (Texas) on Genic Adaptation in Aquatic Organisms", Aug. 1980, Department of Biological Sciences, North Texas State University, Denton, Texas.

CHAPTER 2

ENVIRONMENTAL IMPACT STUDIES FOR DAMS AND RESERVOIRS

Environmental impact studies are typically required on large dams and hydroelectric projects. These studies are an outgrowth of the requirements of international funding agencies and in-country laws such as the National Environmental Policy Act in the United States. The importance of reservoir projects is illustrated by the fact that by the year 2000 it is estimated that the portion of the world's streamflow regulated by reservoirs will increase from one-tenth to two-thirds (Freeman, 1974). While economic and political decisions are involved in the planning and implementation of reservoir projects, it is imperative that environmental impact studies be conducted from a scientific viewpoint, including the use of current technical information on types of impacts, planning and conduction of baseline studies, impact prediction and assessment, and methodologies for trade-off analyses and decision-making. This chapter provides a summary of key technical references useful in environmental impact studies on large dams and hydroelectric projects.

IMPACT IDENTIFICATION

Identification of potential impacts should be an early activity in an environmental impact study. General knowledge about the types of impacts which could occur can be used in identifying potential impacts for new projects. One of the most-studied reservoirs in terms of resultant environmental impacts is Lake Nasser created by the Aswan High Dam (Abu-Zeid, 1979). Beneficial impacts include protection against floods and droughts; while detrimental impacts have resulted from siltation of lakes, loss of routine silt deposits on agricultural lands, changes in water quality and fish habitats, and public health concerns related to disease transmission.

The environmental effects of dam construction and methods for minimizing undesirable ones have been described in a United Nations Conference (United Nations Environment Program, 1978). Dam construction can provide water supply, hydroelectric power, and flood control, and can greatly improve agriculture, forestry, and livestock management. Dams produce a permanent physical transformation, inundating settled areas and destroying habitats, affecting the ground water regime and water table, possibly increasing seismic tendencies, and often leading to explosive aquatic weed growth and the spread of schistosomiasis and other communicable diseases. Dams in tropical areas tend to favor weed propagation and vectors of parasitic diseases, while temperate-zone dams often interface with fish migration. Resettlement of population displaced by dams often leads to housing, disease, and social problems. Due to the potential significance of these effects, UNEP has called for careful social, economic, and ecological evaluation of proposed projects prior to construction.

Systematic approaches to aid in identifying potential impacts include the use of checklists, case studies, interaction matrices, and networks. A checklist of potential impacts and issues from impoundment projects is in Table 18 (U.S.

Environmental Protection Agency, 1976b). A general checklist based on a series of questions has been prepared based on a study of Volta Lake in Ghana (Freeman, 1974). Freeman (1974) also provides guidelines for (1) environmental assessment of alternative impoundment sites--seismic and hydrological impacts, effects on natural rivers and forests, effects on archaeological remains, effects on human settlements and farmland, and impacts below the damsite; (2) assessment of hydrological and biological impacts following dam closure; and (3) assessment of environmental and ecological impacts of a tropical reservoir, including lake geology, ground water geology, lake topography, mineral cycling, plankton, aquatic plants, fish nutrition and production, water supply and sewage, and control of disease.

Table 18: Checklist of Potential Impacts from Impoundment Projects (U.S. Environmental Protection Agency, 1976b)

Category	Potential Impacts and Issues
Construction Phase	Sediment pollution and stream siltation
	Pesticides, petrochemicals, and other potential pollutants
	Quantification of erosion and sediment generation
	Relevant criteria for sediment pollution
	Protection of water quality during construction--general
	Erosion and sediment control techniques
	Treatment of polluted water from construction site
	Activity scheduling
	Components of solid waste from construction operations
	Disposal of chemicals and containers
	Summary of solid waste impacts
	Air pollution sources at construction sites
	Noise generators at impoundment construction site
	Typical construction noise levels
	Rough estimation of noise impacts
	Damaging effects of noise

Table 18: (continued)

Category	Potential Impacts and Issues
Impoundment Area	Probable land use impacts
	General methodology for evaluating land use changes and impacts
	Loss of stream and bottom land
	Relocation impacts
	Recreational development--general
	Secondary air pollution impacts (parking facilities)
	Solid waste generation at recreational areas
	Impact of land inundation on impoundment water quality
	Organic decomposition and dissolved oxygen deficiency
	Solution of iron and manganese
	Loss of wildlife habitat
	Assimilative capacity changes--general
	Primary determinants
	Critical water quality conditions
	Effects of stratification and density currents
	Eutrophication and associated impacts
	Consideration of evaporation
	Shift from river to lake environment and reduction of species diversity
	Sedimentation in impoundment
	Modeling of impoundment water quality
	Estimating significance of site conditions with respect to impoundment water quality
	Potential for erosion in reservoir

Table 18: (continued)

Category	Potential Impacts and Issues
	Relationship of morphometry to potential eutrophication and weed problems
	Nutrient sources and loadings
	Quantification of influent water quality
	Changes in point and nonpoint pollution sources
	Probability of water quality problems in stratified reservoirs
	Evaluation of reservoir fisheries
	Summary of water quality parameters that may be affected by impoundment and relevant criteria
	Thermal criteria for fisheries
Downstream and Areas of Water Use	Influence of land acquisition policy on reservoir development
	Induced development in region
	Land use impacts due to increased flood protection
	Land use impacts of irrigation impoundments
	Evaluation of water pollution from irrigation
	Policy concerning use of flood plains
	Prevention of water quality degradation from irrigation projects
	Impacts of water quality changes on downstream biota
	Impact of dam as barrier
	Flow regime changes--general
	Quantification of hydrographic modification
	Seasonal and diurnal flow variations
	Minimum release requirements

Table 18: (continued)

Category	Potential Impacts and Issues
	Low-flow augmentation analysis
	Effects on riparian vegetation
	Flow requirements for salmon and other species
	Temperature changes--general
	Important categories of fish species
	Effects of outlet location and impoundment operation
	Possible thermal effects on downstream species composition
	Thermal criteria for fisheries
	Effects on downstream uses

One valuable source of information for impact identification is to review reports on environmental impact studies for specific projects. Several illustrations exist for hydroelectric projects in both the United States (Federal Energy Regulatory Commission, 1981) and Canada (Berkes, 1981; Kelly, Underwood and Thirumurthi, 1980; Kenyon, 1981; and Reynolds and Ujjainwalla, 1981).

The development of a simple interaction matrix, or the use of an impact network can be an aid in identifying and summarizing information on anticipated impacts. The simple matrix refers to a display of project actions or activities along one axis, with appropriate environmental factors listed along the other axis of the matrix. When a given action or activity is anticipated to cause a change in an environmental factor, this is noted at the intersection point in the matrix and further described in terms of magnitude and importance considerations. Networks refer to those methodologies which attempt to integrate impact causes and consequences through identifying interrelationships between causal actions and the environmental factors affected, including those representing secondary and tertiary effects. A linear network display for an impoundment project is shown in Figure 1 (U.S. Soil Conservation Service, 1977).

To serve as an example of an approach to develop a matrix/network for a project, a binational Technical Commission (CTM) was formed to evaluate the construction of a dam at Salto Grande on the Uruguay River (Gallopin, Lee and Nelson, 1980). A workshop attended by CTM members defined the system to be controlled and gave as the main objective of the system the maintenance of the quality of life. To gather information as quickly as possible and enhance inter-

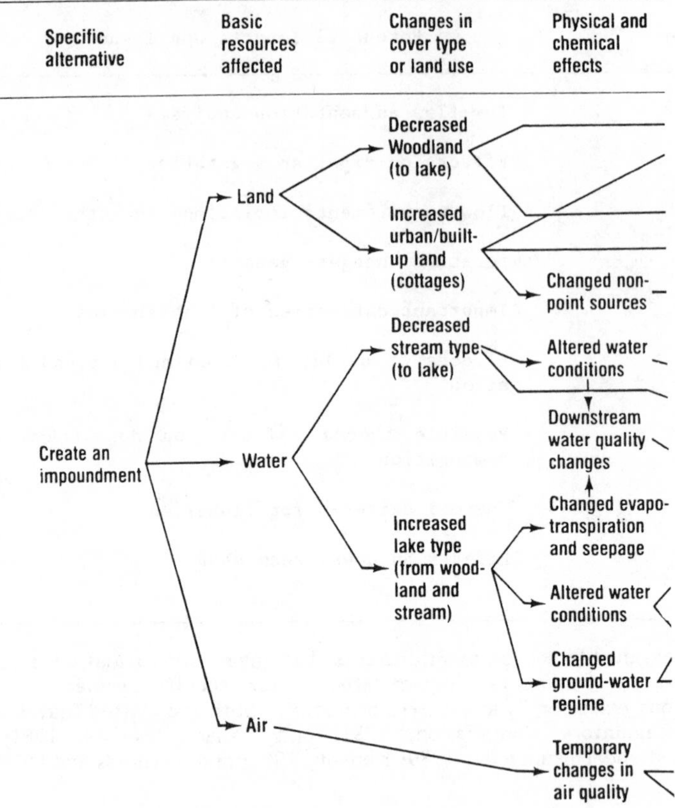

Figure 1: An Example of a Network Diagram for Analyzing Probable Environmental Impacts (U.S. Soil Conservation Service, 1977)

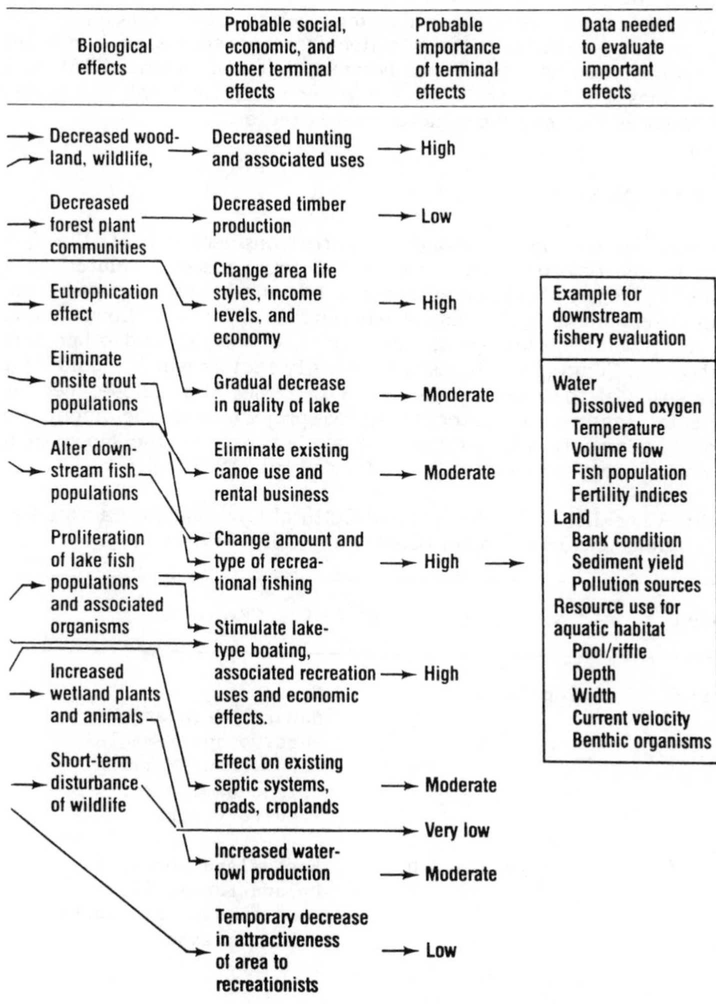
Biological effects
Probable social, economic, and other terminal effects
Probable importance of terminal effects
Data needed to evaluate important effects
Decreased wood-land, wildlife,
Decreased hunting and associated uses
High
Decreased forest plant communities
Decreased timber production
Low
Eutrophication effect
Change area life styles, income levels, and economy
High
Eliminate onsite trout populations
Gradual decrease in quality of lake
Moderate
Alter down-stream fish populations
Eliminate existing canoe use and rental business
Moderate
Proliferation of lake fish populations and associated organisms
Change amount and type of recrea-tional fishing
High
Stimulate lake-type boating, associated recreation uses and economic effects.
High
Increased wetland plants and animals
Effect on existing septic systems, roads, croplands
Moderate
Short-term disturbance of wildlife
Very low
Increased water-fowl production
Moderate
Temporary decrease in attractiveness of area to recreationists
Low
Example for downstream fishery evaluation
Water
Dissolved oxygen
Temperature
Volume flow
Fish population
Fertility indices
Land
Bank condition
Sediment yield
Pollution sources
Resource use for aquatic habitat
Pool/riffle
Depth
Width
Current velocity
Benthic organisms

disciplinary communications, a questionnaire approach was adopted. The questionnaire focused on indicators of the state of each sector and attempted to highlight significant associations that might exist between sectors. A second workshop compiled data from the questionnaire and generated flow diagrams for each sector, showing interactions between systems. A final CTM workshop clarified actions which impacted on the water sector, the soil and terrestrial flora and fauna sector, and the socio-economic sector.

BASELINE STUDIES

Environmental impact studies require consideration of the baseline conditions of the environment. One of the first issues is related to which environmental factors should be addressed. Information on anticipated impacts can aid in developing an initial list of environmental factors. For example, the physical and chemical, biological, and social, economic, and other terminal effects shown in Figure 1 can be used to identify factors which should be a part of a description of the existing environment. Other aids include checklists of environmental factors for water resources projects and the application of professional judgment. An example of a checklist for the biophysical and cultural environment is in Table 19 (Canter and Hill, 1979).

Table 19: Checklist of Biophysical and Cultural Environment Factors for Impoundment Project (Canter and Hill, 1979)

Category	Subcategory	Factor
Terrestrial	Population	Crops
		Natural Vegetation
		Herbivorous Mammals
		Carnivorous Mammals
		Upland Game Birds
		Predatory Birds
	Habitat/Land Use	Bottomland Forest (1)
		Upland Forest (2)
		Open (nonforest) Lands (3)
		Drawdown Zone
		Land Use
	Land Quality/ Soil Erosion	Soil Erosion
		Soil Chemistry
		Mineral Extraction
	Critical Community Relationships	Species Diversity
Aquatic	Populations	Natural Vegetation
		Wetland Vegetation
		Zooplankton
		Phytoplankton
		Sport Fish

Table 19: (continued)

Category	Subcategory	Factor
		Commercial Fisheries Intertidal Organisms Benthos/Epibenthos Waterfowl
	Habitats	Stream (4) Freshwater Lake (5) River Swamp (6) Nonriver Swamp (7)
	Water Quality	pH Turbidity Suspended Solids Water Temperature Dissolved Oxygen Biochemical Oxygen Demand Dissolved Solids Inorganic Nitrogen Inorganic Phosphate Salinity Iron and Manganese Toxic Substances Pesticides Fecal Coliforms Stream Assimilative Capacity
	Water Quantity	Stream Flow Variation Basin Hydrologic Loss
	Critical Community Relationships	Species Diversity
Air	Quality	Carbon Monoxide Hydrocarbons Oxides of Nitrogen Particulates
	Climatology	Diffusion Factor
Human Interface	Noise	Noise
	Esthetics	Width and Alignment Variety within Vegetation Type Animals--Domestic Native Fauna Appearance of Water Odor and Floating Materials

Table 19: (continued)

Category	Subcategory	Factor
		Odor and Visual Quality Sound
	Historical	Historical Internal and External Packages
	Archaeological	Archaeological Internal and External Packages

Footnotes:

(1) Bottomland forest represents a composite consideration of the following 11 parameters: species associations, percent mast-bearing trees, percent coverage by understory, diversity of understory, percent coverage by groundcover, diversity of groundcover, number of trees greater than 16 in (or 18 in) dbh/acre, percent of trees greater than 16 in (or 18 in) dbh, frequency of inundation, edge (quantity) and edge (quality).

(2) Upland forest represents a composite consideration of the following 10 parameters: species associations, percent mast-bearing trees, percent coverage of understory, diversity of understory, percent coverage of groundcover, diversity of groundcover, number of trees greater than or equal to 16 in dbh/acre, percent of trees greater than or equal to 16 in dbh, quantity of edge and mean distance to edge.

(3) Open (nonforest) lands represent a composite consideration of the following 4 parameters: land use, diversity of land use, quantity of edge, mean distance to edge.

(4) Stream represents a composite consideration of the following 8 parameters: sinuosity, dominant centrarchids, mean low water width, turbidity, total dissolved solids, chemical type, diversity of fishes and diversity of benthos.

(5) Freshwater lake represents a composite consideration of the following 10 parameters: mean depth, turbidity, total dissolved solids, chemical type, shore development, spring flooding above vegetation line, standing crop of fishes, standing crop of sport fish, diversity of fishes, and diversity of benthos.

(6) River swamp represents a composite consideration of the following 6 parameters: species associations, percent forest cover, percent flooded annually, groundcover diversity, percent coverage of groundcover, and days subject to river overflow.

Table 19: (continued)

---

(7) Nonriver swamp represents a composite consideration of the following 5 parameters: species associations, percent forest cover, percent flooded annually, groundcover diversity and percent coverage by groundcover.

Professional knowledge and judgment is based on the general effects of impoundments on water quality (Canter, 1977). Considering water quality only, it is known that impoundment of water will lead to beneficial effects in terms of turbidity reduction, hardness reduction, oxidation of organic material, coliform reduction, and flow equalization. Detrimental effects include lower reaeration, buildup of inorganics, algae blooms, stratified flow, and thermal stratification. Perhaps the most significant impact is due to thermal stratification with the following additional changes in water quality: decreased dissolved oxygen in hypolimnion; anaerobic conditions in hypolimnion; and dissolution of iron and manganese from bottom deposits. In addition to changes in water quality resulting from thermal stratification, changes in mixing patterns also occur. Thermal stratification can result in overflow (warmer water flowing over the surface of colder water), interflow (cool water flowing between upper layers of warmer water and lower layers of colder water), or underflow (cooler water flowing underneath warmer surface water). An additional concern of water impoundment is the reduction in waste assimilative capacity of the body of water being impounded. In general, water impoundment decreases the reaeration ability of a body of water, thus reducing the waste loading that the body of water can receive without having the dissolved oxygen concentration decreased below a prescribed water quality standard.

Information for describing baseline conditions can be procured from existing data collected by governmental agencies or private groups, or it can be collected via specific baseline studies. Field studies can be time-consuming and costly, thus they must be carefully planned and conducted. Due to the potential impacts of large dams and hydroelectric projects on the water environment, it may be necessary to conduct studies on water quality and aquatic biology. Table 20 summarizes 12 steps which can be used in the design of a monitoring program (Sanders, 1980). As noted in Step 5, a critical component is related to the selection of parameters which should be monitored (IHD-WHO Working Group on the Quality of Water, 1978). Water quality parameters which are expected to be altered by project construction and operation should be included. An aquatic ecosystem includes algae, zooplankton, fish, and benthic organisms, thus a monitoring program should include consideration of sampling of various planktonic and benthic forms, as well as fish (Hellawell, 1978; Jacobs and Grant, 1978; and Stofan and Grant, 1978).

Table 20: Steps Related to Sampling Network Design (Sanders, 1980)

---

1. Determine monitoring objectives and relative importance of each.

2. Express objectives in statistical terms.

Table 20: (continued)

3. Determine budget available for monitoring and amount to be allocated for each objective.

4. Define the characteristics of the area in which the monitoring is to take place.

5. Determine water quality and aquatic biology parameters to be monitored.

6. Determine sampling station locations.

7. Determine sampling frequency.

8. Compromise previous objective design results with subjective considerations.

9. Develop operating plans and procedures to implement the network design.

10. Develop data and information reporting formats and procedures.

11. Develop feedback mechanisms to fine tune the network design.

12. Prepare a network design report.

An important aspect of describing the baseline setting is the interpretation of the extant or collected data. For the water environment, interpretation can be based on water quality criteria, water quality standards which have been adopted for a given stream setting, or the quality and quantity requirements of existing or potential water users in the area. One approach which can be used to summarize water quality data is an empirical index which combines data from several parameters into one numeric indicator (Ott, 1978). An example is the Water Quality Index (WQI) developed in 1970 using a formal procedure based on the Delphi technique. The WQI is composed of nine parameters--dissolved oxygen, fecal coliforms, pH, biochemical oxygen demand, nitrates, phosphates, temperature, turbidity, and total solids. Aquatic biological data can be summarized in terms of selected indicator species, diversity indices, system photosynthesis and respiration, or the development of aquatic habitat quality indices (U.S. Army Corps of Engineers, 1980). Professional judgment will have to be exercised in the data interpretation phase, particularly as related to biological parameters and characteristics.

IMPACT PREDICTION

The most important technical activity in an environmental impact study for a dam/reservoir project is the scientific prediction of the effects of project construction and operation. Prediction of the impacts of large dams and hydroelectric projects can be based on (1) a qualitative approach which relies on general knowledge of the impacts of similar projects, or specific results of comprehensive studies of similar projects; (2) a quantitative approach based on the use of simple mass balance and environmental dilution calculations; and (3) a quantitative approach based on the use of mathematical models for multiple environmental factors. A given environmental impact study will probably involve all three approaches to some degree.

Qualitative Approach

General knowledge of impacts has already been mentioned in conjunction with impact identification and describing the environmental setting. Examples of comprehensive studies include Martin, Noel and Federer (1981); Nix (1980); McClellan and Frazer (1980); Entz (1980); Teskey and Hinckley (1977); Ploskey (1982); Miracle and Gardner (1980); Petts (1980); Nelson et al. (1976); Hazel et al. (1976); and Johnston, Benson and King (1981). The effects of forest clear-cutting on stream water chemistry and biology was addressed by Martin, Noel and Federer (1981). This nonpoint source runoff can have major implications upon stream impoundments. Specifically, changes in stream chemistry following clear-cutting were sought in 56 streams at 15 locations throughout New England in the United States. Streams draining clear-cut areas were compared with nearby streams in uncut watersheds over periods of up to two years. In general, concentrations of all elements studied (inorganic N, SO4-S, Cl, Ca, Mg, K, Na), as well as pH and specific conductivity, varied as much among uncut streams at a location as between uncut and cutover streams. At four of the locations the effect of cutting on algae and invertebrates in the streams were also examined. Both algal and invertebrate densities were greater in cutover streams by factors of 2 to 4, probably because of increased light and temperature.

As noted earlier, water impoundment can cause changes in basic chemical parameters; in addition, these changes can also affect the distribution of other chemical constituents within impoundments. The results of a detailed water quality survey conducted over 1959-1975 on DeGray Reservoir in southwest Arkansas in the United States were used to describe physical/chemical processes affecting reservoir and river system constituent concentrations (Nix, 1980). Dissolved oxygen (DO) and turbidity profiles delineated water movement through the reservoir. Development of metalimnetic DO minima appeared to be related to advective transport within the metalimnion (with metalimnetic DO depletion greatest during years when spring rains occurred after stratification onset). Dissolved oxygen data showed new reservoir "aging" (with hypolimnetic DO depletion severe during early impoundment years, but moderated after four years). Iron (Fe) and manganese (Mn) expectedly correlated well with DO depletion; other trace metals studied did not correlate with either Fe or Mn. Cobalt correlated significantly with nickel, probably due to their geochemical similarity. Storm event intervention was the most significant factor determining concentrations of calcium and other conservative constituents. Calcium dropped as storm water entered the reservoir. Other constituents, including phosphorus and suspended solids, increased in storm water. Another U.S. study on the fate of trace metals in reservoirs was reported by McClellan and Frazer (1980). Emphasis was placed on selenium, although several metal ions were determined.

Water impoundment can also cause changes in aquatic and riparian plant and animal communities. Succession of animal and plant communities during the first 10 years' existence of Lakes Nasser and Nubia in Egypt is discussed by

Entz (1980). The development and decline of chironomid swarms is compared to similar phenomena observed on other man-made lakes. Their feeding habits and their role as food for fish, toads, spiders, and other insects indicated that they played an important role as food organisms for many animals and insects. About the time the chironomid swarms diminished, large numbers of water bugs (corixids) became common. Fish in the reservoir seemed to cause the eventual disappearance of the bugs. The succession of underwater plants was: algal epitecton, Chara fields, and submerged macrophytes. Succession in lakeshore vegetation was: Glinus lotoides and other weeds, desert plants, Tamarix nicotica, and the reappearance of river grass. Plants and animals were often the types found in the original river, and Nile flooding caused the reintroduction of several species.

Changing water levels in reservoirs can also cause changes in woody riparian and wetland communities. Teskey and Hinckley (1980) prepared a comprehensive literature review of general plant physiological responses to managed or natural changes in water levels, (i.e., submersion, flooding, soil saturation) and on plant tolerance mechanisms involved in water level changes. The major effect of flooding is the creation of an anaerobic environment surrounding the root system, and the maintenance of proper root functioning is the factor which determines tolerance to flooding. Physical tolerance mechanisms involve processes designed to increase oxygen content in the roots either by transport of oxygen from the stem or from parts of the root system where oxygen is more available. Metabolic mechanisms enable the plant to utilize less toxic end-products. Tolerant species are able to maintain root systems with a minimum of stress by incorporating a variety of tolerance mechanisms. Five water level factors--time of year, flood frequency, duration, water depth, and siltation--are considered critical in determining a plant's physiological responses.

Impacts on fish can also occur as a result of reservoir creation. Ploskey (1982) authored a report containing 367 annotations describing the effects of fluctuating reservoir water levels on fish. Citations on phytoplankton, zooplankton, and water quality effects that pertain to reservoir fisheries are also included. Reservoir operation can cause undesirable impacts on fish populations. For example, Miracle and Gardner (1980) have reviewed the literature on the effects of pumped storage operations on ichthyofauna. They found that pumped storage operations pass large volumes of water from one reservoir to another, ultimately affecting fishes, either by entrainment in withdrawn waters or by modification of the aquatic environment. Mortality caused during entrainment is primarily due to the following physical stresses: abrasion and collision; pressure changes; velocity changes; and acceleration effects. Fish passage or entrainment is influenced by: size and life stage of fish; susceptibility of the fish which directly relates to life history aspects; and physical characteristics of the pumped storage facility. Pumped storage facilities can adversely affect reservoir hydrology through changes in water level, water temperature, dissolved oxygen and water velocity.

Downstream effects can also occur from stream impoundment, including changes in geomorphology, species of fish and wildlife, and coastal ecosystems due to decreased fresh water inflows. Geomorphological data from 14 impounded rivers located throughout Britain have been used to assess the long-term impact of reservoirs upon river channels downstream, and to identify the potential problems for river management (Petts, 1980). Changes of channel size and shape have halved the water conveyance capability of 11 of the 14 rivers studied. Four mechanisms, channel erosion, redistribution of the boundary sediment, channel side deposition, and channel bed aggrandizement, contributed to the morphological changes. Long-term changes downstream from reservoirs can affect land drainage, coastal stability, water quality, navigation, structural stability, fisheries, and aesthetics, as well as the aquatic flora and fauna.

Two examples of studies on downstream effects on fish and wildlife will be mentioned. Nelson et al. (1976) reported on the findings, conclusions and recommendations regarding changed flow regions below dams, the impacts on fisheries, and methodologies used to assess flow requirements for 96 dams and diversions in the Rocky Mountain and Pacific Northwest regions in the United States. Hazel et al. (1976) presented the results and conclusions from 47 case studies of California water projects that altered natural streamflow regimes and causally affected the fish and wildlife. Surveys were conducted on existing conditions below dams and diversions to assess the actual effects of the streamflow characteristics on fish and wildlife and to evaluate the adequacy of the methodologies used to determine necessary flows.

Johnston, Benson and King (1981) describe ecological characterization studies to assess the effects of fresh water inflows on estuaries. Several studies have been completed or are underway in the United States. Data from the Sabine Basin in Texas demonstrated the quantified effects of modifying natural river flow on Sabine Lake and the associated estuarine area. Navigation channel developments, construction of reservoirs on the incoming rivers, and impoundment of marshes were the primary causes of ecological changes. The Mississippi Deltaic Plain study mapped habitat changes from the mid-1950's to 1978. Many of the habitat changes resulted from the modification of fresh-water inflow. The study indicated that over 500,000 acres (202,000 ha) of Louisiana coastal wetlands were lost or altered during the period.

Simple Quantitative Approach

Simple mass balance and environmental dilution calculations can also be used for impact prediction. Due to the large number of calculations which might be involved in an environmental impact study, the focus herein will be on selected examples. Calculations can be made for anticipated nonpoint and point sources of pollution in a reservoir through use of unit waste generation factors. These factors express the rate at which a pollutant is released to a drainage area or watercourse as a result of some activity, such as land clearing or production by industry, divided by that activity (Canter, 1977). Table 21 shows representative rates of erosion from various land uses (U.S. Environmental Protection Agency, 1973). Detailed calculations for erosion can be made via usage of the Universal Soil Loss Equation.

Detailed information is available for identifying and estimating pollutant load generation and transport from major urban stormwater sources (U.S. Environmental Protection Agency, 1976a). A structured model for phosphorus loading in lakes has been developed by Ahmed and Schiller (1981). This model for computing loading estimates from nonpoint sources in a watershed (CLENS) was used to quantify the phosphorus in 16 lakes in Connecticut and Massachusetts as part of the development of preliminary management plans. The model is simple and can be used to develop quantitative estimates on nonpoint sources of pollution and their impact on water bodies.

Table 21: Representative Rates of Erosion from Various Land Uses (U.S. Environmental Protection Agency, 1973)

Land Use	Erosion Rate		Relative to forest = 1
	MT/km^2-yr	Tons/mi^2-yr	
Forest	8.5	24	1
Grassland	85	240	10
Abandoned surface mines	850	2,400	100
Cropland	1,700	4,800	200
Harvested forest	4,250	12,000	500
Active surface mines	17,000	48,000	2,000
Construction	17,000	48,000	2,000

Point sources can also be addressed by using unit waste generation factors. Extensive literature information is available on domestic wastewater characteristics expressed on a per capita basis. The pollutional strength of industrial wastewaters varies considerably depending upon the type of industry. A useful approach for industrial waste loadings is the population equivalent (Canter, 1977):

$$PE = \frac{(A)(B)(8.34)}{0.17}$$

where

PE = population equivalent based on organic constituents in the industrial waste

A = industrial waste flow, mgd

B = industrial waste BOD, mg/l

8.34 = number of lb/gal

0.17 = number of lb BOD/person-day

As noted earlier, many of the impacts of dams and hydroelectric projects occur on the socio-economic environment. A good descriptive checklist for the economic and demographic impacts of water resources projects is contained in Chalmers and Anderson (1977). This methodology addresses environmental impacts in terms of population, employment and income, and specifically relates these to community population, community facilities and services, and community fiscal concerns. Impact calculations are presented in terms of ratios of present and future populations, economic indicators, and multiplier factors.

Predictions related to the transport and fate of pollutants in impoundments can be based on comprehensive studies as described earlier, and assumptions relative to the fractions associated with the water phase, suspended matter, sediments, and aquatic organisms. Calculations for evaporation losses and hydraulic dilution can also be used. Ahlgren (1980) described a simple hydraulic dilution model used to compare predicted nitrogen and phosphorus concentrations with those observed in a chain of four heavily eutrophicated shallow lakes north of Stockholm, Sweden, after all sewage effluents were diverted in 1970. Total P loadings were reduced by 80-90 percent, and total N loadings by 50-90 percent. The model successfully predicted P concentrations using a sediment retention coefficient of zero, but nitrogen fixation and denitrification produced differences between calculated and observed N concentrations. The average annual amplitudes in P levels were correlated with wind fetch over lake surface divided by mean depth, thus showing that differences among the lakes may depend on wind-generated turbulence at the sediment surface. Nitrogen was the limiting factor in chlorophyll-a concentration.

It may be necessary to collect specific information on the transport and fate of pollutants in reservoirs through the conduction of laboratory studies. For example, Vieth, DeFoe and Bergstedt (1979) described a method of estimating the bioconcentration factor of organic chemicals in fathead minnows (Pimephales promelas). Water at 25°C was intermittently dosed with the chemical at a nontoxic concentration in a flow-through aquarium. Thirty minnows were placed in the aquarium, and composite samples of five fish were removed for analysis after 2, 4, 8, 16, 24, and 32 days of exposure. The bioconcentration process was summarized by using the first-order uptake model, and the steady-state bioconcentration factor was calculated from the 32-day exposure. A structure-activity correlation between the bioconcentration factor (BCF) and the n-octanol/water partition coefficient (P) of individual chemicals is summarized by the equation log BCF = 0.85 log P - 0.70, which permits the estimation of the bioconcentration factor of chemicals to within 60 percent before laboratory testing.

As mentioned earlier, thermal stratification occurs in many natural lakes and man-made reservoirs. The amount and duration of this stratification depends on the water body geometry, flow, wind velocity and solar radiation. Weak stratification will occur in shallow "run of the river" reservoirs, or deeper reservoirs, with a small flow to volume ratio. One means of predicting whether or not a reservoir will stratify is through the use of a densimetric Froude number (F). If F is less than $1/\pi$, stratification is expected, with the degree of stratification increasing as F becomes smaller. When F is greater than $1/\pi$, no stratification is expected. The Froude number can be approximated by:

$$F = 320 \frac{L}{D} \frac{Q}{V}$$

where

L = reservoir length (meters)

D = mean reservoir depth (meters

Q = volumetric discharge through the reservoir (cu meters/sec)

V = reservoir volume (cu meters)

Empirical indices can also be used in impact prediction. One example relates to land usage around reservoirs and the potential for undesirable impacts. Sargent and Berke (1979) present an analytical procedure developed to enable planners to classify undeveloped lakeshore areas according to their suitability for public and private uses. The undeveloped lakeshore evaluation system rated the physical characteristics of the lakeshore for five of the most common uses: (1) public beaches; (2) camping and picnicking areas; (3) boat access areas; (4) marinas; and (5) development-vacation homes, cottages, and motels. Standards for rating the lakeshore potential have been developed for the following site requirements: size, slope, soil suitability, shoreline type (sand, rock, muck, or gravel), water quality, site location, scenery, and road access. Each of these items are measured and rated on a numerical scale from 1 to 5; the basis of the rating is a comparison with other sites for a similar use in the town or region. Data sources for the ratings include air photos, U.S. Geological Survey topographical maps, U.S. Soil Conservation Service maps, and other previous studies, plus a site visit.

Habitat index approaches have been applied to water resources projects for describing the baseline environment. Flood et al. (1977), described a habitat evaluation procedure for measuring the effects of water development projects on fish, wildlife and related resources. In the procedure developed for usage in the Meramec Park Lake Project in Missouri, six groups of animals were characterized in as many as six habitat types. The animal groups included forest game, upland game, tree squirrels, terrestrial fur bearers, aquatic fur bearers, and waterfowl. The habitat types included bottomland hardwood, upland hardwood, old field, pasture, small grain and row crops, and the Meramec River and riverine habitat. The procedure could be used for impact prediction if projections could be made of changes in habitat quantity and quality.

Another habitat index approach has been developed for projects in the lower Mississippi Valley area in the United States (U.S. Army Corps of Engineers, 1980). Seven habitat types are defined (freshwater stream, freshwater lake, bottomland hardwood forest, upland hardwood forest, open lands, freshwater river swamp, and freshwater nonriver swamp) in terms of the factors listed in the footnotes to Table 19. Functional relationships are used to transform objective information on the factors into a subjective quality scale ranging from 0.0 (undesirable) to 1.0 (desirable). Each factor is assigned a weight which reflects its relative importance in describing habitat quality. The importance weights were assigned through a joint effort by 20 biologists. Again, this habitat index approach is useful for describing the environment. Quantification of the changes in each factor for each relevant habitat type, as well as the quantities of the types, is necessary for impact prediction.

Simple statistical models can also be useful for impact prediction. One example is a study to assess the effects of motorboat usage on the long-term lead concentration of a multipurpose reservoir, Turlock Lake in Stanislaus County, California (Byrd and Perona, 1979). An apparent correlation was found between the lead level at the boat dock and the boats per-unit-volume. The lake inlet and outlet lead data obtained were treated in terms of a simple plug-flow model. The increase in the lead level of the water as it passes through the lake is greater, by at least a factor of 15, than would be expected on the basis of the number of boats on the lake. Experiments were done to find if an alternate source of lead in the lake is the sediment. The mechanism linking lake volume and lead concentrations is unclear but is likely to involve migration of lead from the sediment, which was found to contain exchangeable lead, and/or biotic processes.

Driscoll, DiToro and Thomann (1979) developed a simplified statistical-based methodology which can be used to assess the impact of urban stormloads on the quality of receiving waters. The methodology is particularly appropriate for use at the planning level where preliminary assessments are made to define problems, establish the relative significance of contributing sources, assess feasibility of control, and determine the need for the focus of additional evaluations. It can also be used effectively in conjunction with detailed studies by providing cost-effective screening of an array of alternatives, so that the more detailed and sophisticated techniques can examine only the more attractive alternatives. The methodology is based on the determination of certain statistical properties of the rainfall history of an area. From these statistics, the desired information on loads, performance of controls, and receiving water impacts is generated directly.

A third example of a statistical model was that developed between phosphate phosphorous (PO4-P) and chlorophyll-a at the Roodeplatt Dam in South Africa (Fieterse and Toerien, 1978). Many of the impoundments in South Africa have excessive algae and/or macrophyte growths, which result in poorer water quality. Surface samples were taken at various points in the Roodeplatt impoundments and analyzed for inorganic nitrogen and phosphorous ions. Regression analysis of the averaged data indicated that algal growth was limited by the PO4-P concentration rather than by inorganic nitrogen. A reduction in the PO4-P concentration would then reduce algal growth, and PO4-P concentration data is more applicable to South Africa conditions than Total P values. A model for eutrophication was developed for the Roodeplant Dam which may be applicable to similar bodies of water.

Two other examples of simple models for impact prediction include an input-output model and a steady-state model. An input-output phosphorus lake model has been developed for quantifying the relationship between land use and lake trophic quality (U.S. Environmental Protection Agency, 1980). When the model is employed to predict the impact of projected land use changes, it is necessary to use phosphorus export coefficients extrapolated from other points in time and/or space. These coefficients represent the mass loading of phosphorus to a surface water body per year per unit of source. The model includes an error estimation procedure, and is applicable to a fairly wide range of lake types. A steady state eutrophication model which simplifies assumptions of the kinetic and transport equations has been developed for use in the United States (Schnoor and O'Connor, 1980). Use of the model requires the estimation of the sedimentation, hydrolysis, autocatalytic growth, and death rate constants. The sedimentation rate constant determines the amount of phosphorus lost to the deep sediment, and the total phosphorus levels of the lake. The other trace constants control the partitioning of nutrients among the various organic, inorganic, and phytoplankton fractions.

Quantitative Approach with Multi-factor Models

Numerous sophisticated models are available for usage in predicting the aquatic impacts of dams and hydroelectric projects. Several general reports and literature reviews have been prepared on applicable models. Warner et al. (1974) presents materials intended for use by reviewers of environmental impact statements on major water reservoir projects. The section on water quality impacts contains a detailed comparison of mathematical models for predicting impacts on water temperature, dissolved oxygen, and several chemical constituents. Orlob (1977) presents a review of the state-of-the-art of mathematical modeling of surface water impoundments. Models reviewed included one-dimensional models for simulation of temperature and water quality in stratified reservoirs, two-dimensional circulation and water quality in shallow lakes, two-dimensional stratified flow, circulation in multi-layer large lakes, and eutrophication and ecological responses in lake systems.

Due to numerous types and classifications of aquatic impact models a complete review is beyond the scope of this chapter. However, information on examples of types of models will be presented. Included will be models for storm water impacts, sediment deposition, temperature profiles, and general water chemistry changes. Models for biological productivity, eutrophication, and toxic substances uptake will also be described along with one example of a reservoir operations model.

Freedman, Canale and Pendergast (1980) developed a mathematical model to predict the transient impact of storm loads on phosphorus, fecal coliform, and dissolved oxygen concentrations in Onondage Lake in Canada. Model simulations demonstrated that combined sewer and storm loads have a significant impact on lake fecal coliform but little effect on phosphorus, CBOD, NBOD, and dissolved oxygen in Onondage Lake. Observed variations in lake dissolved oxygen were caused by changes in chlorophyll-a, light, and wind.

An issue of concern in many reservoirs is sediment deposition. A computer-based model called SEDRES has been developed for calculation of the amounts, rates, and spatial distributions of sediment in lakes and reservoirs (Karim, Croley and Kennedy, 1979). The principal components of SEDRES compute the following: sediment entrapment, distribution, and differential settling for three different size classes (clay, silt and sand); compaction of currently and all previously deposited sediments; correction to zero elevation for compaction; sediment slump correction due to compaction at zero elevation and at sediment-type interfaces; and alteration of the elevation-area-capacity relation due to sedimentation. Inputs to the model are water inflows, reservoir operation levels, original reservoir elevation-area-capacity relation, sediment characteristics, type of sediment-entrapment and sediment-distribution methods. The time-interval for simulation may be one week or any multiple thereof.

One-dimensional to three-dimensional temperature models have been developed for predicting thermal stratification and temperature profiles in reservoirs. Ford and Stefan (1980) used a one-dimensional integral energy (mixed-layer) model to simulate the seasonal temperature cycle of three, morphometrically different, temperate lakes. In the model, turbulent kinetic energy supplied by wind shear was used to entrain denser water into the upper mixed layer by working against gravity. The model was calibrated with data from one lake for one year and verified against data from two other lakes and also against data from other years. Predictions of the onset of stratification, surface and hypolimnetic temperatures, mixed layer depths, and periods of turnover were all in agreement with data.

A two-layer mathematical model for water temperature prediction in stratified reservoirs has been presented by Rahman (1979). The model included the nonlinear effects of the physical properties of water and the heat budget which accounts for vertical motion of the water body. Temperature profiles were determined by considering variable density, conductivity, and diffusivity, taking into account the influence of vertical motion of water but neglecting the effects of horizontal currents. An analysis of profiles calculated under various assumptions indicated the importance of the major parameters.

A pseudo three-dimensional time-dependent analytical model was developed by Uzzell and Ozisik (1978) for the prediction of temperature distributions in lakes resulting from thermal discharges. For the special case of lakes having a uniform depth, rectangular geometry, and a constant axial velocity, explicit analytical solutions were presented. Sample calculations were performed to demonstrate the effects of the time variation of the thermal loading, different types of boundary conditions, circulation velocities, and eddy conductivity coefficients on the surface temperature distribution along the lake as a function of time.

A multi-segment deep reservoir water quality simulation model has been developed for the U.S. Environmental Protection Agency (Baca et al., 1974). The model contains a hydrothermal submodel and a water quality submodel. The hydrothermal submodel accurately predicts vertical temperature profiles with little or no subjective effort and requires only standard meteorological data to predict seasonal temperature variations. A water quality submodel simulates natural seasonal patterns of algal growth and death and nutrient cycling, and predicts DO-BOD dynamics in reservoirs and impoundments. Principal environmental variables such as dissolved oxygen, total and benthic BOD, phytoplankton, zooplankton, nitrogen and phosphorus, toxic materials, and coliform bacteria can be predicted.

Formulation, calibration, and verification of a eutrophication model and a limnological model for predicting and simulating water quality changes in municipal water supply reservoirs of Adelaide, Australia, are described by Baca et al. (1977a). These computer models apply to both shallow and deep lakes and reservoirs. The eutrophication model incorporates inflows and outflows, fluctuations of the thermocline, nutrient fixation and mineralization, and sediment-water interactions to simulate monthly changes of four eutrophication indicators: (1) soluble phosphorus, (2) total phosphorus, (3) chlorophyll-a, and (4) Secchi disc depth. The limnological model is based on dynamics of heat and mass transport, hydromechanics, and chemical and biological transformations. The model simulates daily vertical and horizontal variations of: (1) water flow and temperature, (2) phytoplankton and zooplankton biomass, (3) nitrogen and phosphorus forms, (4) BOD, (5) DO, (6) total dissolved solids, and (7) suspended sediments. A user's manual for the two models is available (Baca et al., 1977b).

A numerical model designed for studying the complex relationships that exist between chemical, physical, and biological processes which occur in deep stratified lakes of the temperate zone is described by Walters (1980). Results of a mathematical model of the thermal stratification cycle of a deep lake are combined with a phytoplankton growth and nutrient concentration model to ensure consistency of the vertical eddy diffusion of algal cells and dissolved nutrients with the mixing processes that determine the lake's thermal stratification. Turbulent mixing processes in the thermal model controlled the chlorophyll-a and distribution of nutrients. The thermal model utilizes a heat diffusion equation which is nonlinear and reflects the interaction of wind-induced turbulence and buoyancy gradients related to surface heating and cooling. Changes in surface heat are described by standard meteorological parameters. A pair of coupled, nonlinear partial differential equations are used to form the biological production model.

A comprehensive water quality model for the Nile River Basin in Egypt has been developed (Jorgensen, 1980). Submodels are included for algal growth; zooplankton; fisheries; nutrient exchange between sediment and water; mass balances of P, N, Si, O, and C; snails; DDT; Cu; and the water and salt balance. Flow charts for each submodel and calibration data are available to simplify the understanding of the main model. The water quality model will be an important management tool and will also help in the decision-making process and in the planning of environmental policy.

A mathematical simulation model describing shoreline vegetative succession in response to reservoir water level fluctuations has been developed by Austin, Riddle and Landers (1979). Plant species are grouped into ecologically similar compartments. Differential equations describing compartment intrinsic growth, intraspecies competition, interspecies competition, and other growth limiting factors are solved numerically. The model has been used to evaluate the impacts of various operating policies on plant succession for a new reservoir in central Iowa.

Thomann (1979) has discussed the development of models for simulating the distribution and dynamics of toxic substances within an ecosystem. In order to incorporate both bioaccumulation of toxic substances directly from the water and subsequent transfer up the food chain, a mass balance model is constructed that introduces organism size as an additional independent variable. The principal factors that influence the total toxicant concentration in various regions of the food chain include excretion and uptake rates, the rate of decrease of biomass density with organism size and the food chain transfer velocity, a parameter reflecting average predation along the food chain. The analysis of some PCB data from Lake Ontario in Canada is used as an illustration of the theory. The introduction of organism size as an independent variable in the mass balance of a toxicant provides a generalized analysis framework; this permits the integrated use of diverse laboratory experiments on uptake and excretion as well as an interpretive framework for field data of toxicant concentrations.

In addition to using models to predict the impacts of the construction and basic operation of dams and hydroelectric projects, models can also be used to predict environmental changes from various operational procedures. For example, Austin, Landers and Dougal (1978) have developed mathematical models to simulate the effects of fluctuating water levels in multipurpose reservoirs in Iowa. The models are designed as management and operational tools to evaluate trade-offs between the environmental impacts in the flood pool upstream of the dam and the economic benefits downstream of the dam. A dynamic programming optimization model was developed to select an optimal operating policy. The model is forward looking and has only a single decision variable, the release rate from the reservoir.

IMPACT ASSESSMENT

Assessment or interpretation of predicted impacts represents a vital activity in an environmental impact study. Decisions based on this activity include whether or not to approve a proposed project, whether or not to prepare an environmental impact statement, and the necessity for identification and inclusion of mitigation measures. Impact assessment requires the considerable exercise of professional judgment along with environmental standards or criteria and other scientific information. Warner et al. (1974) contains extensive citations to relevant literature on impact assessment for water quality and ecological impacts; and economic, social and aesthetic impacts. Interpretation of water environment impacts involved the use of water quality standards or criteria which give consideration to multiple uses of water resources and the quality requirements associated with the uses. Within the United States, water quality standards have been established for stream segments, river basins, lakes, estuaries, and coastal areas. A sound approach for impact assessment is to evaluate the calculated impacts relative to existing and resultant water quality if the project is implemented.

One of the difficult areas is related to interpretation of anticipated changes on the aquatic ecosystem. There are some laws and executive orders within the United States which address aquatic biological features, including Executive Order 11990 (Protection of Wetlands); Coastal Zone Management Act of 1972; Deep Water Port Act of 1974; Endangered Species Act amendments of 1978; Fish and Wildlife Coordination Act of 1966; Marine Mammal Protection Act of 1972; Marine Protection, Research and Sanctuaries Act of 1972; and the Clean Water Act of 1977. Most of these regulatory documents provide general guidance for protection of the aquatic environment; however, specific standards for aquatic species or species diversity are not included. The most appropriate technical approach to utilize in aquatic ecosystem impact assessment involves the application of specific biological principles and recommended criteria. Two examples will be cited--one for instream flow requirements and the other for temperature criteria for fish.

It is well known that fish and other aquatic forms are dependent on stream flow and quality. Construction and operation of dams or hydroelectric projects may alter both of these; therefore, it may be necessary to determine the minimum instream flow requirements for various aquatic forms. Stalnaker and Arnette (1976) summarize and evaluate techniques for determining instream flow requirements and assessing the effects of changing stream flows on fish, terrestrial wildlife, and water quality. Bovee and Cochnauer (1977) document the methods and procedures used in the construction of probability criteria curves. Weighted criteria are used to assess the impacts of altered streamflow regimes on a stream habitat. They are developed primarily for those habitat parameters most closely related to stream hydraulics; depth, velocity, substrate, and temperature. Guidelines for data collection, analysis and curve development are presented.

Dams and hydroelectric projects can also alter water temperatures, thus a relevant concern is related to temperature requirements for freshwater fish. Brungs and Jones (1977) summarize temperature criteria for 34 freshwater fish species expressed as mean and maximum temperatures; the means control functions such as embryogenesis, growth, maturation, and reproductivity; and maxima provide protection for all life stages against lethal conditions. The presented criteria are based on numerous field and laboratory studies.

Impact assessment may aid in identifying impacts of sufficient concern that mitigation measures would be warranted. Mitigation may include avoiding the impact altogether by not taking a certain action or parts of an action; minimizing impacts by limiting the degree or magnitude of the action and its implementation; rectifying the impact by repairing, rehabilitating, or restoring the affected environment; reducing or eliminating the impact over time by preservation and maintenance operations during the life of the actions; and/or compensating for the impact by replacing or providing substitute resources or environments. Appropriate mitigation measures must be identified on a project specific basis. However, two general examples can be cited for minimizing water environment impacts. One appropriate approach is to attempt to minimize the nonpoint source pollution that would occur during the construction phase of a project. Ripken, Killen and Gulliver (1977) discuss the nature and amount of solids that may be transported by runoff at construction sites. They also review and evaluate potential control methods. Anton and Bunnell (1976) present general guidelines for minimizing erosion from construction sites.

Aquatic weeds can also be a problem in reservoir operation. A workshop was recently held to review the state-of-the-art of chemical, biological, mechanical, and integrated control of aquatic weeds (U.S. Environmental Protection Agency, 1981). Chemical control is the dominant control method while biological controls have shown some success. The cost of harvesting and disposal of weeds after mechanical control is often prohibitive, but the method allows rapid removal of plants. Integrated control, using two or more of the above methods can achieve more precise vegetation management. Possible combinations include: herbicide treatment followed by stocking with fish, or pathogen application; mechanical harvesting followed by fish or pathogens; treatment with insects followed by pathogens; and mechanical or chemical control followed by competitive plants.

METHODOLOGIES FOR TRADE-OFF ANALYSES AND DECISION-MAKING

Environmental impact studies typically address a minimum of 2 alternatives, and they can include upwards of 50 alternatives. Typical studies address 3 to 5 alternatives. The minimum number typically represents a choice between construction and operation of a project versus project nonapproval. Alternatives for dams and hydroelectric projects may also include project construction and operation at different sites; differences in design and operational procedures, including the incorporation of various mitigation measures; and timing options for the construction and operational phases. Depending upon the project need, still other alternatives for flood control, water supply, recreation, and energy supply could be included in the analysis. Therefore, a systematic environmental impact study should incorporate methodologies for trade-off analyses and decision-making between alternatives.

Several types of methodologies have been used in environmental impact studies on dams and hydroelectric projects, including matrices, scaling or ranking checklists, and scaling-weighting or ranking-weighting checklists. These methodologies provide a structure to the analysis and form the basis for comparisons of alternatives. King (1978) describes the use of matrices to examine the environmental effects of hydraulic structures. An example is included on the effects of an impoundment on downstream water quality. Yorke (1978) discusses a dual-matrix system for reviewing and evaluating the impacts of water development projects on fish and wildlife resources. The generalized matrix presented consists of summary statements of the impacts of common water development projects on selected physical and chemical characteristics of streams.

Davos (1977) discusses the use of three matrices for trade-off analysis and decision-making. He calls this the priority-tradeoff-scanning (PTS) approach. Three objectives are identified: (1) to record the impact of each option on all goals and interest groups; (2) to scan all feasible goal-priority trade-offs for each interest group; and (3) to scan priority trade-offs that are actually acceptable to those groups. On the decision-making evaluation level, the objectives of the PTS approach are: (1) to identify decision choices which will maximize consensus on goal priorities; (2) to identify choices which will maximize the satisfaction of individual interests of competing groups; and (3) to scan priority trade-offs for maximizing achievement of goals, satisfying the aspirations of planners, and maximizing consensus. An approach to evaluation is suggested, rather than an evaluation methodology, in providing a framework for interpreting and synthesizing all information inputs pertinent to evaluation, instead of generating particular information inputs. In PTS, pertinent information for each of the three evaluation objectives is synthesized in the form of a matrix: (1) goals-achievement matrix (GAM); (2) goal-priority-tradeoff matrix (GPTM); and (3) interest-priority-tradeoff matrix (IPTM).

Budge (1981) and Whitlatch (1976) suggest the conjunctive use of matrices with other methodologies. Budge (1981) outlines a method that has been used in conjunction with engineering, economic and hydrological criteria. It consists of the identification of environmental impacts through the use of a matrix, the collection and collation of information on each impact and the comparison of options, using a combination of ordinal rankings in preference to cost-benefit techniques. The methodology is illustrated with reference to water resource developments. Whitlatch (1976) suggests the use of matrix or stepped matrix techniques in conjunction with linear vector or nonlinear evaluation systems.

Checklist methodologies can involve the scaling or ranking of the impacts of alternatives on each of the environmental factors under consideration. Scaling techniques include the use of numerical scores, letter assignments, or linear proportioning. Alternatives can be ranked from best to worst in terms of potential impacts on each environmental factor. Duke et al. (1977) described a scaling checklist for evaluation of water resources projects. Pertinent environmental factors are identified through use of a simple interaction matrix. Scaling is accomplished following the establishment of an evaluation guideline for each environmental factor. An evaluation guideline is defined as the smallest change in the highest existing quality in the region that would be considered significant. For example, assuming that the highest existing quality for dissolved oxygen in a region is 8 mg/l, if a reduction of 1.5 mg/l is considered as significant, then the evaluation guideline is 1.5 mg/l irrespective of the existing quality in a given regional stream. Scaling of impacts is accomplished by quantifying the impact of each alternative relative to each environmental factor, and if the net change is less than the evaluation guideline it is considered to be insignificant. If the net change is greater than the evaluation guideline and moves the environmental factor toward its highest quality, then it is considered to be a beneficial impact; the reverse exists for those impacts that move the measure of the environmental factor away from its highest existing quality. Sondheim (1978) describes a methodology devised in response to a problem involving whether or not a dam should be constructed at a given site. The method uses interval or ratio rating schemes instead of ordinal ones for the impacts of alternatives on environmental factors.

Scaling-weighting or ranking-weighting checklist methodologies involve the assignment of importance weights to environmental factors and the scaling or ranking of the impacts for each alternative on each factor. Resultant comparisons can be made through the development of a product matrix which consists of multiplying importance weights by the scale or rank for each alternative. An early scaling-weighting methodology for water resources projects was developed by Dee et al. (1972). Included are 78 environmental factors defined within the categories of ecology, environmental pollution, aesthetics, and human interest. Weighting is accomplished through use of the ranked pairwise comparison technique, while scaling is achieved via the use of functional curves. As noted earlier, functional relationships are used to relate the objective evaluation of an environmental factor to a subjective judgment regarding its quality based on a range from high quality to low quality. Final product matrices developed for each alternative can be used as a basis for trade-off analyses and decision-making.

Yapijakas and Molof (1981) presented a decision-making method for evaluating alternatives for multinational river basin development. Factors considered are benefit-cost ratio, capital outlay, environmental and social impacts, and manageability/technology level. Each government creates a coordinating group, a weighting panel, and a rating panel, each composed of specialists and experts. The coordinating group lists project alternatives, defines the factors in terms of its government's policies, and chooses the rating and weighting panels. The weighting panel establishes a weighting scheme for each of the aspects. Each member of the rating panel judges the alternatives in their field of expertise by any method--model construction, experiments, point assignment procedures, and others. Final selection of an alternative is done by one of two methods: ranking the alternatives from all countries in a single matrix on an equal weight basis; or producing a single list from the combined project ratings from each country.

The conceptual basis for scaling-weighting and ranking-weighting checklists is from multiattribute utility theory. Environmental problems usually involve multiple conflicting objectives, large uncertainties concerning the possible environmental impact, and several individuals or groups whose preferences are very different, but yet very important in choosing an alternative. If one wishes to influence the decision-making process using analysis, the above issues should be addressed. One critical aspect, which is usually conducted informally, involves considering the advantages and disadvantages of the possible impacts of the various alternatives by each of the interested parties. Keeney (1976) presents the basic ideas and examples of the usage of multiattribute utility in the context of a decision-making framework.

Despite the availability of many scientific tools, determination of the environmental impacts of dams and hydroelectric projects has many uncertain components. Duckstein et al. (1977) have developed a conceptual framework to address uncertainty in environmental impact studies. Current procedures used to assess environmental impact do not take proper cognizance of uncertainty and the problems of trade-off among multiple objectives in the analysis and evaluation of actions affecting the environment. Some specific and legal concepts such as proof and model validation must be thought of in terms of probability, if optimal decisions are to be made affecting the future environment. The methodology developed by Duckstein et al. (1977) shows how uncertain information and scientific models can be used to advantage in planning the future.

Two other types of methodologies which can be useful include simulation and computer graphics. Hodgins, Wisner and McBean (1977) describe a computer simulation model which was used to screen the most promising alternatives for a series of eastern Canadian reservoirs. The model, with both hydrologic and water quality components, simulates and thereby indicates probable changes in downstream flows, reservoir surface fluctuations, and temperature and dissolved oxygen changes in the reservoirs and streams. Through easily adjusted operating policies, reservoir sizes, and other factors, the model can rapidly determine the potential impact of alternative possible developments. However, computer simulation and optimization models that are used to assist in multipurpose, multiobjective water resource planning often suffer from the lack of an easy yet comprehensive means of interpreting and communicating the results of model studies to others. These deficiencies may be minimized with the help of computer graphic input and display methods. French et al. (1980) applied interactive computer graphics to four planning problems, which included the prediction and management of water quality, multireservoir simulation for water supply, multiobjective analyses for reservoir sizing, cost and yields, and flood management. Tablet digitizing routines were frequently used to input spatial and other data, while the graphical output was accomplished by vector display methods. Visual feedback was obtained at all stages of the procedures.

SUMMARY

Construction and operation of large dams and hydroelectric projects can cause significant environmental impacts. Systematic environmental impact studies are needed to predict and assess these impacts and identify appropriate mitigation measures. Extensive technical literature is available to aid in impact identification and planning and conduction of appropriate studies of the baseline environment. Impact prediction can be achieved via (1) a qualitative approach which relies on general knowledge of the impacts of similar projects, or specific results of comprehensive studies of similar projects; (2) a quantitative approach based on the use of simple mass balance and environmental dilution calculations; and (3) a quantitative approach based on the use of mathematical models for multiple environmental factors. A given environmental impact study will probably involve all three approaches. Impact assessment involves the use of water quality standards and criteria, appropriate laws and regulations related to the biological environment, and professional knowledge and judgment. Systematic methods are also available for comparing alternatives for large dams and hydroelectric projects.

Despite the availability of technical information, many environmental impact studies have not used available scientific methods and approaches. Some possible reasons for this lack of extensive use include lack of knowledge about available approaches on the part of many practitioners, nonexistence of current technology during the early years following the initiation of environmental impact studies, and general reluctance to use approaches perceived to be difficult and time- and cost-consuming. However, usage of scientific methods and techniques is expected to increase as a result of the expanding knowledge base for conducting studies, and the emphasis being given to public justification and accountability in project planning and decision-making.

SELECTED REFERENCES

Abu-Zeid, M., "Short and Long-Term Impacts of the River Nile Projects", Water Supply and Management, Vol. 3, No. 4, 1979, pp. 275-283.

Ahlgren, I., "A Dilution Model Applied to a System of Shallow Eutrophic Lakes After Diversion of Sewage Effluents", Archive fur Hydrobiologie, Vol. 89, No. 1/2, June 1980, pp. 17-32.

Ahmed, R. and Schiller, R.W., "A Methodology for Estimating the Loads and Impacts of Non-Point Sources on Lake and Stream Water Quality", Proceedings of a Technical Symposium on Non-Point Pollution Control--Tools and Techniques for the Future, Technical Publication 81-1, Jan. 1981, Interstate Commission on the Potomac River Basin, Rockville, Maryland, pp. 154-162.

Anton, W.F. and Bunnel, J.L., "Environmental Protection Guidelines for Construction Projects", Journal of American Water Works Association, Vol. 68, No. 12, Dec. 1976, pp. 643-646.

Austin, T.A., Landers, R.Q. and Dougal, M.D., "Environmental Management of Multipurpose Reservoirs Subject to Fluctuating Flood Pools", Technical Completion Report No. ISWRRI-84, June 1978, Water Resources Research Institute, Iowa State University, Ames, Iowa.

Austin, T.A., Riddle, W.F. and Landers, Jr., R.Q., "Mathematical Modeling of Vegetative Impacts from Fluctuating Flood Pools", Water Resources Bulletin, Vol. 15, No. 5, Oct. 1979, pp. 1265-1280.

Baca, R.G. et al., "A Generalized Water Quality Model for Eutrophic Lakes and Reservoirs", Nov. 1974, Battelle Pacific Northwest Laboratory, Richland, Washington.

Baca, R.G. et al., "Water Quality Models for Municipal Water Supply Reservoirs. Part 2. Model Formulation, Calibration and Verification", Jan. 1977a, Battelle Pacific Northwest Laboratory, Richland, Washington.

Baca, R.G. et al., "Water Quality Models for Municipal Water Supply Reservoirs. Part 3. User's Manual", Jan. 1977b, Battelle Pacific Northwest Laboratory, Richland, Washington.

Berkes, F., "Some Environmental and Social Impacts of the James Bay Hydroelectric Project, Canada", Journal of Environmental Management, Vol. 12, No. 2, Mar. 1981, pp. 157-172.

Bovee, K.D. and Cochnauer, T., "Development and Evaluation of Weighted-Criteria, Probability-of-Use Curve for Instream Flow Assessments: Fisheries", Report No. FWS/OBS-77/63, IFIP-3, Dec. 1977, U.S. Fish and Wildlife Service, Fort Collins, Colorado.

Brungs, W.A. and Jones, B.R., "Temperature Criteria for Freshwater Fish: Protocol and Procedures", EPA/600/3-77/061, May 1977, U.S. Environmental

Protection Agency, Duluth, Minnesota.

Budge, A.L., "Environmental Input to Water Resources Selection", Water Science and Technology, Vol. 13, No. 6, 1981, pp. 39-46.

Byrd, J.E. and Perona, M.J., "The Effect of Recreation on Water Quality", Technical Completion Report, Jan. 1979, California Water Resources Center, University of California, Davis, California.

Canter, L.W., Environmental Impact Assessment, McGraw-Hill Book Company, New York, New York, 1977, 331 pages.

Canter, L.W. and Hill, L.G., Handbook of Variables for Environmental Impact Assessment, Ann Arbor Science Publishers, Inc., Ann Arbor, Michigan, 1979, 203 pages.

Chalmers, J.A. and Anderson, E.J., "Economic/Demographic Assessment Manual: Current Practices, Procedural Recommendations, and a Test Case", Nov. 1977, Engineering and Research Center, U.S. Bureau of Reclamation, Denver, Colorado, (prepared for Bureau of Reclamation by Mountain West Research, Inc., Tempe, Arizona).

Davos, C.A., "A Priority-Tradeoff-Scanning Approach to Evaluation in Environmental Management", Journal of Environmental Management, Vol. 5, No. 3, 1977, pp. 259-273.

Dee, N. et al., "Environmental Evaluation System for Water Resources Planning", Final Report, 1972, Battelle-Columbus Laboratories, Columbus, Ohio (prepared for the Bureau of Reclamation, U.S. Department of the Interior, Washington, D.C.).

Driscoll, E.D., DiToro, D.M. and Thomann, R.V., "A Statistical Method for Assessment of Urban Stormwater", Final Report, May 1979, Water Planning Division, U.S. Environmental Protection Agency, Washington, D.C.

Duckstein, L. et al., "Practical Use of Decision Theory to Assess Uncertainties About Actions Affecting the Environment", Completion Report, Feb. 1977, Department of Systems and Industrial Engineering, Arizona University, Tucson, Arizona.

Duke, K.M. et al., "Environmental Quality Assessment in Multi-objective Planning", Nov. 1977, Final Report to U.S. Bureau of Reclamation, Denver, Colorado (prepared for BuRec by Battelle-Columbus Laboratories, Columbus, Ohio).

Entz, B., "Ecological Aspects of Lake Nasser-Nubia", Water Supply and Management, Vol. 4, No. 1-2, 1980, pp. 67-72.

Federal Energy Regulatory Commission, "Final Environmental Impact Statement: North Fork Payette River Project Number 2930-Idaho", FERC/EIS-0027, Oct. 1981, Washington, D.C.

Fieterse, A.J.H. and Toerien, D.F., "The Phosphorus-Chlorophyll Relationshp in Roodeplaat Dam", Water SA (Pretoria), Vol. 4, No. 3, 1978, pp. 105-112.

Flood, B.S. et al., "A Handbook for Habitat Evaluation Procedures", Resource Publ. 132, 1977, 77 pages, U.S. Fish and Wildlife Service, Washington, D.C.

Ford, D.E. and Stefan, H.G., "Thermal Predictions Using Integral Energy Model", Journal of the Hydraulics Division, American Society of Civil Engineers, Vol. 106, No. HY1, Jan. 1980, pp. 39-55.

Freedman, P.L., Canale, R.P. and Pendergast, J.F., "Modeling Storm Overflow Impacts on a Eutrophic Lake", Journal of Environmental Engineering Division, American Society of Civil Engineers, Vol. 106, No. EE2, Apr. 1980, pp. 335-349.

Freeman, P.H., "The Environmental Impact of a Large Tropical Reservoir: Guidelines for Policy and Planning, Based Upon a Case Study of Lake Volta, Ghana, in 1973 and 1974", 1974, Smithsonian Institution, Washington, D.C.

French, P.N. et al., "Water Resources Planning Using Computer Graphics", Journal of the Water Resources Planning and Management Division, American Society of Civil Engineers, Vol. 106, No. WR1, Mar. 1980, pp. 21-42.

Gallopin, G., Lee, T.R. and Nelson, M., "The Environmental Dimension in Water Management: The Case of the Dam at Salto Grande", Water Supply and Management, Vol. 4, No. 4, 1980, pp. 221-241.

Hagan, R.M. and Roberts, E.B., "Energy Impact Analysis in Water Project Planning", Journal of the Water Resources Planning and Management Division, American Society of Civil Engineers, Vol. 106, No. WR1, Mar. 1980, pp. 289-302.

Hazel, C. et al., "Assessment of Effects of Altered Stream Flow Characteristics on Fish and Wildlife, Part B: California, Case Studies", FWS/OBS-76/34, Dec. 1976, U.S. Fish and Wildlife Service, Washington, D.C.

Hellawell, J.M., Biological Surveillance of Rivers: A Biological Monitoring Handbook, 1978, 332 pp., Water Research Center, Stevenage, England.

Hodgins, D.B., Wisner, P.E. and McBean, E.A., "A Simulation Model for Screening a System of Reservoirs for Environmental Impact", Canadian Journal of Civil Engineering, Vol. 4, No. 1, Mar. 1977, pp. 1-9.

IHD-WHO Working Group on the Quality of Water, "Water Quality Surveys", Studies and Reports in Hydrology - 23, 1978, 350 pp., United Nations Educational Scientific and Cultural Organization, Paris, France, and World Health Organization, Geneva, Switzerland.

Jacobs, F. and Grant, G.C., "Guidelines for Zooplankton Sampling in Quantitative Baseline and Monitoring Programs", EPA/600-3-78/026, Feb. 1978, 62 pp., Virginia Institute of Marine Science, Gloucester Point, Virginia.

Johnston, J.B., Benson, N.G. and King, III, B.D., "Values of Ecological Characterization Studies to Assess Effects of Fresh Water Inflow to Estuaries", Proceedings of the National Symposium on Fresh Water Inflow to Estuaries, FWS/OBS-81-04, Vol. II, Oct. 1981, U.S. Fish and Wildlife Service, Washington, D.C., pp. 155-164.

Jorgensen, S.E., "Water Quality and Environmental Impact Model of the Upper Nile Basin", Water Supply and Management, Vol. 4, No. 3, 1980, pp. 147-153.

Karim, F., Croley, II, T.E. and Kennedy, J.F., "A Numerical Model for Computation of Sedimentation in Lakes and Reservoirs", Completion Report No. 105, 1979, Water Resources Research Institute, Iowa State University, Ames, Iowa.

Keeney, R.L., "Preference Models of Environmental Impact", IIASA-RM-76-4, Jan. 1976, 23 pp., International Institute for Applied Systems Analysis, Laxenburg, Austria.

Kelly, D.M., Underwood, J.K. and Thirumurthi, D., "Impact of Construction of a Hydroelectric Project on the Water Quality of Five Lakes in Nova Quality Surveys", Studies and Reports in Hydrology - 23, 1978, 350 pp., United Nation Scotia, Canadian Journal of Civil Engineering, Vol. 7, No. 1, 1980, pp. 173-184.

Kenyon, G.F., "The Environmental Effects of Hydroelectric Projects", Canadian Water Resources Journal, Vol. 6, No. 3, 1981, pp. 309-314.

King, D.L., "Environmental Effects of Hydraulic Structures", Journal of Hydraulics Division, American Society of Civil Engineers, Vol. 104, No. 2, Feb. 1978, pp. 203-221.

Martin, C.W., Noel, D.S. and Federer, C.A., "The Effect of Forest Clear-Cutting in New England on Stream Water Chemistry and Biology", Research Report 34, July 1981, Water Resources Research Center, University of New Hampshire, Durham, New Hampshire.

McClellan, B.E. and Frazer, K.J., "An Environmental Study of the Origin, Distribution, and Bioaccumulation of Selenium in Kentucky and Barkley Lakes", Research Report No. 122, 1980, Water Resources Research Institute, University of Kentucky, Lexington, Kentucky.

Miracle, R.D. and Gardner, Jr., J.A., "Review of the Literature on the Effects of Pumped Storage Operations on Ichthyofauna", Proceedings of the Clemson Workshop on Environmental Impacts of Pumped Storage Hydroelectric Operations, FWS/OBS-80/28, Apr. 1980, U.S. Fish and Wildlife Service, Washington, D.C., pp. 40-53.

Nelson, W. et al., "Assessment of Effects of Stored Stream Flow Characteristics on Fish and Wildlife, Part A: Rocky Mountains and Pacific Northwest (Executive Summary)", Publication No. FWS/OBS-76/28, Aug. 1976, Environmental Control, Inc., Rockville, Maryland.

Nix, J., "Distribution of Trace Elements in a Warm Water Release Impoundment", Oct. 1980, Water Resources Research Center, University of Arkansas, Fayetteville, Arkansas.

Orlob, G.T., "Mathematical Modeling of Surface Water Impoundments, Volume I and II", 1977, Resource Management Associates, Lafayette, California.

Ott, W.R., Environmental Indices--Theory and Practice, 1978, Ann Arbor Science Publishers, Inc., Ann Arbor, Michigan, pp. 202-213.

Petts, G.E., "Morphological Changes of River Channels Consequent Upon Headwater Impoundment", Journal of the Institution of Water Engineers and Scientists, Vol. 34, No. 4, July 1980, pp. 374-382.

Ploskey, G.R., "Fluctuating Water Levels in Reservoirs: An Annotated Bibliography on Environmental Effects and Management for Fisheries", Tech. Report E-82-5, May 1982, U.S. Army Engineer Waterways Experiment Station, Vicksburg, Mississippi.

Rahman, M., "Temperature Structure in Large Bodies of Water: Analytical Investigation of Temperature Structure in Large Bodies of Stratified Water", Journal of Hydraulic Research, Vol. 17, No. 3, 1979, pp. 207-215.

Reynolds, P.J. and Ujjainwalla, S.H., "Environmental Implications and Assessments of Hydroelectric Projects", Canadian Water Resources Journal, Vol. 6, No. 3, 1981, pp. 5-19.

Ripken, J.F., Killen, J.M. and Gulliver, J.S., "Methods for Separation of Sediment from Storm Water at Construction Sites", EPA-600/2-77-033, 1977, U.S. Environmental Protection Agency, Washington, D.C.

Sanders, T.G., editor, "Principles of Network Design for Water Quality Monitoring", July 1980, 312 pp., Colorado State University, Ft. Collins, Colorado.

Sargent, F.O. and Berke, P.R., "Planning Undeveloped Lakeshore: A Case Study on Lake Champlain, Ferrisburg, Vermont", Water Resources Bulletin, Vol. 15, No. 3, June 1979, pp. 826-837.

Schnoor, J.L. and O'Connor, D.J., "A Steady State Eutrophication Model for Lakes", Water Research, Vol. 14, No. 11, Nov. 1980, pp. 1651-1665.

Sondheim, M.W., "A Comprehensive Methodology for Assessing Environmental Impact", Journal of Environmental Management, Vol. 6, No. 1, Jan. 1978, pp. 27-42.

Stalnaker, C.B. and Arnette, J.L., "Methodologies for the Determination of Stream Resource Flow Requirements: An Assessment", FWS/OBS-76/03, Apr. 1976, U.S. Fish and Wildlife Service, Washington, D.C.

Stofan, P.E. and Grant, G.C., "Phytoplankton Sampling in Quantitative Baseline and Monitoring Programs", EPA/600/3-78-025, Feb. 1978, Virginia Institute of Marine Science, Gloucester Point, Virginia.

Teskey, R.O. and Hinckley, T.M., "Impact of Water Level Changes on Woody Riparian and Wetland Communities. Volume I: Plant and Soil Responses to Flooding", Office of Biological Services Report 77/58, Dec. 1977, U.S. Fish and Wildlife Service, Washington, D.C.

Thomann, R.V., "An Analysis of PCB in Lake Ontario Using a Size-Dependent Food Chain Model", Perspectives in Lake Ecosystem Modeling, 1979, Manhattan College, Bronx, New York, pp. 293-320.

United Nations Environment Program, "Environmental Issues in River Basin Development", Proceedings of the United Nations Water Conference on Water

Management and Development, 1978, Vol. 1, Part 3, Pergamon Press, New York, New York, pp. 1163-1172.

U.S. Army Corps of Engineers, "A Habitat Evaluation System for Water Resources Planning", 1980, Vicksburg, Mississippi.

U.S. Environmental Protection Agency, "Methods for Identifying and Evaluating the Nature and Extent of Non-Point Sources of Pollutants", EPA-430/9-73-014, Oct. 1973, Washington, D.C.

U.S. Environmental Protection Agency, "Areawide Assessment Procedures Manual Volume I", EPA/600/9-76/14-1, July 1976a, Municipal Environmental Research Laboratory, Cincinnati, Ohio.

U.S. Environmental Protection Agency, "Guidelines for Review of Environmental Impact Statements, Vol. 3, Impoundment Projects", Interim Final Report, July 1976b, 147 pp., Washington, D.C. (prepared for EPA by Curren Associates, Inc., Northampton, Massachusetts).

U.S. Environmental Protection Agency, "Modeling Phosphorus Loading and Lake Response Under Uncertainty: A Manual and Compilation of Export Coefficients", EPA-4405-80-011, June 1980, Washington, D.C.

U.S. Environmental Protection Agency, Proceedings of the Workshop on Aquatic Weeds, Control and Its Environmental Consequences, EPA-600/9-81-010, Feb. 1981, Washington, D.C.

U.S. Soil Conservation Service, "Guide for Environmental Assessment", Mar. 1977, Washington, D.C.

Uzzell, Jr., J.C. and Ozisik, M.N., "Three-Dimensional Temperature Model for Shallow Lakes", Journal of the Hydraulics Division, American Society of Civil Engineers, Vol. 104, No. HY12, Dec. 1978, pp. 1635-1645.

Veith, G.D., DeFoe, D.L. and Bergstedt, B.V., "Measuring and Estimating the Bio-Concentration Factor of Chemicals in Fish", Journal of the Fisheries Research Board of Canada, Vol. 36, 1979, pp. 1040-1048.

Walters, R.A., "A Time- and Depth-Dependent Model for Physical, Chemical and Biological Cycles in Temperate Lakes", Ecological Modeling, Vol. 8, Jan. 1980, pp. 79-96.

Warner, M.L. et al., "An Assessment Methodology for the Environmental Impact of Water Resource Projects", Report No. EPA-600/5-74-016, July 1974, Battelle-Columbus Laboratories, Columbus, Ohio (report prepared for Office of Research and Development, U.S. Environmental Protection Agency, Washington, D.C.).

Whitlatch, Jr., E.E., "Systematic Approaches to Environmental Impact Assessment: An Evaluation", Water Resources Bulletin, Vol. 12, No. 1, Feb. 1976, pp. 123-137.

Yapijakas, C. and Molof, A.H., "A Comprehensive Methodology for Project Appraisal and Environmental Protection in Multinational River Basin Development", Water Science and Technology, Vol. 13, No. 7, 1981, pp. 425-436.

Yorke, T.H., "Impact Assessment of Water Resource Development Activities: A Dual Matrix Approach", FWS/OBS-78/82, Sept. 1978, 37 pp., U.S. Fish and Wildlife Service, Kearneysville, West Virginia.

CHAPTER 3

ENVIRONMENTAL IMPACT STUDIES FOR CHANNELIZATION PROJECTS

A number of environmental impact studies have been conducted on channelization projects. The physical features of these projects may include the clearing of snags and other physical blockages of flow, and stream channel deepening and realignment. In some cases, natural channels are converted to concrete-lined channels. This chapter provides a summary of key technical references useful in environmental impact studies on channelization projects.

IMPACT IDENTIFICATION

As noted in Chapter 2, the identification of potential impacts should be an early activity in an environmental impact study. Numerous follow-on studies have been made of actual impacts caused by channelization projects; these impacts can be considered in terms of stream flows, water quality, and aquatic and terrestrial biology.

Impacts on Stream Flows

Stream channelization is typically done to provide drainage and improve river basin flow patterns. Inadvertent impacts can occur on area wetlands; for example, stream channelization under the Small Watershed Program (P.L. 83-566) was the major influence on wetland drainage in the Wild Rice Creek Watershed in North and South Dakota (Erickson, Linder and Harmon, 1979). Drainage rates were 2.6 times higher in the channeled area than in the unchanneled area during project planning, and 5.3 times higher during and following construction. Although the channel's claimed benefits were watershed protection and flood control, the channel permitted and stimulated wetland drainage deleterious to wildlife.

In a study of the effects of stream channelization on bottomland and swamp forest ecosystems, 16 well lines were established, representing 8 channelized situations and 8 nonchannelized situations in 10 stream bottoms from Craven County northward to Gates County, North Carolina, covering over a hundred miles of Coastal Plain (Maki, Hazel and Weber, 1975). One of the conclusions from a year's study was that channelized streams maintain perennial and clear flow during drought periods in sharp contrast to the murky water in natural streams during low flow.

While not directly involving channelization per se, Stone, Bahr and Way (1978) noted that water flow and quality determine and control species composition and function in the freshwater marshes of coastal Louisiana; however, man's activities alter this when he disrupts or removes the marsh. For example, man-made canals can change the hydrologic regime, depending on its alignment and local elevations, from -1 percent to -35 percent of normal flow. This in turn likely accelerates land loss from increased wave action. It is estimated that perhaps 172 hectares per year of freshwater marsh in coastal Louisiana is being lost due to man's activities. Canals also tend to divert runoff water away from the marsh (where it would be purged of pollutants) to open water bodies, thereby probably causing eutrophication.

Impacts on Water Quality

Channelization projects can cause stream water quality impacts. For example, a three-year statewide study was made of the occurrence and consequences of channelization in Hawaiian streams (Parrish et al., 1978). The 366 perennial streams of the state were inventoried for the first time, and some basic information was catalogued on their physical characteristics, complete status of channel alteration, and macrofaunal communities. Fifteen percent of the state's streams have channels altered in at least 1 of 6 forms. Forty percent of the modified channel length is concrete lined--the form of alteration found to be most ecologically damaging. Field measurements showed that channel alterations commonly caused large changes in pH values, conductivity, dissolved oxygen and daily temperatures.

Impacts on Aquatic Ecosystems

A number of studies have been conducted on the aquatic biological effects of stream channelization projects. Impacts can occur on a range of aquatic floral and faunal species. For example, habitat diversity and invertebrate drift were studied in a group of natural and channelized tributaries of the upper Des Moines River during 1974 and 1975 (Zimmer and Bachmann, 1978). Channelized streams in this region had lower sinuosity index values than natural channel segments. There were significant (P=0.05) positive correlations between channel sinuosity and the variability of water depth and current velocity. Invertebrate drift density, expressed as biomass and total numbers, also was correlated with channel sinuosity. Channelization has decreased habitat variability and invertebrate drift density in streams of the upper Des Moines River Basin, and probably has reduced the quantity of water stored in streams during periods of low flow.

Benke, Gillespie and Parrish (1979) noted that invertebrate production dynamics in the Satilla River in Georgia were suited to determine the effect of channelization or other river alterations on animal diversity, productivity, and general river ecology. The 362 km river has a drainage basin of 9,143 sq km with good to excellent water quality. Invertebrates were studied in three major habitats: snags, submerged wooden substrates; the main channel, sandy benthic conditions; and the muddy benthic habitat of backwater areas. Sampling was conducted at two sites on the river, one near Waycross 290 km from the Atlantic Ocean, and one at Atkinson 129 km from the ocean. Samples taken to determine the relative importance of the various invertebrate habitats included: (1) quantitative sampling to estimate invertebrate production; (2) invertebrate drift sampling with notation of habitat of origin for drift organisms; and (3) sampling of the feeding habitats of major fish species to determine trophic pathways and the habitats of origin of the fish. Results show that the snag habitat had the greatest species diversity, standing stock biomass, and total production with weights of 57 to 72 g dry wt/sq m. Drift samples revealed that approximately 80 percent of the animals found in the drift originated from the snags. Drift densities were high compared to other rivers with roughly 3 invertebrates/cu m. The major insectivorous fish species, redbreast, bluegills, and large-mouth bass, are largely dependent upon snags as sources of food. Due to this importance to invertebrates, the removal of snag habitats and/or channelization of the river would cause a significant decline in animal diversity and productivity.

Several studies have also been made on the impacts of channelization projects on fish populations. Two examples will be cited (Duvel et al., 1979; and Headrick, 1976). Duvel et al. (1979) undertook geologic, engineering, and biological investigations of six Pennsylvania coldwater streams to determine the impact of channel modifications instituted both prior to and following Hurricane Agnes. The primary focus of the study was on the ecological changes brought about by stream channelization. No long-term deleterious effects on water quality, attached algae, benthic fauna, or forage fish populations were found. Trout, however, were found to be greater in number and weight in natural than in channelized stream reaches. Lack of suitable physical habitat appeared to be the primary cause of reduced trout populations in stream reaches which have been channelized.

Fish populations from ditches 6 to 8 years old and 52 to 62 years old were compared with populations in adjacent portions of natural streams in Portage County, Wisconsin (Headrick, 1976). Two study areas were selected: an upstream zone of good brook trout habitat and a downstream zone of marginal trout habitat. Loss of year-round instream cover through channelization limited brook trout density, which reduced the annual brook trout production to 28.8 kg/stream km in the upstream new ditch, compared to 72.2 kg/km and 65.5 kg/km in the upstream old ditch and the upstream natural stream, respectively. Angler success was also reduced in the upstream new ditch. Midsummer water temperatures reached upper lethal levels for brook trout in the downstream ditches where current velocity was reduced and white sucker were abundant. Mottled sculpin were consistently absent from the upstream new ditch and scarce in the downstream new ditch. The natural stream had the greatest number of fish species in both study areas; the new ditch had the fewest species. Recovery was more rapid in ditches where spoil was spread on adjacent fields and bank vegetation left in place than where spoil was left on the banks.

Channelization can also create an impact on aquatic insect populations. White and Fox (1980) found that recovery, defined as development of an aquatic insect fauna similar to that of the control streams, does not occur in South Carolina streams following channelization. Five channelized streams and two natural streams were sampled over the 1975 to 1979 period to determine channelization effects on species composition and diversity of aquatic insects in the Piedmont/Coastal Plain regions. Stream samples revealed that channelization yields a fauna composed principally of very tolerant or normally pond-inhabitating species. Taxa preferring fast currents, vegetation, and low turbidity were rarely found in the channelized streams. Surber square foot and core samplings afforded the greatest utility for characterizing stream aquatic insect diversity; kick-screen sampling was found preferable in determining population composition.

Two other types of biological impacts which could be associated with channelization project construction or use activities can be noted. First, in channel straightening the construction process may create cutoff bendways.

Second, use of channelized sections by navigation traffic has the potential to cause impacts. Examples of studies related to both of these issues will be cited (Pennington and Baker, 1982; and Wright, 1982).

Biological and physical data were collected from four bendways within the river portion of the Tennessee-Tombigbee Waterway (TTW) from Columbus, Mississippi, to Demopolis, Alabama: Rattlesnake Bend, Cooks Bend, Big Creek Bendway, and Hairston Bend (Pennington and Baker, 1982). During this study, the four bendways had not all been cut off and had been impounded for various lengths of time. At the completion of the TTW project, all four of the bendways will be severed from the main navigation channel. Four distinct areas within each bendway were compared: above the bendway, within the bendway, below the bendway, and within the cut. Sampling was conducted from January 1979 to September 1980 to coincide with four different river stage/water temperature regimes.

Sediment analysis and bottom profiles indicated that the substrate composition of some of the bendways was changing (Pennington and Baker, 1982). Overall, the substrate of the study area is changing from a sand-gravel-fines mixture to one of predominantly sand and fines. Areas of some bendways, in particular the upper areas, were accumulating sediments. Phytoplankton composition and chlorophyll concentrations showed only small differences among the four bendways. Aquatic macrophytes were scatted and uncommon in the four bendways. Based upon total collections, a consistent family assemblage of macroinvertebrates characterized the four bendways. Although 60 family-level taxa were collected, 9 families of macroinvertebrates accounted for between 93.5 and 97.2 percent of the benthos. The importance of these families varied among bendways and appeared to reflect differences in physical bendway conditions, particularly substrate type and current velocities. Eighteen species of Unionid mollusks, plus the Asian clam Corbicula, were collected during the surveys.

Based on overall ichthyofaunas, two groups of bendways were delineated that corresponded to impoundment and riverine habitats (Pennington and Baker, 1982). Rattlesnake Bend and Cooks Bend were located in lower pool sections, where impoundment conditions prevailed, and their icthyofaunas were dominated by clupeids (shad) and centrarchids (sunfishes, crappies, and basses). Hairston Bend, essentially a riverine reach during this study, was dominated by cyprinids (minnows), ictalurids (catfishes), and catostomids (suckers). Big Creek Bendway, unique in having both riverine and lacustrine habitats, was faunistically most similar to Hairston Bend, but also showed moderate similarities to the other bendways.

Wright (1982) conducted a literature search to identify research relating physical and chemical changes associated with navigation traffic to potential biological impacts. It was found that, although some information on physical and chemical changes was available, documentation to demonstrate biological impacts was generally lacking. Where possible impacts were identified, they were observed at the organism level, rather than at the population, community, or ecosystem level. A particular problem was encountered in attempting to separate impacts of navigation from those caused by natural and/or anthropogenic perturbations.

Impacts on Terrestrial Ecosystems

In addition to aquatic ecosystems effects, channelization projects can also cause impacts on terrestrial ecosystems. For example, Frederickson (1979) reported on the floral and faunal changes in low land hardwood forests in Missouri resulting from channelization, drainage, and impoundment. The study was designed to gather data on the effects of decreasing forest cover in the lower Mississippi Valley. The objectives were to identify the effects of channelization, impoundments, and drainage on plant and animal communities, and to develop reliable techniques to monitor and predict changes in these communities as a result of such actions. The Mingo Swamp, along a portion of the St. Francis River in southeastern Missouri, was the study area. Photo interpretation provided overall information on the nature of the tree cover; data on species composition and density were gathered along line transects selected randomly along the river. Among the pertinent conclusions reached from the study were that stream channelization reduced or changed riparian habitat, decreasing the forest area by as much as 78 percent as compared to no more than 7 percent in unchannelized areas; bird populations tended to avoid channelized streams; and channelization does reduce flooding and benefit agriculture.

A study of the effects of stream channelization on terrestrial wildlife and their habitats was conducted in the Buena Vista Marsh in Wisconsin (Prellwitz, 1976). Stream channelization affected wildlife in the Buena Vista Marsh by draining wetlands, setting back plant succession, and decreasing habitat diversity along streambanks by removing or burying plants. Plant and animal species composition and abundance were studied in a continuum of plant successional stages from grassland to mature woods on streambanks adjacent to recently dredged (6 years), old dredged (50 years), and natural streams. Sheet-water area and longevity and wildlife use of three sheet-water areas with various degrees of drainage were compared. Bird and mammal species diversity and bird abundance increased as streambank plant succession advanced, until a mature wooded stage was reached. The abundance of small mammals was related to the amount of ground cover and diversity of habitats along the streambanks. Sheet-water area and longevity were greatest on undrained wetlands, and least near recently dredged channels. Waterfowl use, bird nesting, and reptile and amphibian abundance were also greatest on undrained areas.

The effects of stream channelization on songbirds and small mammals were documented for the White River watershed in Vermont (Possardt and Dodge, 1978). Stream channelization effects were documented in the watershed during the first and second years after channelization. Birds were mist-netted during four sampling periods. Percentages of birds collected from channelized areas as compared to nonchannelized areas were 33%, 27%, 38%, and 46% for fall 1974, spring 1975, summer 1975, and late summer 1975, respectively. Species diversity was significantly less in channelized areas for fall 1974 and early summer 1975; highly significantly less for spring 1975; however, no significant difference existed for later summer 1975. Swallows and spotted sandpipers were more abundant in channelized areas; thrushes, vireos, and particularly warblers were more abundant in nonchannelized areas. Small mammals were live-trapped during three sampling periods. Percentages of mammals collected from channelized areas as compared to nonchannelized areas were 28%, 39%, and 39% for fall 1974, early summer 1975, and late summer 1975, respectively. Shrews and jumping mice were most adversely affected; the white-footed mouse, the most abundant small mammal collected, recovered rapidly in channelized areas. Impacts on small mammals and songbird populations was most dramatic where streamside vegetation had been extensively destroyed.

One other terrestrial ecosystem effect which could occur from channelization is a result of raised water tables in the vicinity of the channel. Raised water tables can detrimentally affect trees primarily by causing oxygen depletion in the root zone. The ability of an individual tree to withstand this type of stress may be related to a variety of factors, including the tree species, rooting depth, soil conditions, and the timing, duration, and frequency of encroachment of the water table into the root zone (Klimas, 1982). In some situations, trees may show improved growth as a result of raised water tables, although this response may be short-lived. The forest communities likely to be most sensitive to partial root-zone saturation are those typically found on well-drained upland sites. Where such communities are subjected to drastic changes in the root environment, mortality may be widespread and rapid. In contrast, swamp forest communities (cypress-tupelo) are generally much less sensitive to increases in site moisture status, and may be affected primarily with respect to long-term growth and reproduction, or not at all. The complex flood plain forest (bottomland hardwood) communities present the greatest difficulty in predicting potential impacts resulting from raised water tables. Possible effects range from near-complete mortality in areas where permanent surface saturation occurs to no effect or growth rate increases in response to very minor, subsurface water table rises. The complexity of this issue suggests that only very general impact predictions may be possible except in areas where extreme stresses are anticipated.

BASELINE STUDIES

Information on the anticipated impacts of projects can be used to delineate which environmental factors should be addressed as part of the baseline conditions description. For example, descriptions of the baseline conditions for stream flows, certain water quality constituents, and selected features of the aquatic and terrestrial ecosystems would be appropriate. Specific information to enable the preparation of these descriptions can be procured from existing data collected by governmental agencies or private groups, or it can be collected via specific baseline studies. Some planning considerations for baseline field studies are summarized in Chapter 2.

IMPACT PREDICTION

As noted in Chapter 2 for dam and reservoir projects, the most important technical activity in an environmental impact study for a channelization project is the scientific prediction of the effects of project construction and operation. Examples of prediction approaches which could be used include: (1) a qualitative approach which relies on general knowledge of the impacts of similar projects, or specific results of comprehensive studies of similar projects; (2) a quantitative approach based on the use of mathematical models; and (3) a quantitative approach based on the use of physical models. Mathematical or physical modeling can be used for physical and chemical impacts such as those on stream flows and water quality; the qualitative approach will probably be needed for aquatic and terrestrial ecosystem impacts.

Qualitative Approach

A general knowledge of the types of impacts anticipated from channelization projects can be useful in identifying impacts, describing the environmental setting, and preparing qualitative predictions of the anticipated impacts. The earlier section on Impact Identification contains detailed information from several studies of the impacts of channelization projects.

Quantitative Mathematical Model Approach

Mathematical models can be used to predict the effects of channelization projects on stream flows and water quality. An example of a mathematical model is a self-calibrating watershed model for predicting the effect of channel improvements on downstream flows (Huang and Gaynor, 1977). The model is called MOPSET because it is a modified version of OPSET developed several years ago at the University of Kentucky. OPSET is a computerized procedure for determining an optimum set of parameter values by matching synthesized flows with recorded flows. Major modifications include the replacement of the modified Muskingum method of channel routing by a kinematic finite difference method, the division of the watershed into a number of segments, and the inclusion of a storage routing procedure to take care of any reservoirs or flood control structures located in the watershed. The computer program is well documented and can be used not only as a flood predicting model but also as a general model for hydrologic simulations. The model was applied to three different watersheds in Kentucky. It was found that an optimum set of parameter values obtained automatically by the model was not unique and might not yield the most desirable solution. For this reason, new features were added so that the user can exercise his judgment in selecting the most desirable parameter values. The synthesized flows obtained from these watersheds are presented and compared with the recorded flows. The effects of channel improvements, flood control structures, and routing procedures on downstream flows are discussed.

Although not specifically related to channelization projects per se, Stone and McHugh (1977) reported on a computer simulation of the hydrography of the Barataria Basin in Louisiana. The simulation indicated significant hydrologic changes due to navigation and transportation canals. The simulations compared hydrologic parameters in the Basin before and after the construction of the Barataria and Intracoastal Waterways, and the canals associated with eight oil and gas fields. The waterways accounted for about 90 percent of the simulated changes; the remaining 10 percent was due to the canals of the eight oil and gas fields.

Quantitative Physical Model Approach

Physical scale-models can also be used for predicting the impacts of channelization projects on stream flows. One example involved tests on the Chesapeake Bay Model, the world's largest estuarine model (Bastian, 1980). The model was used to assess the effects of increasing the approach channels to Baltimore, Maryland from 13 to 15 meters. While the main project effort involved dredging, it is addressed herein as an example of a physical model. In the proposed project there are four sections of dredged channels comprising 55 km of the 277 km distance from the bay mouth to Baltimore. The increased depth of channel would extend the length of dredged channels to 79 km. Base tests using the existing 13-m channels were conducted to determine the synoptic velocity, salinity, and tidal conditions at a number of locations throughout the bay, but primarily in the dredged channels. A 2½ year hydrographic period was simulated in the model to enhance the evaluation by adding a variable discharge as a parameter. A 12-constituent harmonic tide was used, giving a 28 lunar day tidal sequence which simulates a lunar month. The entire test was repeated, but with the 15-m channel installed. The tide is far more significant than discharge in affecting the velocities for the stations observed. No shift in flow predominance was identified between the base and the plan which could be used for conclusions about the effects of depth change on sediment transport. The deepened plan depths appear to induce higher salinities in the Patapsco and surrounding bay area. Salinity differences, base to plan, in this area attenuate with distance from the deepened channels.

IMPACT ASSESSMENT

As noted in Chapter 2, the assessment or interpretation of predicted impacts represents an important activity in an environmental impact study. Impact assessment for channelization projects should be based on the application of professional judgment in conjunction with environmental standards and criteria and other pertinent scientific information. Public input can also be used in this activity. Several relevant laws and executive orders, along with some key technical references, are summarized in the Impact Assessment section of Chapter 2.

Mitigation of undesirable impacts should also be considered within the context of impact assessment. Mitigation may include avoiding, minimizing, rectifying, reducing or eliminating impacts, and/or compensating for them. Examples of mitigation measures are included in several channelization studies (Frederickson, 1979; Parrish et al., 1978; and White and Fox, 1980).

METHODOLOGIES FOR TRADE-OFF ANALYSES AND DECISION-MAKING

Environmental impact studies for channelization projects typically address 3 to 5 alternatives. In their literature review, Thackston and Sneed (1982) addressed a number of alternatives to traditional channel modification. Structural alternatives to channel modification include levees, floodways, reservoirs, and land treatment measures. Additional alternatives include various forms of flood plain management; flood plain zoning; construction of bypass channels around sensitive wetland areas; construction of numerous, very small, water-retention structures; and substitution of clearing and snagging, or only snagging, for complete channelization.

An environmental impact study for a channelization project should include a systematic evaluation of alternatives via the application of methodologies for trade-off analyses and decision-making. Potentially useful methodologies include matrices, scaling or ranking checklists, and scaling-weighting or ranking-weighting checklists. Examples of these methodologies are contained in the Chapter 2 section on Methodologies for Trade-off Analyses and Decision-making.

SUMMARY

Construction and usage of channelization projects can cause significant environmental impacts. Systematic environmental impact studies are needed to predict and assess these impacts and identify appropriate mitigation measures. There is considerable technical literature to aid in impact identification and the planning and conduction of appropriate studies of the baseline environment. Impact prediction can be achieved via (1) a qualitative approach which relies on general knowledge of the impacts of similar projects, or specific results of comprehensive studies of similar projects; (2) a quantitative approach based on the use of mathematical models; and (3) a quantitative approach based on the use of physical scale-models. A given environmental impact study could involve all three approaches. Impact assessment involves the use of water quality standards and criteria, appropriate laws and regulations related to aquatic and terrestrial ecosystems, and professional knowledge and judgment. Systematic methods such as matrices and checklists are available for comparing alternatives for channelization projects.

SELECTED REFERENCES

Bastian, D.F., "The Salinity Effects of Deepening the Dredged Channels in the Chesapeake Bay", Report NWS-81-S1, Dec. 1980, U.S. Army Institute for Water Resources, Fort Belvoir, Virginia.

Benke, A.C., Gillespie, D.M. and Parrish, F.K., "Biological Basis for Assessing Impacts of Channel Modification: Invertebrate Production, Drift, and Fish Feeding in a Southeastern Blackwater River", Report No. ERC 06-79, 1979, Environmental Resources Center, Georgia Institute of Technology, Atlanta, Georgia.

Duvel, W.A. et al., "Environmental Impact of Stream Channelization", Water Resources Bulletin, Vol. 12, No. 4, Aug. 1979, pp. 799-812.

Erickson, R.E., Linder, R.L. and Harmon, K.W., "Stream Channelization (PL 83-566) Increased Wetland Losses in the Dakotas", Wildlife Society Bulletin, Vol. 7, No. 2, Summer 1979, pp. 71-78.

Frederickson, L.H., "Floral and Faunal Changes in Low Land Hardwood Forests in Missouri Resulting from Channelization, Drainage, and Impoundment", FWS/OBS-78-91, Jan. 1979, U.S. Fish and Wildlife Service, Washington, D.C.

Headrick, M.R., "Effects of Stream Channelization on Fish Populations in the Buena Vista Marsh, Portage County, Wisconsin", Sept. 1976, U.S. Fish and Wildlife Service, Stevens Point, Wisconsin.

Huang, Y.H. and Gaynor, R.K., "Effects of Stream Channel Improvements on Downstream Floods", Research Report No. 102, Jan. 1977, Kentucky Water Resources Research Institute, University of Kentucky, Lexington, Kentucky.

Klimas, C.V., "Effects of Permanently Raised Water Tables on Forest Overstory Vegetation in the Vicinity of the Tennessee-Tombigbee Waterway", Misc. Paper E-82-5, Aug. 1982, U.S. Army Engineer Waterways Experiment Station, Vicksburg, Mississippi.

Maki, T.E., Hazel, W. and Weber, A.J., "Effects of Stream Channelization on Bottomland and Swamp Forest Ecosystems", Completion Report, Oct. 1975, School of Forest Resources, North Carolina State University, Raleigh, North Carolina.

Parrish, J.D. et al., "Stream Channelization Modification in Hawaii, Part D: Summary Report", Report No. FWS/OBS-78/19, Oct. 1978, U.S. Fish and Wildlife Service, Hawaii Cooperative Fishery Research Unit, Honolulu, Hawaii.

Pennington, C.H. and Baker, J.A., "Environmental Effects of Tennessee-Tombigbee Project Cutoff Bendways", Misc. Paper E-82-4, Aug. 1982, U.S. Army Engineer Waterways Experiment Station, Vicksburg, Mississippi.

Possardt, E.E. and Dodge, W.E., "Stream Channelization Impacts on Songbirds and Small Mammals in Vermont", Wildlife Society Bulletin, Vol. 6, No. 1, Spring 1978, pp. 18-24.

Prellwitz, D.M., "Effects of Stream Channelization on Terrestrial Wildlife and Their Habitats in the Buena Vista Marsh, Wisconsin", Report FWS/OBS-76-25, Dec. 1976, Wisconsin Cooperative Fishery Research Unit, Stevens Point, Wisconsin.

Stone, J.H. and McHugh, G.F., "Simulated Hydrologic Effects of Canals in Barataria Basin: A Preliminary Study of Cumulative Impacts", Final Report, June 1977, Louisiana State University Center for Wetland Resources, Baton Rouge, Louisiana.

Stone, J.H., Bahr, Jr., L.M. and Way, Jr., J.W. "Effects of Canals on Freshwater Marshes in Coastal Louisiana and Implications for Management", Freshwater Wetlands: Ecological Processes and Management Potential, 1978, Academic Press, New York, New York, pp. 299-320.

Thackston, E.L. and Sneed, R.B., "Review of Environmental Consequences of Waterway Design and Construction Practices as of 1979", Tech. Report E-82-4, Apr. 1982, U.S. Army Engineer Waterways Experiment Station, Vicksburg, Mississippi.

White, T.R. and Fox, R.C., "Recolonization of Streams by Aquatic Insects Following Channelization", Technical Report 87, Vol. I, May 1980, Water Resources Research Institute, Clemson University, Clemson, South Carolina.

Wright, T.D., "Potential Biological Impacts of Navigation Traffic", Misc. Paper E-82-2, June 1982, U.S. Army Engineer Waterways Experiment Station, Vicksburg, Mississippi.

Zimmer, D.W. and Bachmann, R.W., "Channelization and Invertebrates in Some Low Streams", Water Resources Bulletin, Vol. 14, No. 4, Aug. 1978, pp. 868-883.

CHAPTER 4

ENVIRONMENTAL IMPACT STUDIES FOR DREDGING PROJECTS

A number of environmental impact studies have been conducted on dredging projects. Dredging involves the removal of bottom materials from rivers, lakes or estuaries, and the subsequent disposal of the dredged material in open water or on land. Dredging projects are typically done to improve flow characteristics and/or to enhance navigational opportunities. This chapter provides a summary of key technical references useful in environmental impact studies on dredging projects.

Several general references on the environmental impacts of dredging can be cited. For example, Lehmann (1979) generated a bibliography of 174 abstracts on worldwide research on dredging related to biology, nutrient composition, sedimentation, and water pollution. Abstracts on the effects of disposal of dredge material in containment areas, landfills, or oceans are cited, as is dredge material reuse in land reclamation. Massoglia (1977) prepared guidelines to assist the U.S. Army Corps of Engineers, the U.S. Environmental Protection Agency, the Maritime Administration, estuary managers, and individuals in evaluating and preparing environmental impact statements for dredging in estuaries. Finally, Slotta et al. (1974) reported on a research project dealing with the impacts of dredging and marine traffic, impacts of sediment turnover, impacts of sediment physical changes, impacts of turbidity, and impacts of release of toxins from sediments.

IMPACT IDENTIFICATION

As noted in Chapters 2 and 3, the identification of potential impacts should be an early activity in an environmental impact study. Many studies have been conducted on the actual impacts experienced from dredging project activities. Figure 2 shows a network analysis of dredging relative to potential environmental impacts (Sorensen, 1971). The reasons for doing dredging are shown on the left; both initial and related impacts are delineated across the right portion of the network. The impacts of dredging projects are influenced by the type of dredging, and they can be considered in terms of impacts in the dredged area and resulting from either open water on land disposal of dredged material.

Dredging Technology

Dredging can be accomplished with a variety of types of equipment, including grab, bucket, clamshell, cutterhead, and specialized dredges such as mud cat, bucket wheel, and Japanese special purpose dredges (Peterson, 1979). The environmental concerns associated with dredging include resuspension of bottom sediments, toxic substances, oxygen depletion, reduced primary production, temperature alteration, increased nutrient levels, benthic community alteration, and sediment disposal sites. Negative and positive

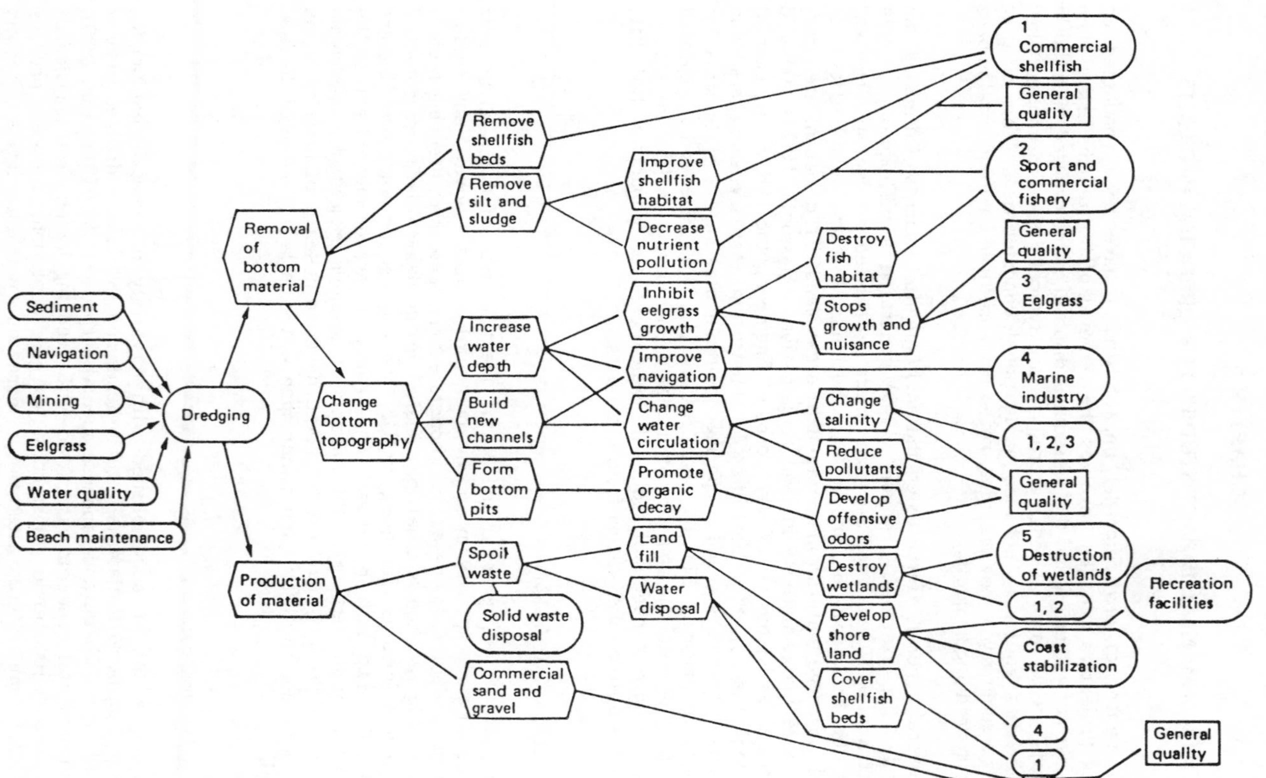

Figure 2: Network Analysis of Dredging (Sorensen, 1971)

aspects are discussed by Peterson (1979) for various types of dredging equipment.

Impacts on the Area Being Dredged

Allen and Hardy (1980) prepared a review of the impacts to fish, other aquatic organisms and wildlife, as well as habitat enhancement opportunities, resulting from construction of new navigational channels and maintenance dredging of existing channels. As noted earlier, the type of dredging equipment used determines, to a great extent, the viable disposal alternatives, the type and magnitude of potential impacts, and the potential for habitat development. About 99 percent of the dredging volume in the United States is accomplished by hydraulic dredges. The composition of dredged material from a particular site affects its pollution potential as well as the potential for habitat development or other beneficial uses. Dredged material from new channels or new works projects often has chemical and engineering properties which create fewer environmental problems than material from maintenance projects. Material removed during maintenance dredging of navigational channels is an accumulation of detached soil particles which have been transported by wind and water, and may contain a variety of contaminants.

Two case studies of environmental impacts which have occurred in dredged areas will be cited (Bohlen, Cundy and Tramontano, 1979; and Conner and Simon, 1979). Bohlen, Cundy and Tramontano (1979) described the sampling of the suspended material downstream of a large volume bucket dredge operating in the lower Thames River estuary near New London, Connecticut. The collected data indicated that approximately 1.5 to 3 percent of the sediment volume in each bucket-load is introduced into the water column producing suspended material concentrations adjacent to the dredge of 200 mg/l to 400 mg/l. These values exceeded background levels by two orders of magnitude. Analysis of particulate organic carbon and grain size characteristics indicated that resuspension also alters suspended load composition increasing the percentage of inorganic materials and the median grain size. Proceeding downstream, material concentrations along the centerline of the dredge-induced plume decreased rapidly, approaching background within approximately 700 m. Compositional variations display similar trends with the major perturbations confined to the area within 300 m of the dredge. The observed spatial distributions indicated that the dredge-induced resuspension is primarily a near-field phenomenon producing relatively minor variations as compared to those caused by naturally occurring storm events. Previous work has shown that these latter systems can produce estuary-wide variations in suspended material concentrations, increasing the mass of material in suspension by at least a factor of two. This increase in total suspended load is nearly an order of magnitude larger than that produced by the dredge.

The second case study determined the extent and nature of the effects on the benthos of physical disruptions associated with dredging fossil oyster shell (Conner and Simon, 1979). Two dredged areas and one undisturbed control area in Tampa Bay, Florida, were quantitatively sampled before dredging, and for one year after dredging. The immediate effects of dredging on the soft-bottom community were reductions in numbers of species (40 percent loss), densities of macroinfauna (65 percent loss), and total biomass of invertebrates (90 percent loss). During months 6 to 12 after dredging, the analysis used (Mann-Whitney U

Test, alpha = 0.05) showed no difference between dredged and control areas in number of species, densities, or biomass (except for one area). Community overlap (Czeckanowski's coefficient) between dredged and control areas was reduced directly after dredging, but after six months the predredging level of similarity was regained.

Impacts from Open Water Disposal

Engler (1978) summarized some of the key impacts associated with open water disposal of dredged material. He noted that, with few exceptions, the impacts of aquatic disposal are mainly associated with the physical effects. These possible effects are persistent, often irreversible, and compounding. Geochemically, releases are limited to nutrients, with negligible release of toxic metals and hydrocarbons. Biochemical interactions are infrequent, with no clear trends; elevated uptake of toxic metals and hydrocarbons are negligible to nonexistent in typical open water disposal operations. Three case studies on the environmental impacts of open water disposal will be cited; one for a lake (Flint, 1979), one for a river and estuary (National Marine Fisheries Service, 1977), and one for an estuary (Pavlou et al., 1980).

Temporal variations in the natural benthic macroinvertebrate community of areas subjected to open-lake dredged spoil disposal were examined in Lake Erie near Ashtabula (Flint, 1979). The two types of dredged spoil used were sandy-silt sediment from the Ashtabula harbor and coarse sand sediments from the Ashtabula River. Two untreated reference sites were also sampled for comparison. Grab samples were used to characterize the communities at the sources of dredging to determine the types of material that would be deposited at the spoil sites. Samples were taken at the disposal and control sites at intervals of 5 days, 90 days, 256 days, and 340 days after disposal. For each sample taken, a modified Spade box corer was used to collect four replicate bottom samples. Macroinvertebrates were identified to the lowest taxa possible. Results showed that for the sandy silt site the density of macroinvertebrates increased while diversity decreased. Species brought in with the river and harbor sediments contributed to the increased density. Also, while the untreated sites had communities of 60 to 80 percent oligochaetes, both disposal sites had communities of 98 percent oligochaetes. The river spoils disposal sites showed a more rapid recovery to predisposal conditions. The opportunistic nature of the Oligochaeta life history pattern allows them to take advantage of and thrive in the environment existing after disposal. This dominance of Oligochaeta along with the increased population density form unstable communities at the disposal sites.

The National Marine Fisheries Service (1977) reported on a study of the impacts of dredging in the Thames River in Connecticut, and the impacts of dredged material disposal at the New London dumping ground in the Thames River estuary. The impacts of dredging operations on suspended material transport in the lower Thames River estuary were confined to an area within 300 to 500 yards of the operating dredge and barge, produced an increase in total suspended load within the estuary that was small in comparison to that produced by typical aperiodic storm events, and caused no major alterations in mass transport within the estuary. Field surveys of the Thames River hydrography, phytoplankton, and trace metal concentrations in water, sediment, and shellfish suggested that effects of dredging on primary production were spatially and temporally limited. The highest concentrations of nickel, lead, cadmium and mercury in water samples were observed before or during dredging, while copper was highest after dredging but were generally higher upriver. Sediment levels of these five metals, plus zinc and organic carbon, increased in an upriver direction. Dredging-related changes in trace metal body burdens in shellfish were difficult to separate from normal seasonal variations. No gross pathology was detected in the shellfish. The physical oceanography of the disposal area showed that turbidity was higher in bottom waters than near the surface, and was not restricted to the vicinity of the disposal pile. Maximum transport was in the east/west direction with highest values occurring during the ebbtide.

Short-term and long-term ecological impact studies were initiated at the time of open-water disposal of contaminated dredged material from the Duwamish River in Elliott Bay, Puget Sound, Washington (Pavlou et al., 1980). The experimental disposal site was located at a depth of 60 m in a marine estuary with generally weak circulation. Approximately 114,000 cubic meters of dredged material contaminated with PCB's were dumped at the site from split-hull barges. Replicate water, suspended particulate matter, and sediment samples were taken at intervals up to nine months after disposal and investigated for PCB content. The PCB concentration in water and suspended matter generally declined over time. The sediments at the disposal site were slumping from the center of the site to the periphery during the monitoring period, and there was a slight but not significant reduction in PCB concentrations. Long-term (3-years post-disposal) PCB concentrations showed further declines, with the highest concentrations in the vicinity of the disposal site. Macrofauna, including marine worms, clams and crustaceans, was collected by grab sampler. Statistical analysis of the abundance figures suggest that some taxa may be more abundant within and close to the disposal site, and that stations in close proximity to the grid are more similar to each other than to more distant stations. Nonparametric Wilcoxon two-sample tests revealed significant differences in abundances for some taxa that were grouped as stations within the grid site versus more distant stations. All but one of the taxa exhibited greater abundances at the disposal site.

Impacts from Land Disposal

One of the key concerns related to land disposal of dredged material is related to the potential release and transport characteristics of material components such as heavy metals and organics. Brannon (1978) synthesized data from seven research projects that investigated the pollution properties of dredged material, and procedures for determining their potential for effects on water quality and aquatic organisms. The results can be applied to either open water or land disposal activities. Short-term impacts of dredged material on water quality and aquatic organisms are related to the concentration of chemically mobile, readily available contaminants rather than total concentration. The Elutriate Test, which measures the concentrations of contaminants released from dredged material, can be used to evaluate short-term impacts on water quality. The only constituents generally released from dredged material are manganese and ammonium-N. Elevated concentrations of these constituents, however, are of short duration because of rapid mixing and are of low frequency due to the intermittent nature of most disposal operations. Short-term chemical and biological impacts of dredging and disposal have generally been minimal. Longer-term impacts of dredged material on water quality have generally been slight and can be evaluated by means of the

Elutriate Test and analysis of mobile forms of sediment contaminants. The greatest hazard of dredged material disposal is the potential effect of material on benthic organisms. Most dredged material has not proven particularly toxic. Some dredged material, however, can be extremely toxic or of unknown toxicological character.

Hoeppel (1980) reported on the monitoring conducted at nine dredged material land containment areas. These areas were monitored during hydraulic dredging operations in fresh and brackish-water riverine, lake, and estuarine environments. Influent-effluent sampling at the diked disposal areas showed that, with proper retention of suspended solids, most chemical constituents could be removed to near background water levels. Most heavy metals, oil and grease, chlorinated pesticides, and PCB's were almost totally associated with solids in influent and effluent samples. Ammonia, manganese, total mercury and possibly iron, copper and zinc are the only contaminants which may occasionally exceed background water levels and present water quality standards after suspended solids removal. Actively growing vegetation appeared to be efficient in reducing ammonium nitrogen to low levels and for filtering out suspended solids. Therefore, it was concluded that confined land disposal of dredged material seems to be a viable method for the containment and treatment of most contaminated sediments.

One of the growing areas of environmental concern is related to the potential for ground water quality impacts at land disposal sites. Morrison and Yu (1981) described a study in which ground water was monitored over a 2-year period at three dredge disposal sites: Grandhaven, Michigan (fresh water); Mobile, Alabama; and Sayreville, New Jersey (both estuarine). Leachate samples collected from monitoring wells and vacuum/pressure lysimeters were analyzed for 27 water quality parameters. The Na, Cl, and K concentrations were high (maximum 4310, 8330, and 250 mg/l, respectively) in ground water sites where saline dredge material was deposited. Dilution is the major mechanism regulating transport of these ions. The Mg and Ca maxima were 720 and 380 mg/l, respectively, representing significant increases in water hardness over background levels. Mechanisms for transport of Mg and Ca are dilution, precipitation/dissolution, and ion exchange. Alkalinity ranged from undetectable to 1000 mg/l, and was controlled by dissolution of carbonates, weathering, and oxidation-reduction reactions. Total organic carbon was highly correlated with alkalinity. Maximum iron and manganese levels were 538 and 12 mg/l, respectively, both above recommended drinking water quality standards. Sulfides were generally below 20 mg/kg. Phosphate, Cd, Cu, Pb, Ni, Hg, and Zn occurred in concentrations too small to have significant effects on ground water quality. The major controlling factors for these ions include complexation, adsorption, and precipitation/dissolution.

BASELINE STUDIES

Information on the anticipated impacts of dredging projects can be used to delineate which environmental factors should be addressed as part of the baseline conditions description. For example, descriptions of the baseline conditions for certain water quality parameters and aquatic biological species would be appropriate for open water disposal areas. Examples of the type of baseline information necessary for land disposal environmental impact studies can be surmised from a report by Chen et al. (1978). This report summarized the findings of five work units concerned with the impacts of dredged material disposal in confined land disposal areas. Three work units dealt with active disposal operations at 11 sites; impact was assessed by comparing the quality of influents and effluents at each site with background surface receiving water. Two work units evaluated the impact of confined disposal area leachates on ground waters. Leachate studies included laboratory column elutions (3-9 month period) of each of five types of dredged material overlying one of two different soils; four field sites were also monitored for changes in leachate and ground water quality in a 9-month study. Field sites included fresh and brackish water dredging environments in geographical areas where contamination problems were anticipated. Dredged material and environmental features varied greatly at different sites. In most cases soluble concentrations of most chemical constituents were very low. Only soluble manganese and ammonia nitrogen levels failed to meet most criteria. Leachate studies suggested that the disposal of brackish water dredged material in upland disposal areas may render subsurface water unsuitable for public water supply or irrigation purposes. Guidelines for evaluation of potential disposal sites should be developed in steps not requiring complete execution of the total program to determine site suitability.

Specific information to enable the preparation of a description of baseline conditions can be procured from existing data collected by governmental agencies or private groups, or it can be collected via specific laboratory and field studies. Some planning considerations for baseline field studies are summarized in Chapter 2.

IMPACT PREDICTION

The most important technical activity in an environmental impact study for a dredging project is the scientific prediction of the effects of removal of bottom material and its subsequent disposal. Examples of prediction approaches which could be used include: (1) a qualitative approach which relies on general knowledge of the impacts of similar projects, or specific results of comprehensive studies of similar projects; (2) a quantitative approach using mathematical models; and (3) a quantitative/qualitative approach based on laboratory and/or field studies. A given environmental impact study could involve all three approaches depending upon the impacts being addressed.

Qualitative Approach

A general knowledge of the types of impacts anticipated from dredging projects can be useful in identifying impacts, describing the environmental setting, and preparing qualitative predictions of the anticipated impacts. The earlier section on Impact Identification contains detailed information from several studies of the impacts of dredging projects.

The general type of information necessary for an impact study of disposal sites was described by Holliday (1978). Holliday (1978) noted that determination of the fate of dredged material placed on the bottom of an ocean, lake, estuary, or river is an environmental concern that requires consideration and adequate prediction in the planning of a dredging project. In the selection process for a disposal site, consideration must be given to the eventual disposition of dredged material in order that adequate determination of the site capacity can be made. Prediction of the fate of dredged material at a disposal site requires knowledge of: (1) currents, (2) waves, (3) tides, (4) suspended sediment concentrations, (5) seasonal energy fluctuations, (6) storms, (7) dredging/disposal operations, (8) shipping traffic, (9) fisheries activities, (10) bathymetry, (11) sedimentology, and (12) biological activity.

Quantitative Mathematical Model Approach

Mathematical models can be used to predict certain physical, chemical, and biological impacts of dredging projects. For example, a methodology for the prediction of effluent water quality from the disposal of dredged materials in confined areas was presented by Eichenberger and Chen (1980). The computation method was based on the use of laboratory determination of contaminants in different particle size fractions together with the application of Hazen's theory of sedimentation for longitudinal basins. Hazen's equation was applied to each particle size fraction. The removal efficiency of each parameter can be obtained from the data of suspended solids removal and its concentration in that size fraction. The validity of this methodology has been verified in three active disposal sites.

Holliday, Johnson and Thomas (1978) summarized some research efforts concerned with predicting the short-term fate of dredged material discharged to open water. Extant short-term fate models were evaluated and calibrated using field data collected at several disposal sites, including the Duwamish, New York Bight, and Lake Ontario sites. Two two-dimensional finite element models for the long-term prediction of sediment transport in estuaries were also evaluated.

A compartmental model addressing nutrient and heavy metal cycling in marsh-estuarine ecosystems has been developed to enhance the understanding of these complex phenomena (Gunnison, 1978). The model was developed based on both literature surveys and discussions with authorities in marsh-estuarine ecology. However, one of the conclusions was that present knowledge is inadequate to permit accurate predictions of environmental impacts resulting from use of heavily contaminated materials in marsh creation. Only those dredged materials containing nutrients and heavy metals in quantities no greater than those in marsh soils at the creation site should be used for marsh development.

A final example of a mathematical approach is a methodology for determining land values and associated benefits from the productive use of dredged material containment sites (Conrad and Pack, 1978). A discussion of productive uses of dredged material sites, their physical characteristics, institutional and legal constraints, and local land demand is included in the methodology, as well as an overview of property valuation. Site description, establishment of use potential, value estimation, and associated benefits and impacts are discussed, and working tables are presented. Fifteen case studies of productively used dredged material containment sites were conducted to validate and refine the methodology. One of the case studies was used as a site-specific example of how the methodology can be applied.

Quantitative/Qualitative Approach Based on Laboratory/Field Studies

Laboratory and/or field studies can be used to develop quantitative and/or qualitative information useful for predicting the impacts of dredging projects (Johanson, Bowen and Henry, 1976). Laboratory testing may include the use of a bulk sediment-chemistry evaluation procedure, bioassays, and/or the standard elutriate test.

A bulk sediment-chemistry evaluation procedure and the elutriate test were compared with a bioassay technique as a means of evaluating disposal methods for dredged materials (Laskowski-Hoke and Prater, 1981). Forty sediment samples collected in 1977 from Lake Michigan were analyzed. The elutriate process required analysis of ammonia, COD, total P, total Kjeldahl N, nitrate, nitrite, chloride, sulfate, and eight heavy metals. The sediment-chemistry technique relied on analysis of total Kjeldahl N, total P, COD, sieve size, oil, grease, cyanide, and 11 metals. Test organisms for the bioassay were Pimephales promelas Rafinesque, Hexagenia limbata Walsh, Lirceus fontinalis Rafinesque, and Daphnia magna Straus. Statistical analysis of results indicated that a combination of bulk sediment-chemistry and sediment bioassays would be the most efficient and ecologically sound approach to evaluating the effects of dredged materials on benthic communities.

The standard elutriate test has been frequently used to simulate the quality of the effluent during dredging operations, and is generally reasonably accurate. Grimwood and McGhee (1979) reported on data collected at five southern Louisiana locations to assess the predictive value of the test as a tool for estimating pollutant release. The Kolmogorov-Smirnov two-sample test was used to test the correspondence of pollutant release in the dredge effluent and the standard elutriate test for total Kjeldahl nitrogen, chemical oxygen demand, cyanide, arsenic, cadmium, copper, chromium, lead, mercury, nickel, zinc and phenol. It was found that for chemical oxygen demand, chromium, cadmium, and phenol--and to a lesser degree for total Kjeldahl nitrogen, arsenic, nickel, and zinc--the test afforded a useful prediction of release. The relative failure of the test with respect to lead, mercury, and copper, was a result, in part, of adsorption of these materials by the suspended sediment.

IMPACT ASSESSMENT

As noted in Chapter 2 for dams and reservoir projects, and in Chapter 3 for channelization projects, the assessment or interpretation of predicted impacts represents an important activity in an environmental impact study. Impact assessment for dredging projects should be based on the application of professional judgment in conjunction with environmental standards and criteria and other pertinent scientific information. Public input can also be sought and utilized. Several relevant laws and executive orders, along with some key technical references, are summarized in the Impact Assessment section of Chapter 2. Mitigation of undesirable impacts should also be considered within the context of impact assessment. Mitigation may include avoiding, minimizing, rectifying, reducing or eliminating impacts, and/or compensating for them.

In considering mitigation measures, it is important to recognize natural characteristics of the environmental system. Pertinent characteristics of benthic invertebrates can be used as an example since a major concern in dredging and disposal projects is the effect of burial on the survival of benthic invertebrates. Maurer et al. (1981) reported on a study to determine the ability of estuarine benthos, and in particular three species of mollusks (Mercenaria

mercenaria, Nucula proxima, and Illyanassa obsoleta), to migrate vertically in natural and exotic sediments and to determine the survival of benthos when exposed to particular amounts of simulated dredged material. Mortalities generally increased with increased sediment depth, with increased burial time, and with overlying sediments whose particle size distribution differed from that of the species' native sediment. Temperature affected mortalities and vertical migration, with both being greater in summer temperatures than in winter temperatures. It was concluded that vertical migration is a viable process which can significantly affect rehabilitation of a dredged disposal area. Under certain conditions, vertical migration should be considered, together with larval settling and immigration from outside impacted areas, as a mechanism of recruiting a dredge-dump site.

Another example related to natural characteristics is associated with the potential impacts of dredged material disposal on colonial nesting waterbirds. Landin (1978) developed an extensive bibliography on the life requirements of colonial nesting waterbirds in the United States. An additional bibliography pertaining to the vegetation and soils on dredged material islands and environmental impacts of dredged material deposition on waterbird habitats was also presented. Selected references from Canada, Europe, and Africa that pertain to related waterbirds, or those introduced to the United States, are also included. Usage of the information in this report would enable the planning of a dredged material disposal activity which should have minimal negative impacts on nesting waterbird habitats.

Enhancement of environmental conditions at dredged material disposal sites can be achieved by a terrestrial wildlife habitat development program. Ocean Data Systems, Inc. (1978) compiled a user-oriented handbook of existing published and unpublished data on terrestrial wildlife habitat development on dredged material within the contiguous United States. A general list of 250 plant species with food and cover value for wildlife is indexed by life form and state; a synopsis is given for each of 100 plant species chosen from the general list on the basis of their importance to wildlife, ease of establishment, and geographic distribution. Each synopsis includes a description and discussion of habitat, soil requirements, establishment and maintenance, disease and insect problems, and wildlife value. A range map and illustrations are given along with appropriate miscellaneous comments. The handbook also outlines a suggested approach for developing terrestrial wildlife habitat on dredged material; discusses wildlife species inhabitating dredged material areas; and recommends techniques for propagation, establishment, and maintenance of plantings.

Mitigation measures can also be used for reducing some of the physical effects of dredging, with this in turn reducing some of the chemical and biological impacts (JBF Scientific Corporation, 1975). For example, the turbidity created in the vicinity of dredging operations can be reduced through the use of a turbidity screen or diaper (Johnston, 1981). This measure is desirable since the biological effects of turbidity may cause reduced visibility and reductions in the availability of food for fish. High levels of suspended solids can reduce oyster growth and may have toxic effects on various larvae.

METHODOLOGIES FOR TRADE-OFF ANALYSES AND DECISION-MAKING

Environmental impact studies for dredging projects typically address several alternatives, particularly those related to dredging technology (Peterson, 1979), and dredged material disposal. Disposal options may be related to open water versus land disposal, and choices of specific sites within each option. Additional options for land disposal can include minimal management versus planned development and use.

Raster et al. (1978) described a logical, step-by-step methodology for site selection and design of reusable dredged material disposal sites. The methodology is capable of handling anything from a single disposal site serving a single dredging location to an entire dredging program involving several dredging locations and disposal sites. Pertinent factors--legal, environmental, and technological--which influence selection of candidate disposal sites and determine their suitability as reusable and nonreusable sites are identified. Site design and operating recommendations are presented, along with a preliminary costing procedure to enable evaluation of alternative disposal options for each site and cost modifications of the entire dredging program. Numerical examples are provided to assist the user in applying the procedures to particular cases. Although this methodology is focused on reusable disposal sites, nonreusable sites of a nonconventional nature are also discussed for situations where reusable sites are inappropriate or economically unsound.

Gushue and Kreutziger (1977) evaluated 12 selected cases where dredged material from navigation projects was used to create productive land. The principal output was the development of an overall set of "implementation factors" for disposal-productive use projects. Thirty-seven factors were identified and categorized as environmental, technical, economic/financial, legal, institutional, or planning/implementation. These factors provide a framework for ensuring that project planners address the full range of substantive and procedural considerations that are important to successful project implementation. The case studies provide documented proof that disposal-productive use project success is as much affected by procedural factors as by substantive factors.

An environmental impact study for a dredging project should include a systematic evaluation of alternatives via the application of methodologies for trade-off analyses and decision-making. Potentially useful methodologies include matrices, scaling or ranking checklists, and scaling-weighting or ranking-weighting checklists. Examples of these methodologies are described in the Chapter 2 section on Methodologies for Trade-off Analyses and Decision-making.

SUMMARY

Dredging projects involving the removal of bottom material and its subsequent open water or land disposal can cause significant environmental impacts. Systematic environmental impact studies are needed to predict and assess these impacts and identify appropriate mitigation measures. There is considerable technical literature on dredging projects to aid in impact identification and the planning and conduction of appropriate studies of the baseline environment. Impact prediction can be achieved via (1) a qualitative approach which relies on general knowledge of the impacts of similar projects, or specific results of comprehensive studies of similar projects; (2) a quantitative approach based on the use of mathematical models; and (3) a quantitative/qualitative approach based on laboratory and/or field studies. A given environmental impact study could involve all three approaches. Impact assessment involves the use of water quality standards and criteria, appropriate laws and regulations related to aquatic and terrestrial ecosystems, and professional knowledge and judgment. Several mitigation measures are available for reducing the undesirable impacts of dredging projects. Finally, systematic methods such as matrices and checklists are available for comparing alternatives for dredging project decisions.

SELECTED REFERENCES

Allen, K.O. and Hardy, J.W., "Impacts of Navigational Dredging on Fish and Wildlife: A Literature Review", FWS/OBS-80/07, Sept. 1980, U.S. Fish and Wildlife Service, Washington, D.C.

Bohlen, W.F., Cundy, D.F. and Tramontano, J.M., "Suspended Material Distributions in the Wake of Estuarine Channel Dredging Operations", Estuarine and Coastal Marine Science, Vol. 9, No. 6, Dec. 1979, pp. 699-711.

Brannon, J.M., "Evaluation of Dredged Material Pollution Potential", Technical Report No. DS-78-6, Aug. 1978, U.S. Army Engineer Waterways Experiment Station, Vicksburg, Mississippi.

Chen, K.Y. et al., "Confined Disposal Area Effluent and Leachate Control Laboratory and Field Investigations", Technical Report No. DS-78-7, Oct. 1978, U.S. Army Engineer Waterways Experiment Station, Vicksburg, Mississippi.

Conner, W.G. and Simon, J.L., "The Effects of Oyster Shell Dredging on an Estuarine Benthic Community", Estuarine and Coastal Marine Science, Vol. 9, No. 6, Dec. 1979, pp. 749-758.

Conrad, E.T. and Pack, A.J., "A Methodology for Determining Land Value and Associated Benefits Created from Dredged Material Containment", Technical Report No. D-78-19, June 1978, U.S. Army Engineer Waterways Experiment Station, Vicksburg, Mississippi.

Eichenberger, B.A. and Chen, K.Y., "Methodology for Effluent Water Quality Prediction", Journal of the Environmental Engineering Division, American Society of Civil Engineers, Vol. 106, No. EE1, Feb. 1980, pp. 197-209.

Engler, R.M., "Impacts Associated with the Discharge of Dredged Material Into Open Water", Proceedings of the Third U.S.-Japan Expert's Meeting on Management of Bottom Sediments Containing Toxic Substances, Report No. EPA-600/3-78-084, 1978, U.S. Environmental Protection Agency, Washington, D.C., pp. 213-223.

Flint, R.W., "Responses of Freshwater Benthos to Open-Lake Dredged Spoils Disposal in Lake Erie", Journal of Great Lakes Research, Vol. 5, No. 3-4, 1979, pp. 264-275.

Grimwood, C. and McGhee, T.J., "Prediction of Pollutant Release Resulting from Dredging", Journal of the Water Pollution Control Federation, Vol. 51, No. 7, July 1979, pp. 1811-1815.

Gunnison, D., "Mineral Cycling in Salt Marsh-Estuarine Ecosystems; Ecosystem Structure, Function, and General Compartmental Model Describing Mineral Cycles", Technical Report No. D-78-3, Jan. 1978, U.S. Army Engineer Waterways Experiment Station, Vicksburg, Mississippi.

Gushue, J.J. and Kreutziger, K.M., "Case Studies and Comparative Analyses of Issues Associated with Productive Land Use at Dredged Material Disposal Sites", Technical Report No. D-77-43, Dec. 1977, Two Volumes, U.S. Army Engineer Waterways Experiment Station, Vicksburg, Mississippi.

Hoeppel, R.E., "Contaminant Mobility in Diked Containment Areas", Proceedings of the 5th United States-Japan Experts Meeting on Management of Bottom Sediments Containing Toxic Substances, EPA-600/9-80-044, Sept. 1980, U.S. Environmental Protection Agency, Washington, D.C., pp. 175-207.

Holliday, B.W., "Processes Affecting the Fate of Dredged Material", Technical Report No. DS-78-2, Aug. 1978, U.S. Army Engineer Waterways Experiment Station, Vicksburg, Mississippi.

Holliday, B.W., Johnson, B.H. and Thomas, W.A., "Predicting and Monitoring Dredged Material Movement", Technical Report No. DS-78-3, Dec. 1978, U.S. Army Engineer Waterways Experiment Station, Vicksburg, Mississippi.

JBF Scientific Corporation, "Dredge Disposal Study, San Francisco Bay and Estuary. Appendix M--Dredging Technology", Sept. 1975, U.S. Army Corps of Engineers, San Francisco, California.

Johanson, E.E., Bowen, S.P. and Henry, G., "State-of-the-Art Survey and Evaluation of Open-Water Dredged Material Placement Methodology", Contract Report No. D-76-3, Apr. 1976, U.S. Army Engineer Waterways Experiment Station, Vicksburg, Mississippi.

Johnston, Jr., S.A., "Estuarine Dredge and Fill Activities: A Review of Impacts", Environmental Management, Vol. 5, No. 5, Sept. 1981, pp. 427-440.

Landin, M.C., "A Selected Bibliography of the Life Requirements of Colonial Nesting Waterbirds and Their Relationship to Dredged Material Islands", Misc. Paper No. D-78-5, 1978, U.S. Army Engineer Waterways Experiment Station, Vicksburg, Mississippi.

Laskowski-Hoke, R.A. and Prater, B.L., "Dredged Material Evaluations: Correlations Between Chemical and Biological Evaluation Procedures", Journal of the Water Pollution Control Federation, Vol. 53, No. 7, July 1981, pp. 1260-1262.

Lehmann, E.J., "Dredging: Environmental and Biological Effects (Citations from the Engineering Index Data Base)", Dec. 1979, National Technical Information Service, U.S. Department of Commerce, Springfield, Virginia.

Massoglia, M.F., "Dredging in Estuaries--A Guide for Review of Environmental Impact Statements", Report No. NSF/RA-770284, 1977, Oregon State University, Corvallis, Oregon.

Maurer, D. et al., "Vertical Migration and Mortality of Benthos in Dredged Material, Part I: Mollusca", Marine Environmental Research, Vol. 4, No. 4, 1981, pp. 299-319.

Morrison, R.D. and Yu, K.Y., "Impact of Dredged Material Disposal Upon Ground Water Quality", Ground Water, Vol. 19, No. 3, May/June 1981, pp. 265-270.

National Marine Fisheries Service, "Physical, Chemical and Biological Effects of Dredging in the Thames River (CT) and Spoil Disposal at the New London (CT) Dumping Ground", Final Report, Apr. 1977, Division of Environmental Assessment, Highlands, New Jersey.

Ocean Data Systems, Inc., "Handbook for Terrestrial Wildlife Habitat Development of Dredged Material", Technical Report No. D-78-37, July 1978, U.S. Army Engineer Waterways Experiment Station, Vicksburg, Mississippi.

Pavlou, S.P. et al., "Release, Distribution, and Impacts of Polychlorinated Biphenyls (PCB) Induced by Dredged Material Disposal Activities at a Deep Water Estuarine Site", Proceedings of the 5th United States-Japan Experts Meeting on Management of Bottom Sediments Containing Toxic Substances, EPA-600/9-80-044, Sept. 1980, U.S. Environmental Protection Agency, Washington, D.C., pp. 129-174.

Peterson, S.A., "Dredging and Lake Restoration", Report No. EPA 440/5-79-001, 1979, U.S. Environmental Protection Agency, Corvallis Environmental Research Laboratory, Corvallis, Oregon, pp. 105-144.

Raster, T.E. et al., "Development of Procedures for Selecting and Designing Reusable Dredged Material Disposal Sites", Technical Report No. D-78-22, June 1978, U.S. Army Engineer Waterways Experiment Station, Vicksburg, Mississippi.

Slotta, L.S. et al., "An Examination of Some Physical and Biological Impacts of Dredging in Estuaries", Interim Progress Report to the National Science Foundation, Dec. 1974, Oregon State University, School of Oceanography, Corvallis, Oregon.

Sorensen, J.C., "A Framework for Identification and Control of Resource Degradation and Conflict in the Multiple Use of the Coastal Zone", June 1971, University of California, Berkeley, California.

CHAPTER 5

ENVIRONMENTAL IMPACT STUDIES FOR OTHER WATER RESOURCES PROJECTS

In addition to the traditional categorization of water resources projects into dams and reservoirs, channelization projects, and dredging projects, still other categories can be considered. The additional categories include irrigation projects, major flow-altering projects, land creation projects, and shoreline projects. This chapter provides a summary of some key technical references related to the environmental impacts resulting from these four additional categories of water resources projects. Complete discussions for each of these additional categories of projects in terms of impact identification, baseline studies, impact prediction, impact assessment, and methodologies for trade-off analyses and decision-making will not be presented.

IRRIGATION PROJECTS

Irrigation projects have been developed throughout the world in semi-arid and arid regions. A semi-arid region can be defined as one having a dry season of three to four months, regardless of the total annual precipitation. An arid region can be taken as one in which the usual dry season will last longer than four months, and sometimes the whole year. Semi-arid regions need supplementary or seasonal irrigation, whereas arid regions must be designated for perennial irrigation. Ahmad (1982) noted that the impact of irrigation projects is not exclusively beneficial, and when assessing the viability of such a project the adverse consequences must also be taken into consideration. The consequences can be most damaging to the fragile and delicately balanced ecosystems that characterize the arid and semi-arid regions.

Some countries have developed extensive irrigation schemes, including the People's Republic of China (Smil, 1981). Regional lack of moisture is a critical environmental constraint to China's agro-ecosystem. The climate ranges from subtropical in the southeast to arid in the north and northwest. Mean annual precipitation varies from 577 mm to 1720 mm. In some areas evaporation exceeds precipitation by up to 4 times. Water resources development and control have been an integral part of China's agriculture for centuries. As of 1979, the irrigation system included 80,000 reservoirs, 5000 canal networks, and 2 million power-operated wells. Several large multipurpose water projects are under construction. The capacity of agricultural pumps for drainage and irrigation was 47 GW in 1978, a large increase from 0.13 GW in 1950. The mechanical irrigation systems are not efficient. Leaks and seepage losses are 50-60 percent, power shortages idle the pumps for long periods, and sprinkler irrigation is in an early stage of development. Finally, soil erosion is a widespread and serious environmental problem caused by land mismanagement.

One of the chief concerns associated with irrigation projects is the potential for spreading water-borne diseases. For example, schistosomiasis, malaria, and yellow fever have occurred in the Gezira region in Sudan as a result of water resources projects developed for the irrigation of cotton (Pollard, 1981).

A specific example of water-borne disease is schistosomiasis. Many water resource development projects in tropical areas have had the unintended consequence of markedly increasing the prevalence of schistosomiasis in the local population (Tucker, 1983). Water contact is the most critical variable in the transmission of schistosomiasis. Irrigation of formerly arid regions creates additional habitats for the snail vectors beyond those already present in ponds and rivers. Defects in the design or engineering of water projects may also provide new habitats for the snail hosts. The water projects have provided new opportunities for villagers to come into contact with the infested water, particularly when the projects are located close to villages in which no alternative water sources are available. The advent of perennial irrigation, in which the irrigation canals are in use year-round, has facilitated the multiplication of the snail vectors of schistosomiasis. Perennial irrigation has removed natural checks on the snail population, enabling them to multiply out of control. Once snails have been introduced into an irrigation system it is impossible to eradicate them completely. However, if irrigation projects are supplied with efficient drainage, good water management and regular maintenance, it is possible to avoid an increase in the incidence of schistosomiasis.

There are opportunities for design and operation schemes to minimize the potential health impacts of irrigation projects. For example, environmental health aspects should be included in the planning, construction, and operation of large-scale irrigation projects (Diamant, 1980). Vector control measures, which break the three-part cycle (sick carrier, transmitting vector, and recipient), can be accomplished by proper engineering to produce a habitat unfit for snails and mosquitos. Some suggestions for destroying habitats for disease vectors are: controlling vegetation, straightening banks, producing frequent changes in water levels, reducing evaporation, eliminating wastewater influx, installing snail screens, using closed channels, sprinkling irrigation water, and choosing crops which do not need flooding. Operation and maintenance of the water projects should be regularly surveyed by teams of ecologists, environmental engineers, entomologists, and health workers.

Environmental Resources Limited (1983) issued a report which systematically addresses environmental health problems associated with irrigated agricultural development projects. This report provides outline guidance, based on past experience, on how environmental health impact assessment may be carried out. The report is organized into four main chapters, with additional background information provided in the form of annexes to the report. In Chapter 2, guidance is provided on identification of impacts on the environment; Chapter 3 focuses on prediction, and in particular on prediction of environmental health impacts; Chapter 4 focuses on mitigation of health impacts; while Chapter 5 discusses the organization and presentation of information for the decision-maker, that is, the individual or agency who must take this information into account in deciding whether the development should proceed.

MAJOR FLOW-ALTERING PROJECTS

Major flow-altering projects include those which are designed to change normal river flow patterns via the use of storm surge barriers, reversal of the direction of flow, or inter-basin water transfers. Smies and Huiskes (1981) discussed the possible environmental consequences of construction and use of the Eastern Scheldt Barrier which is being built to protect the Netherlands from flood waters. The project includes the construction of a storm-surge barrier across the mouth of the estuary along with two secondary dams to separate the tidal basin of the Eastern Scheldt from the brackish-fresh water of the Rhine Scheldt shipping route. Turbulence and turbidity will decrease with decreasing mean tidal current velocities, and mean water residence time will increase. Chlorinity and nutrient concentrations depend on water residence time, so these values will be changing. There will probably be increases in net particulate carbon, which may result in increased sedimentation and resulting changes in depth. None of the effects of the barrier will be extreme. If the barrier should be completely closed for over a week, for example, to protect the estuaries from a tanker spill, problems may occur. The estuary may become anoxic, and severe ecological damage could result from lack of oxygen and flooding of the intertidal zone. Elgershuizen (1981) noted similar environmental impact findings for the Eastern Scheldt Barrier as those by Smies and Huiskes (1981).

Relative to flow reversal projects, proposals are being considered in the Soviet Union for transferring waters from northward-flowing rivers to the south. Territorial redistribution of water resources in the USSR has become a matter of national policy. One of the two major plans being considered calls for diversion of waters from the Sukhona and Onega Rivers and Lake Onega (possibly also from Lakes Lacha and Vozhe) into the Rybinsk reservoir and then to the upper reaches of the Volga River. About 31 cu km a year would be withdrawn by 1990. Later stages of the plan envisage additional withdrawals variously from the Severnaya Dvina, Vychegda, Pinega, Mezen and Yug, as well as from the basin of Lake Ladoga. The second plan differs primarily in the degree of probable impact on the environment. The first plan would involve drawing water from a reservoir to be built at the confluence of the Tobol and Irtysh and lifting it about 75 m, passing it through the Turgay Gate into Kazakhstan, after which it would flow by gravity towards the Aral region. In the second stage, water would be drawn from a reservoir in the Khanty-Mansiysk area and be transferred either by a special canal or along the bed of the Irtysh. In the second variant, the main difference is that in the first stage use would be made of the existing Novosibirsk reservoir on the Ob, and it would involve transferring the water by a canal starting at Kamen-on-Ob which would cross the Kulunda Steppe to a proposed reservoir on the Irtysh above Pavlodar, then by canal to Kurgan and on south. The second stage envisages drawing water from the Chulym and Tom Rivers, transferring it to a reservoir to be built on the Irtysh and then on to Kurgan. Yet a third plan combined features from the other two, withdrawing water from the Novosibirsk reservoir and the Irtysh and Tobol in the first stage, and from the Chulym and Tom in addition to these in the second stage. Other water transport projects include the Irtysh-Karaganda canal, the Ob-Kulunda canal and the Volga-Ural canal. Impacts on the environment include changes in the levels of the Caspian and Aral Seas. There are international implications associated with potential impacts in the Caspian Sea (Micklin, 1977).

Major water transfer schemes are also being considered in England and the People's Republic of China. Of the alternative proposed water resource development schemes to meet future demands in the southeastern part of England, one contender is a strategy to transfer water from the River Severn to the River Thames at appropriate times (Birtles and Brown, 1978). Severn flows would be augmented when necessary from surface storages in Wales. Simulation experiments have been performed to estimate the water quality effects of interbasin transfers on the recipient system. A range of nearly conservative river quality determinants has been investigated using river quality models. The results obtained are presented in a form which facilitates estimating (1) the size and cost of any treatment plant which may be necessary to make transferred water compatible with receiving waters, and (2) the adverse effects of transfers on organisms with known concentration tolerance levels.

China's south-to-north water project is a broad concept employing numerous pipelines for diverting water of the Yangtze River in the south to the north of the Yellow River (Yiqui, 1981). In 1975 the North China Plain suffered an acute water shortage, and the eastern line was proposed by the relevant agencies in 1976 as the first phase of the plan to be established. Since the water supply is interrelated with the regional economy and agricultural production, these factors should be considered conjunctively. An environmental impact assessment must be performed to assess the impacts of the proposed projects on the environment, and to select alternatives which might achieve the same goal as the proposed project, while minimizing any harmful effects.

LAND CREATION PROJECTS

Artificial islands represent one category of land creation projects. The impact of the construction and operation of an artificial island on the marine ecosystem of the northeastern United States was examined by Watling, Pembroke and Lind (1975). Information needs and criteria for assessing the impacts of artificial islands have been identified by Watling (1975). To serve as a specific example, the building of artificial islands is contemplated in the coastal waters of the Netherlands (deGroot, 1979). The most suitable sand mining methods, the effect on water quality turbidity, the disturbance of the balance between bottom sediment and the water column in relation to heavy metals, nutrients, and PCB's are issues, as well as the effect of dredging upon plankton, bottom fauna, fishes and on larvae and young stages of sea animals in general. The recovery time for the bottom fauna is estimated to be about three years. The consequences for the fisheries are important as the Netherlands coastal zone plays a very important role in the life of the North Sea species. This is related to the Dutch shrimp, plaice and sole fishery. An estimation of the yearly damage to the fisheries during the building phase is made since the construction time for the island would be about eight years.

Another category of land creation projects is associated with the filling of wetland areas to create new land for residential and/or commercial development. Pressures for development and economic growth are often such that the potential ecological impacts become obscured (Elkington, 1977). Darnell (1977) noted that the primary categories of environmental impact of construction activities in wetlands, in order of importance, include habitat modification and loss, increase in suspended sediments, bottom sedimentation, and modification of the water flow regime.

One example of a coastal land creation project is at Marco Island in southwestern Florida. This project includes filling for new land and the excavation of a large group of interconnected lakes (Huber and Brezonik, 1981). The quality of the proposed lakes is of considerable importance, both to the riparian owners and to the nearby estuarine areas that will receive surface discharges. The deep lakes will be stratified due to the influx of hypersaline ground water below a depth of 2.0 m. Density differences are so great across the chemocline that the possibility of overturn is nil. The lakes receive water from surface runoff from the various land uses; from inflow, from regional ground water flow in the shallow fresh water layer, and from direct rainfall. Water is lost by surface runoff, ground water outflows, and evaporation. Annual precipitation in the area is about 50 inches. Residence times of 0.42 and 0.65 years have been calculated for two units of the development. On the basis of phosphorus loading rates, the lakes are expected to be mesotrophic with fair to good water quality. On the whole, predicted water quality is good with Secchi disk transparencies on the order of 1.2 to 1.5 m, and total nitrogen of about 1.3 mg/l. On the basis of nutrient loads, the urban development is expected to have little impact on the estuaries.

Several procedures and methodologies have been developed for evaluating projects in coastal areas. One example is a report from the National Oceanic and Atmospheric Administration (1976) which provides state coastal zone management (CZM) agencies with information and recommendations for developing guidelines for facility development in the coastal zone. Section A of the report presents a methodology for identifying and initiating implementation procedures for management recommendations for specific facility types. Sections B and C apply the methodology to marinas and power plants in the states of Florida and Maryland, respectively. The two case studies from Florida and Maryland serve the dual purpose of (1) providing a useful CZM reference source on environmental mitigation techniques and relevant Federal authorities for the two facility types; and (2) further clarifying the format, intended information content, and applications envisioned in the methodology.

SHORELINE PROJECTS

Shoreline projects can include breakwaters, jetties, groins, bulkheads, and many others. These types of projects can cause environmental impacts as illustrated by the following two case studies.

Witten and Bulkley (1975) studied four types of bank stabilization structures (revetments, retards, permeable jetties, and impermeable jetties) to determine their impact upon game fish habitat in Iowa. Permeable jetties and retards deepened the channel near the structures 7 to 110 percent greater than the maximum depth in control sections. No other significant differences in either physical parameters or in mean body length or abundance of game fish were found between structured and nonstructured stream sections. Unusually high water during the sampling period may have allowed fish to remain dispersed throughout the streams rather than concentrating in the deep pools at or near stabilization structures. Rock revetments and impermeable jetties fostered the growth of some invertebrates, primarily mayflies and caddisflies. Revetments, which presented the most rock surface for invertebrate colonization, had the greatest impact on invertebrate abundance. A long rock jetty, extending far enough into the stream to produce a scour hole, would combine most of the advantages noted in the structures studied. For habitat improvement, rock was superior to steel as a construction material, and structures which cause the formation of scour holes superior to those that do not deepen the stream.

Bradt and Wieland (1978) reported on a study to evaluate the effect of a gabion installation and stream reconstruction in a 2 km section of a rechanneled stream named Bushkill Creek. This creek supports a naturally reproducing brown trout population in Northampton County, Pennsylvania. The creek was sampled bi-weekly biologically, chemically and physically for 16 months. Prior to the sampling, stream reconstruction efforts included both a gabion (rock current deflectors) installation to narrow and deepen the stream bed, and tree and shrub planting to cover bare banks and provide eventual shade. The stream bed was open to sunlight and primary productivity, as evidenced by larger algae populations, increased in the rechanneled area. The following benthic macroinvertebrate parameters significantly increased also through the rechanneled area: density index, biomass, total numbers, and number of taxa. The following chemical parameters increased significantly through the rechanneled area: conductivity, dissolved oxygen, percent oxygen saturation and alkalinity. Orthophosphate decreased significantly and flow velocity increased significantly. Limestone springs contributed to the increase in conductivity and alkalinity. Increased photosynthesis and turbulence also contributed to the increase in conductivity and alkalinity, and to the increase in dissolved oxygen and oxygen saturation. Finally, the gabions deepened and narrowed the stream channel resulting in a cooler stream in the summer.

A generic study of the biological impacts of minor shoreline structures has been conducted by Mulvihill et al. (1980). Information from 555 sources located in an information search was used to develop a computer data base for the analysis of the biological impacts. Structures included were: breakwaters, jetties, groins, bulkheads, revetments, ramps, piers and other support structures, buoys and floating platforms, small craft harbors, bridges, and causeways. Data were compiled by type of structure and by coastal region including the following: structure functions; site characteristics; geographic prevalence; engineering, socio-economic, and biological placement constraints; construction materials; expected life span; environmental conditions; methodology of impact studies; physical and biological impacts; and alternatives. Results show that structure impacts on the environment are site-specific. Fourteen case studies are included in the report. Small boat harbors, bridges and causeways, bulkheads, breakwaters, and jetties were found to have the most potential for coastal environment impact. Revetments, groins, and ramps have moderate impact potential, while buoys and floating platforms, piers, and other support structures have low impact potentials. Little information relative to the potential impacts of bridges, causeways, and small boat harbors was identified. Also, very little information on the quantitative impacts of specific structures was located.

SUMMARY

Environmental impact studies are needed on irrigation projects, major flow-altering projects, land creation projects, and shoreline projects. These types of water resources projects have not been studied as extensively as dam and reservoir projects, channelization projects, and dredging projects. However, by systematically assembling information on the types of experienced impacts, scientifically based environmental impact studies can be planned and conducted. Elements which should be included in these studies are a description of the baseline environment, impact prediction and assessment, and a systematic evaluation of the impact features of alternative design and operational schemes.

SELECTED REFERENCES

Ahmad, Y.J., "Irrigation in Arid and Semi-Arid Areas", 1982, United Nations Environment Programme, Nairobi, Kenya.

Birtles, A.B. and Brown, S.R.A., "Computer Prediction of the Changes in River Quality Regimes Following Large Scale Inter Basin Transfers", Proceedings: Baden Symposium on Modeling the Water Quality of the Hydrological Cycle, IIASA (Laxenburg, Austria) and IAHS (United Kingdom), IAHS-AISH Publication No. 125, Sept. 1978, Reading, England, pp. 288-298.

Bradt, P.T. and Wieland, III, G.E., "The Impact of Stream Reconstruction and a Gabion Installation on the Biology and Chemistry of a Trout Stream", Completion Report, Jan. 1978, Department of Biology, Lehigh University, Bethlehem, Pennsylvania.

Darnell, R.M., "Overview of Major Development Impacts on Wetlands", Proceedings of the National Wetland Protection Symposium, June 6-8, 1977, Reston, Virginia, Department of Oceanography, Texas A and M University, College Station, Texas.

deGroot, S.J., "An Assessment of the Potential Environmental Impact of Large-Scale Sand-Dredging for the Building of Artificial Islands in the North Sea", Ocean Management, Vol. 5, No. 3, Oct. 1979, pp. 211-232.

Diamant, B.Z., "Environmental Repercussions of Irrigation Development in Hot Climates", Environmental Conservation, Vol. 7, No. 1, Spring 1980, pp. 53-58.

Elgershuizen, J.H., "Some Environmental Impacts of a Storm Surge Barrier", Marine Pollution Bulletin, Vol. 12, No. 8, Aug. 1981, pp. 265-271.

Elkington, J.B., "The Impact of Development Projects on Estuarine and Other Wetland Ecosystems", Environmental Conservation, Vol. 4, No. 2, Summer 1977, pp. 135-144.

Environmental Resources Limited, "Environmental Health Impact Assessment of Irrigated Agricultural Development Projects", Dec. 1983, World Health Organization Regional Office for Europe, Copenhagen, Denmark.

Huber, W.C. and Brezonik, P.L., "Water Budget and Projected Water Quality and Proposed Man-Made Lakes Near Estuaries in the Marco Island Area, Florida", Proceedings of the National Symposium on Fresh Water Inflow to Estuaries, FWS/OBS-81-04, Vol. I, Oct. 1981, U.S. Fish and Wildlife Service, Washington, D.C., pp. 241-251.

Micklin, P.P., "International Environmental Implications of Soviet Development of the Volga River", Human Ecology, Vol. 5, No. 2, June 1977, pp. 113-135.

Mulvihill, E.L. et al., "Biological Impacts of Minor Shoreline Structures on the Coastal Environment: State-of-the-Art Review, Volume I", FWS/OBS-77-51, Mar. 1980, U.S. Fish and Wildlife Service, Washington, D.C.

National Oceanic and Atmospheric Administration, "Coastal Facility Guidelines: A Methodology for Development with Environmental Case Studies on Marinas and Power Plants", Working Paper, Aug. 1976, Rockville, Maryland.

Pollard, N., "The Gezira Scheme--A Study in Failure", Ecologist, Vol. 11, No. 1, Jan.-Feb. 1981, pp. 21-31.

Smies, M. and Huiskes, A.H., "Holland's Eastern Scheldt Estuary Barrier Scheme: Some Ecological Considerations", Ambio, Vol. 10, No. 4, 1981, pp. 158-165.

Smil, V., "China's Agro-Ecosystem", Agro-Ecosystems, Vol. 7, No. 1, 1981, pp. 27-46.

Tucker, J.B., "Schistosomiasis and Water Projects: Breaking the Link", Environment, Vol. 25, No. 7, Sept. 1983, pp. 17-20.

Watling, L., "Artificial Islands: Information Needs and Impact Criteria", Marine Pollution Bulletin, Vol. 6, No. 9, Sept. 1975, pp. 139-141.

Watling, L., Pembroke, A. and Lind, H., "Environmental Assessment", Final Technical Report No. NSF-RE-E-054A, 1975, College of Marine Studies, University of Delaware, Lewes, Delaware, pp. 294-431.

Witten, A.L. and Bulkley, R.V., "A Study of the Effects of Stream Channelization and Bank Stabilization on Warmwater Sport Fish in Iowa. Subproject No. 2. A Study of the Impact of Selected Bank Stabilization Structures on Game Fish and Associated Organisms", Report No. 76/12, May 1975, Iowa Cooperative Fishery Research Unit, Ames, Iowa.

Yiqui, C., "Environmental Impact Assessment of China's Water Transfer Project", Water Supply and Management, Vol. 5, No. 3, 1981, pp. 253-260.

APPENDIX A

IMPACTS OF IMPOUNDMENT PROJECTS

Abu-Zeid, M., "Short and Long-Term Impacts of the River Nile Projects", Water Supply and Management, Vol. 3, No. 4, 1979, pp. 275-283.

A detailed examination of Egypt's supplies and demands for water indicates that the Nile River has been and still is the principal factor in the development of this arid country. Full control of Nile water depends, however, on the Aswan High Dam, completed in 1968, which has created the huge long-term, man-made reservoir, Lake Nasser. The execution of this dam, which was preceded by extensive feasibility studies, has had both positive and negative effects. These effects which include protection against high floods and droughts, and such agricultural and environmental impacts as siltation of lakes, degradation downstream, loss of silt on agriculture, impacts on fisheries, water quality changes and public health impacts, form the basis of this paper. It is felt that the negative effects are far outweighed by benefits so far accrued.

Armaly, B.F. and Lepper, S.P., "Diurnal Stratification of Deep Water Impoundments", Report No. 75-HT-35, 1975, American Society of Mechanical Engineers, New York, New York.

The diurnal temperature distribution in a deep water impoundment which is exposed to heating and cooling loads by convection, evaporation, and radiation is examined. The diurnal behavior of the solar load, directional and spectral, is used with the selective attenuation and reflection characteristics of the water to account for the internal solar energy absorption rate. Simple atmospheric and water attenuation models were used to examine and simulate the diurnal behavior. The results demonstrate the stratification developments of the reservoir and indicate that in some cases the deep layer responds continuously to the diurnal fluctuations of surface energy loads.

Berkes, F., "Some Environmental and Social Impacts of the James Bay Hydroelectric Project, Canada", Journal of Environmental Management, Vol. 12, No. 2, Mar. 1981, pp. 157-172.

The James Bay hydroelectric project, which consists of a series of dams affecting the watersheds of several rivers in northern Quebec, is the most ambitious energy project ever attempted in Canada. Construction of the project, located in subarctic Canada, began in 1972, shortly after the project was announced. However, environmental information needed to plan mitigation measures was not available until about 1975. This pattern appears to be characteristic of large-scale development projects in remote areas, where the time lags involved in obtaining detailed environmental data are so lengthy that engineering plans proceed, for economic reasons, without the benefit of this environmental data as planning input. None of the 1975 recommendations for environmental protection measures with respect to the estuarine subsystem were adopted by the project. An environmental impact case study of the effects of the LaGrande Complex phase of the James Bay development examines the estuarine fisheries subsystem and the effects on this subsystem of changes in the flow regime of the LaGrande River, the relocation of the first dam on the LaGrande, saltwater encroachment up the river during the filling of the second dam, and the changes in the thermal regime of the river. A social impact case study discusses the effects of the road network associated with the hydroelectric project on the land tenure system of the native Cree-Indians of the area. These analyses show that it is not easy for large projects in frontier areas to accommodate environmental considerations, since the lead times are too great and too little is known about the environment when the engineering decisions are being made. In addition, it is often impossible to assess the impacts of these projects after the fact due to the absence of baseline data and to the existence of a cumulative effect of relatively insignificant incremental impacts.

Biswas, A.K., "Environment and Water Development in Third World", Journal of the Water Resources Planning and Management Division, American Society of Civil Engineers, Vol. 106, No. WR1, Mar. 1980, pp. 319-332.

This paper discusses the social and environmental effects of water development in the third world under three categories--physical, biological, and human effects. The positive and negative environmental effects of several water resource development projects are discussed--Egypt's Aswan Dam, India's Koyna Dam, and irrigation systems and agriculture throughout the world. On a global scale 20 million square km of soil, or 35 percent of the currently used arable land, has been destroyed or degraded by poor management. Similarly, water development projects can affect the biological subsystem in beneficial or adverse ways. Although safe drinking water dramatically reduces the incidence of many water-borne diseases, irrigation projects on tropical regions can spread diseases such as schistosomiasis, liver fluke and malaria into new areas and promote growth of undesirable aquatic weeds in waterways. Potable water made easily available to a rural population can improve the life of the water-gatherers, 90 percent of whom are women. It is not unusual for a person to spend 5 hours every day hauling water. This wastes productive time, uses up dietary calories, and exposes the carriers to disease vectors. However, severe social problems may result when a large population is forced to move out of the path of a dam inundation. The planning process must be sensitive to social and environmental guidelines, or the overall strategy may be self-defeating.

Biswas, A.K., "Environmental Implications of Water Development for Developing Countries", Water Supply and Management, Vol. 2, No. 4, 1978, pp. 283-297.

The environmental implications of water resources development can be divided into three categories of subsystems: physical, biological and human. The physical subsystem can be further divided into the hydrologic, atmospheric and crustal systems. Water development projects invariably change river and ecosystem regimes; therefore the issue is not whether the environment will be affected but rather how much change is acceptable to society as a whole, and what countermeasures should be taken to keep the adverse changes to a minimum, at a reasonable economic cost. The biological subsystem is divided into the aquatic ecosystem and terrestrial ecosystem. Changing the biological subsystem affects the quality and availability of potable water. This in turn either reduces or increases health hazards such as cholera, typhoid, infectious hepatitis and other water-borne diseases. Water resources developments do not only bring unmitigated benefits, they also often are responsible for unanticipated social costs. Thus in irrigation developments in tropical and semi-tropical regions a secondary effect is the spreading of parasite water-borne diseases because of the creation of favorable ecological environments for these diseases. The impacts of water developments on the human subsystem could be direct or indirect, stemming from direct effects on physical and biological subsystems. This addition of environmental quality to the traditional objectives of water resources development of economic efficiency and regional income redistribution has made the planning process more complex than ever before. Inclusion of environmental quality as an objective recognizes the fact that the welfare of society has dimensions besides economics; the issue is how it should be incorporated objectively within the planning framework. Planners must give emphasis to the social and environmental consequences of water development projects.

Bombowna, M., Bucka, H. and Huk, W., "Impoundments and Their Influence on the Rivers Studied by Bioassays", Proceedings: Congress in Denmark 1977, Part 3; Internationale Vereingung fur Theoretische und Angewandte Limnologie, Vol. 20, 1978, Polish Academy of Sciences, Krakow, Poland, pp. 1629-1633.

Water flowing in and out of the Goczalkowice, Roznow, and Tresna reservoirs in Poland showed higher eutrophication levels than the Solina impoundment in 1973-74. Carpathian mountain rivers drain one-tenth of Poland; however, their impoundments supply over 30 percent of total water resources. Agricultural wastes are the primary cause of eutrophication, with municipal and industrial waste effects secondary. In the spring, the greatest loads flow into the impoundments. Phosphates are the limiting factor with nitrates in excess. Diatoms such as Cyabella ventricosa, Cyclotella, Navicula, and Diatoma vulgare are dominant, with lesser quantities of green algae, blue-green algae, and chrysophytes. The greatest fertility is in the Goczalkowice impoundment. Green algae are more numerous above impoundments; whereas diatoms grow better below impoundments. One positive effect of these impoundments is the decreased number of green algae below impoundments with better diatom development; therefore, impoundments enhance microorganism succession typical for mountain rivers.

Budweg, F.M., "Reservoir Planning for Brazilian Dams", International Water Power and Dam Construction, Vol. 34, No. 5, May 1982, pp. 48-49.

Brazil's Codigo das Aguas, a collection of laws relevant to water resources, was compiled in 1934, before the industrial and population explosion and before large-scale water resources development. Pollution control regulations are still in an initial phase. Hydropower development is the responsibility of Electrobras, the federal holding company. Permits for construction of dams must be obtained from the Ministry of Mines and Energy. To date most reservoirs built in Brazil have been single-purpose. Little regional planning was done until 1979, when the federal government established a special committee on integrated watershed studies. This is responsible for regional planning and multipurpose development of streams and rivers. Increasing numbers of dam owners and designers favor multipurpose uses of reservoirs. Planning is done in three stages: (1) a baseline survey of hydrology, etc.; (2) an appraisal of environmental effects; and (3) the actual planning phase.

Buikema, Jr., A.L. and Loeffelman, P.H., "Effects of Pumpback Storage on Zooplankton Populations", Proceedings of the Clemson Workshop on

Environmental Impacts of Pumped Storage Hydroelectric Operations, FWS/OBS-80-28, Apr. 1980, U.S. Fish and Wildlife Service, Washington, D.C., pp. 109-124.

The impact of pumped storage operations on the aquatic microfauna of impoundments was investigated at Smith Mountain Lake, a mainstream pumped storage complex southeast of Roanoke, Virginia. Smith Mountain Lake was sampled for zooplankton and water temperature at four stations once a month for 13 months, and then at two stations once a day for seven days. The most obvious effect of pumpback on temperature was disruption of thermal stratification in the forebay. Thermal layers fluctuated greatly, and isolated lenses of water of differing temperature were apparent. A total of 1088 pump samples and 104 tow samples were qualitatively and quantitatively analyzed for zooplankton. Forty-nine species of rotifers, 15 species of Cladocera, and 8 species of copepods were identified. It appears that the interaction of pumped storage operations and zooplankton populations in Smith Mountain Lake is very complex. Pumpback appears to stimulate predation, and normal thermal characteristics of the lake were altered to the apparent benefit of the zooplankton.

Burton, Jr., G.A., "Microbiological Water Quality of Impoundments: A Literature Review", Misc. Paper E-82-6, Dec. 1982, U.S. Army Engineer Waterways Experiment Station, Vicksburg, Mississippi.

Assessing the microbiological water quality of impoundments and the potential for waterborne disease outbreaks is a difficult task when using traditional sampling programs. Problems associated with using fecal coliform bacteria as indicators of human pathogen presence complicates assessments of future water quality in preimpoundment areas. Reliable determination of future and present microbiological water quality requires knowledge of how the chemical, physical, and biological characteristics of the watershed and impoundment interrelate to influence microbial indicator and pathogen densities. Accurate estimates of microbial indicator and pathogen densities, obtainable by using the enumeration methods and their modifications suggested in this report, will allow monitoring of the proper indicator organisms and estimation of potential sites of pathogen occurrence, density, and survival. Sampling programs must be geared toward critical time periods and areas; that is, summer months, storm flows, feeder streams, agricultural and urban runoff, and swimming areas, including water and sediments. Frequency of sampling should be dictated by variability of water conditions, confidence level of data, and extent of human contact. Choice of proper indicator organisms and enumeration methods and appropriate sampling strategies will allow sound preimpoundment assessment and reservoir management to greatly reduce the risk of waterborne disease outbreaks.

Byrd, J.E. and Perona, M.J., "The Effect of Recreation on Water Quality", Technical Completion Report, Jan. 1979, California Water Resources Center, University of California, Davis, California.

The purpose of this study was to assess the effect of motorboat usage on the long-term lead concentration of a multipurpose reservoir, Turlock Lake, in Stanislaus County, California. As the study was made during two drought years, it was possible to examine the effects of large volume changes in the lake. Samples were taken at weekly intervals at the inlet canal where water enters the lake, at the outlet, and at the boat dock. An apparent correlation was found between the lead level at the boat dock and the boats per-unit-volume. Also, at the boat dock, the largest lead concentration correlated with the highest boat concentration. The inlet and outlet lead data obtained were treated in terms of a simple plug-flow model. The increase in the lead level of the water as it passes through the lake is greater, by at least a factor of 15, than would be expected on the basis of the number of boats on the lake. The lead concentration was found to correlate with lake volume. Experiments were done to find if an alternate source of lead in the lake is the sediment. The mechanism linking lake volume and lead concentrations is unclear, but is likely to involve migration of lead from the sediment, which was found to contain exchangeable lead, and/or biotic processes.

Byrd, J.E. and Perona, M.J., "Water Quality Effects of Lead from Recreational Boating", Technical Completion Report, Dec. 1979, California Water Resources Center, University of California, Davis, California.

The temporal and spatial variations in the lead concentration of a freshwater recreational lake were determined, and the results compared with daily records of lake volume, residence time and number of boats launched. In addition, laboratory studies were carried out to establish the influence of sediment-water interactions on the lead concentration of the lake water. The variation in the lead concentration in the main body of the lake was found to correlate with the lake volume. This fact, together with the laboratory studies and calculations based on a plug flow model, suggest that sediment-water interactions are significant in controlling the lead concentration in the main body of the lake. On the other hand, boating was found to be important in controlling the lead concentration in the boat dock area. This water is subjected to both poor mixing and heavy boat traffic.

Byrd, J.E. and Perona, M.J., "The Temporal Variations of Lead Concentration in a Fresh Water Lake", Water, Air, and Soil Pollution, Vol. 13, No. 2, June 1980, pp. 207-220.

The temporal and spatial variations in the Pb concentration of a freshwater recreational lake were determined, and the results compared with daily records of lake volume, residence time, and number of boats launched. In addition, laboratory studies were carried out to establish the influence of sediment-water interactions on the Pb concentration of the lake water. The variation in the Pb concentration in the main body of the lake was found to correlate with the lake volume. This fact, together with the laboratory studies and calculations based on a plug flow model, suggests that sediment-water interactions are significant in controlling the Pb concentration in the main body of the lake. On the other hand, boating was found to be important in controlling the Pb concentration in the boat dock area. The water in this area is subjected to both poor mixing and heavy boat traffic.

Deudney, D., "Hydropower--An Old Technology for a New Era", Environment, Vol. 23, No. 7, Sept. 1981, pp. 16-20, 37-45.

Hydropower technology has been available since Roman times. Political and social conditions have affected its use, which decreased in periods of plentiful, cheap labor and increased with the industrial revolution and the 1973 oil shock. Many of the small dams abandoned in the era of cheap oil are being retrofitted, and the Middle Ages technique of fastening waterwheels to floating river barges is being resurrected. Total world hydropower, currently 3,207 trillion watt-hours per year, could realistically expand to 4 to 6 times that amount, considering environmental and economic constraints. Africa taps only 5% of its hydropower potential, while Norway receives 99% of its electricity and 50% of its energy from water. Switzerland exports electricity, and Nepal and Peru have this potential. The largest existing dam, Itiapu, in Brazil-Paraguay, has a 12,600 MW capacity, but this is dwarfed by proposed dams on the Yangtze River, China (25,000 MW) and the Amazon River, Brazil (66,000 MW). Ecological damage can occur--waterborne disease transmission, silting, erosion, inundation of prime agricultural land, soil salinization, soil waterlogging, fisheries destruction, and loss of unique plant and animal species. For undeveloped countries, small projects can often provide needed power without the disruptions produced by the large dams. China's policy of building small projects to complement large dams is a model for developing nations. Since the untapped water resources are largely in uninhabited areas, an eventual shift of wealth, population, power, and employment to areas with abundant hydropower production can be expected.

El-Hinnawi, E.E., "The State of the Nile Environment: An Overview", Water Supply and Management, Vol. 4, No. 1-2, 1980, pp. 1-11.

The main environmental changes that have occurred or are anticipated as a result of development of the Nile basin are reviewed. The most important environmental impacts are related to the construction of dams and the creation of man-made lakes. The filling of a new lake causes geochemical and biochemical changes such as stratification, eutrophication, bacterial growth, and deoxygenation. Water hyacinth, water fern and water lettuce increase water loss by evaporation and use nutrients which could be used by fish. Placid lake waters as well as the rapid flows through dam sluices have aided the spread of water-borne human effects through several systems. A model for a viable organizational framework for coordinating research on an array of natural and social sciences is presented.

Entz, B., "Ecological Aspects of Lake Nasser-Nubia", Water Supply and Management, Vol. 4, No. 1-2, 1980, pp. 67-72.

Succession of animal and plant communities during the first 10 years' existence of two man-made lakes, Lake Nasser and Lake Nubia, is discussed. The development and decline of chironomid swarms is compared to similar phenomena observed on other man-made lakes. Their feeding habits and their role as food for fish, toads, spiders, and other insects indicated that they played an important role as food organisms for many animals and insects. About the time the chironomid swarms diminished, large numbers of water bugs (corixids) became common. Fish in the reservoir seemed to cause the eventual disappearance of the bugs. The succession of under water plants was: algal epitecton, Chara fields, and submerged macrophytes. Succession in lake shore vegetation was: Glinus lotcides and other weeds, desert plants, Tamarix nicotica, and the reappearance of river grass. Plants and animals were often the types found in the original river, and Nile flooding caused the re-introduction of several species. The possible role of the microclimate in limiting the dispersion of organisms in the two lakes is discussed.

Environmental Control Technology Corporation, "Analysis of Pollution from Marine Engines and Effects on the Environment", EPA-670/2-75-062, June 1975, Ann Arbor, Michigan.

The objective of this study was to obtain sufficient laboratory and field data to be able to predict the number of outboard engines which can be operated on any particular body of water without causing adverse effects on the aquatic environment. Four small natural bodies of water were subjected to outboard engine emissions three times greater than saturation (maximum) levels occurring under normal boating conditions. These natural water bodies were stressed at these high levels with outboard engine exhaust emission for a period of three years. The study showed that there were no acute changes in the physical, chemical or biological characteristics of the water or sediments of the test lakes. No major adverse effects were found on water quality or the biological communities (periphyton, phytoplankton, zooplankton, benthos, and fish) indigenous to the test lakes. No adverse accumulation of lead or hydrocarbons in the water column was observed. Also, no statistically significant accumulation of lead or hydrocarbons was noticed to be taking place in the lake sediments.

Fast, A.W. and Hulquist, R.G., "Supersaturation of Nitrogen Gas Caused by Artificial Aeration in Reservoirs", Tech. Report E-82-9, Sept. 1982, U.S. Army Engineer Waterways Experiment Station, Vicksburg, Mississippi.

Dissolved nitrogen gas (N_2) contents of 12 southern California reservoirs were assessed during the summer of 1979. Eleven of the reservoirs were artificially destratified by compressed air injection at depths, while one reservoir was not artificially aerated. Destratification ranged from nearly complete to quite incomplete. The artificially aerated reservoirs could be categorized into four types of conditions based on their degrees of destratificiation and their dissolved N_2 depth profiles: (1) Type I reservoir conditions were characterized by nearly uniform oxygen and temperature conditions at all depths and N_2 saturations between 100 and 110 percent; (2) in Type II reservoirs, the air diffuser was not placed at the deepest depth, consequently, a thorough mix occurred above the diffuser resulting in N_2 saturations which were 100 to 110 percent; however, strong thermal and chemical stratification persisted below the difuser where N_2 saturations could exceed 140 percent; (3) Type III reservoir conditions occurred when insufficient air was injected to cause a thorough mix; consequently, there was a shallow thermocline near the lake's surface below which were uniform temperatures; under Type III conditions, N_2 saturations normally ranged from 106 to 120 percent below the thermocline; and (4) Type IV reservoirs had a combination of Type II and III conditions where the air diffuser was placed above the bottom and insufficient air was injected to cause a thorough mix above the diffuser depth; under Type IV conditions, N_2 saturations also represented a combination of Type II and III conditions. With all types, N_2 saturations often increased even though absolute N_2 concentrations decreased. This was caused by an incomplete degassing of the deep waters as it was warmed during destratification. Although there is a general relationship between the types of conditions and N_2 saturations, there are no general procedures for predicting N_2 saturations.

Freeman, P.H., "The Environmental Impact of a Large Tropical Reservoir: Guidelines for Policy and Planning, Based Upon a Case Study of Lake Volta, Ghana, in 1973 and 1974", 1974, Smithsonian Institution, Washington, D.C.

By the year 2000, the portion of the world's streamflow regulated by reservoirs will increase from one-tenth to two-thirds. Many new reservoirs will be located in the tropics where reservoirs have been shown to create problems: ecological--water weeds, evaporation and/or sedimentation, obstruction of fish migration and perturbation of downstream hydrology and aquatic systems; health--impounded water as a growth medium for malaria and schistosomiasis; and social and economic--resettling of people from rich alluvial and flood plain farmlands. Using the Volta Lake case study and other published experience, a brief narrative of likely impacts is given, followed by questions that should be asked to determine impacts. Guidelines are given for: (1) environmental assessment of alternative impoundment sites--seismic and hydrological impacts, effects on natural rivers and forests, effects on archaeological remains, effects on human settlements and farmland, and impacts below the damsite; (2) assessment of hydrological and biological impacts following dam closure; and (3) assessment of environmental and ecological impacts of a tropical reservoir, including lake geology, ground water geology, lake topography, mineral cycling, plankton, aquatic plants, fish nutrition and production, water supply and sewage, and control of disease. Researchers found it was not possible to assess lake effects on ground water movement. The main variable for the lake appears to be runoff. Problems of exposure to schistosomiasis and bacterial contamination complicate use of the lake as a water supply. Lake assimilative capacity for wastewaters is great, and agricultural and industrial wastewaters are not current problems.

Gallopin, G., Lee, T.R. and Nelson, M., "The Environmental Dimension in Water Management: The Case of the Dam at Salto Grande", Water Supply and Management, Vol. 4, No. 4, 1980, pp. 221-241.

A bi-national Technical Commission (CTM) was formed to handle the project of construction of a dam at Salto Grande on the Uruguay River. CTM initiated numerous study programs to develop regulatory and management criteria to maintain environmental quality in the area that the dam would affect, which included land from the nations of Uruguay and Argentina. A workshop attended by CTM members defined the system to be controlled and gave as the main objective of the system the maintenance of the quality of life. This system was subdivided, yielding an economic and socio-cultural subsystem which was in direct interaction with a physical subsystem to maintain the integrity of air, water, soils, and terrestrial and aquatic biota. To gather information as quickly as possible and enhance interdisciplinary communications, a questionnaire approach was adopted. The questionnaire focused on indicators of the state of each sector and attempted to highlight significant associations that might exist between sectors. A second workshop compiled data from the questionnaire and generated flow diagrams for each sector, showing interactions between systems. A final CTM workshop clarified actions which impacted on the water sector, the soil and terrestrial flora and fauna sector, the socio-economic sector, and actions which had chain effects. Water management is so complex it is important to utilize any available tool such as systems analysis to analyze all possible strategies and viable alternatives. Models do not fail because of lack of technical expertise or inadequate computer technology. Failure is usually the result of managerial problems. Models can be used in developing countries, although some models of this type are rather crude and are dependent upon the judgment and expertise of the analyst.

Garzon, C.E., "Water Quality in Hydroelectric Projects--Considerations for Planning in Tropical Forest Regions", Tech. Paper No. 20, 1984, The World Bank, Washington, D.C.

This paper identifies and describes the studies necessary to predict water quality changes, at an early state of planning, in large tropical reservoirs with long retention times. Emphasis is placed on both the reservoir area and the region downstream. The need for defining the "baseline" environment is presented as a requirement for conducting studies associated with the flooding and operating stages. These studies are classified according to the stage of project development. In the reservoir area, aspects such as biomass quantification, reservoir thermal stratification, water circulation, dissolved oxygen consumption and reservoir recovery are of major importance. Downstream from the project, the stress is placed on river recovery capacity, water uses and conflicts, and flow requirements. The results obtained from the studies serve as the basis for deciding the extent of forest clearing and other mitigatory measures. The paper illustrates that biological degradation in tropical reservoirs follows a significantly different path from that in reservoirs in temperate zones, thus, conventional approaches to reservoir clearing and filling may not be adequate for projects in forested tropical regions. Two approaches--for project feasibility and project design--are suggested in order to meet the need for successive refinement in the results, and to take advantage of the increasing availability of project and environmental information.

Godden, G.F., Nicol, S.M. and Venn, A.C., "Environmental Aspects of Rural Development with Particular Reference to the Keiskamma River Basin Study", Civil Engineer in South Africa, Vol. 22, No. 5, May 1980, pp. 111-116.

The Keiskamma River rises at an elevation of 1,500 meters in the Amatole Mountains, at the southern tip of the Drakensberg Range of Southern Africa, where it drains into a number of minor tributaries towards the more arid coastal plateau. Two distinct river valleys are formed at this lower elevation. The total area of the river basin is 2700 square kilometers, of which about 85 percent experiences an annual rainfall less than 700 mm, and about 60 percent in the lower and middle reaches has about 400 to 600 mm per year. The highest precipitation, 1400 mm, falls in the mountainous areas forming the northern periphery of the catchment area. The need arose for a new water supply scheme to serve the developing industrial growth points at Dimbaza and Middledrift. The Keiskamma River was the logical choice. However, the potential inundation of 400 ha of valuable agricultural land by the dam, the effects of the new irrigation scheme and possible further afforestation in the upstream catchment, the effects of the dam runoff and irrigation withdrawals in the lower catchment, and the obvious potential for new irrigation development immediately downstream of the dam were factors to be considered. The Keiskamma River Basin Study should serve as an example of the type of investigation required to ensure that development of rural areas is achieved in the most efficient manner and in harmony with the environment.

Gould, M.S., "A Water Quality Assessment of Development in the Senegal River Basin", Water Resources Bulletin, Vol. 17, No. 3, June 1981, pp. 466-473.

The impacts of proposed water resources development in the Senegal River basin were studied with mathematical modeling and examination of case histories. This paper concerns the water quality aspect of the environmental assessment, which also draws on nine other major disciplines. Although this basin is relatively undeveloped, analytical techniques for more highly developed basins were successfully used. It was concluded that the proposed development plans should have minor negative impact on water quality in the basin. However, the Diama Dam, built to prevent salt water intrusion into the upper river, will adversely affect spawning of marine and estuarine fish. Dissolved oxygen in the river would remain high and more uniform with the aid of flow regulation. Minimum dissolved oxygen concentrations in the estuary would be lower, reaching a projected 1 mg/l by 2028. Other projects under consideration are the Manantali Dam for flow regulation and hydropower, irrigated agriculture, dredging for improved navigation, and industrial and municipal development. The reservoir created by Manantali Dam will initially be eutrophic due to nutrient releases and organic matter from inundation of 480 sq km of savanna forest. For the first few years aquatic vegetation could be a nuisance, and fish productivity would be high. The low dissolved oxygen concentration in the first 20 km of river below the dam would be hazardous to aquatic life. Similar lakes in Africa usually take about 5 years to evolve into a mesotrophic state. Agricultural fertilizers and pesticides will have a minor impact on water quality if proper practices are used. Dissolved solids in the river should not increase beyond 150 mg/l with proper irrigation practices.

Grizzle, J.M., "Effects of Hypolimnetic Discharge on Fish Health Below a Reservoir", Transactions of the American Fisheries Society, Vol. 110, No. 1, Jan. 1981, pp. 29-43.

Dams often cause changes in the impounded stream or river which have important effects on their fish populations. The release of anoxic, hypolimnetic waters, which often contain dissolved manganese, iron, ammonia, hydrogen sulfide, and other chemicals, from impoundments may increase the downstream concentrations of these substances. Rainbow trout, brown trout, brook trout, and yellow perch from two locations in the tailwater from the Buford Dam, in Georgia, and trout from a hatchery which used the tailwater as a water supply were examined for infections, lesions, and liver metal concentrations. The fish were found to be infected by 9 genera of parasites, 19 bacterial species, and 1 virus species. Many of the fish that did not have bacterial infections were found to have microscopic lesions in the gills, liver, spleen, and trunk kidney. The occurrence of lesions indicating exposure to a toxicant was related to increases in iron and manganese concentrations, but not to copper concentrations. Manganese did not indicate exposure to high levels of these metals. Since the tailwater contained copper, zinc, iron, manganese, and dissolved oxygen concentrations considered possibly harmful to fish, acclimation may have allowed survival of the fish in the river. The release of hypolimnetic water from the impoundment resulted in low pH, low dissolved oxygen concentrations, low oxidation-reduction potential, and high metal concentrations. Water characteristics, acclimation by the fish, and possible synergistic activity make most bioassay data unreliable for the prediction of the toxicity of tailwaters.

Grover, B. and Primus, C., "Investigating Whether a Large Hydro Development Can Be Environmentally Compatible: The Slave River Hydro Feasibility Study", Canadian Water Resources Journal, Vol. 6, No. 3, 1981, pp. 47-62.

This paper discusses the principles used in studying the environmental effects of a possible 2000 megawatts development on the Slave River, Alberta. These principles are: (1) environmental considerations should be included from the beginning of project planning and evaluation; (2) environmental consequences can be positive as well as negative; (3) power projects should be evaluated in a systems context, that is, alternative sites must be included in the planning process; (4) the scale of a project is no indication of its efficiency or environmental impact; and (5) preconceived constraints should not enter into planning at the initial stages. The provisional findings of a feasibility study indicate that the Slave River development is technically feasible and economically attractive. A single stage development is more economical than two smaller developments. Four possible environmental concerns are leakage of the reservoir through karstic formations, spread of the Arctic lamprey upstream, increases in the naturally high mercury levels of Lake Athabasca, and the most environmentally acceptable location for a transmission system.

Hafez, M. and Shenouda, W.K., "The Environmental Impacts of the Aswan High Dam", Proceedings of the United Nations Water Conference on Water Management and Development, 1978, Vol. 1, Part 4, Pergamon Press, New York, New York, pp. 1777-1786.

Completion in 1968 of the Aswan High Dam on the Nile River in Egypt has profoundly affected the entire Nile Basin in Egypt and the Sudan, but benefits far outweigh negative effects. The dam created a 500-km lake, known as Lake Nassar in Egypt and Lake Nubia in the Sudan, which provides for "century storage" of water which accumulates during the 50-year cycle of heavier rainfall, and is used during the 50-year period of reduced rainfall. Benefits include: (1) a sufficient water supply for agricultural and industrial growth; (2) doubling of electric power output; (3) greatly augmented fisheries in Lake Nasser; (4) high and low flood protection; (5) improved navigation; and (6) increased tourism. Negative effects include: (1) inundation of vast areas of land requiring relocation of 50,000 people and several historical monuments; (2) a change from a riverine to a lacustrine system; (3) regulation of the once-variable water flow, with reduction of flow to the Mediterranean Sea by as much as 100 cu km/yr; (4) loss of 60 million cu m/yr reservoir capacity due to siltation, with increased river bed and bank erosion; (5) increased evaporation from the reservoir from 6 to 10 cu km/yr; (6) alteration of shoreline ecology; (7) an anaerobic lake hypolimnion; (8) alteration of river morphology, water quality, and ecology, including eutrophication and excessive plant growth; (9) social changes as a result of improved agriculture, fishing, and employment; (10) an increase in schistosomiasis; and (11) increased irrigation canal salinity.

Hagan, R.M. and Roberts, E.B., "Energy Impact Analysis in Water Project Planning", Journal of the Water Resources Planning and Management Division, American Society of Civil Engineers, Vol. 106, No. WR1, Mar. 1980, pp. 289-302.

Since the energy crisis is a relatively recent development compared with water shortages and pollution, this paper illustrates some problems in preparing and interpreting the impact of energy costs and availability on water project planning. Energy analysis should be carried out in energy units separately from the economic aspects. Total energy impact should be considered and should include items such as energy consumed in producing the electricity and energy consumed in producing steel for construction of a dam. Planning for the New Melones, California, water management project produced estimates from a maximum power production of 430 million kilowatt hours per year to a net loss of 3.9 million kilowatt hours per year. The proponents of the higher figure did not include energy costs, line losses incurred in delivery of power, and the flooding of the existing electric power plant. The opponents, who claimed the project would produce loss in available energy, calculated costs in BTU of primary energy, and benefits in BTU of delivered electricity with a benefit-cost ratio of 0.65. When the authors recalculated these figures and considered the savings produced by inactivation of the old power plant, the benefit-cost ratio increased to 2.7. If the reduction of decline in shallow wells is considered, the ratio rises to 4.1.

Hefny, K., "Land Use and Management Problems in the Nile Delta", Nature and Resources, Vol. 18, No. 2, April/June 1981, pp. 22-27.

The Nile River delta and its ecosystem have been affected by water control projects and urban development, especially within the last 100 years. The Aswan High Dam, constructed during the 1960's, has reduced the silt load to 1.5-4 percent of that formerly carried by the river. Unless a new equilibrium is reached, the delta will gradually recede. The estuarine ecosystem has been altered drastically by changes in the salinity regime and nutrient levels. Basin irrigation has given way to perennial irrigation, but drip or sprinkler irrigation could save an additional 30-50 percent of the water. Water hyacinth and other aquatic plants, which have flourished in the irrigation canals, influence biotic relationships, both positively (food sources) and negatively (disease-carriers). The large ground water reservoir underlying the delta is of great economic importance, but present knowledge is insufficient for proper management. Several strips of the seacoast several kilometers wide have been removed by the sea during the last 200 years. The erosion process, especially in the softer western and eastern parts of the delta, is expected to increase with the reduction of sediment deposition. Research and monitoring of the river and delta must be interdisciplinary to provide reliable information for managing water resources and protecting the coastline.

Hirschberg, R.I., Goodling, J.S. and Maples, G., "The Effects of Diurnal Mixing on Thermal Stratification of Static Impoundments", Water Resources Bulletin, Vol. 12, No. 6, Dec. 1976, pp. 1151-1159.

The time variation in the temperature distribution of a static water impoundment was predicted. The body of water was modeled as a discrete member of horizontally isothermal layers, and the energy equation was solved using an implicit numerical scheme. Vertical energy transport mechanisms included were solar absorption, molecular diffusion, and convective mixing due to nocturnal turnover. The latter mechanism, called diurnal mixing, was found to have a profound effect on the stratification, particularly in the epilimnion patterns for a typical deep static impoundment.

Hussong, D. et al., "Microbial Impact of Canada Geese (Branta Canadensis) and Whistling Swans (Cygnus Columbianum Columbianus) on Aquatic Ecosystems", Applied and Environmental Microbiology, Vol. 37, No. 1, Jan. 1979, pp. 14-20.

Healthy whistling swans and Canadian geese, migrating along the eastern U.S. flyway in Chesapeake Bay do not harbor detectable enteric, bacterial pathogens. Freshly collected fecal matter from caged birds and from autopsied wild swan and geese do not yield Salmonella sp. Furthermore, Salmonella sp. was not recovered from the four aquatic roosting sites included in the study. A random selection of 75 E. coli isolates from various waterfowl yielded seven enterotoxin-producing E. coli. Waterfowl harboring these enterotoxigenic E. coli are all confined to Lake Shore Pond. Most of the anaerobic bacteria isolated are Escherichia and Streptococcus sp. Wild birds harbor significantly more fecal coliforms than fecal streptococci, while the reverse is true for captive and fasting birds. In a 24 hour period, a single swan eliminates one billion and a goose 10 million fecal coliforms (FC). At an artificial pond, Lake Shore, 25 captive swans and geese were maintained. Before the birds entered Lake Shore Pond, FC was about 1/100 ml and after 200 bird hours, the count rose to 2,400/ml. Despite waste quality deterioration, neither salmonella nor shigella were isolated from the water or sediment. At Canninghouse Cove, the largest natural waterfowl roosting site, sediment EC counts rose from 13 to 170 FC/100 ml during the bird season. This sufficiently high population index could result in shellfishing restrictions.

Interim Committee for Coordination of Investigations of the Lower Mekong Basin, Environmental Impact Assessment--Guidelines for Application to Tropical River Basin Development, 1982, ESCAP, Bangkok, Thailand.

It is now well recognized that sustainable benefits can only be derived from development projects if environmental factors and parameters are considered in project designs at the planning stage. The role of environment in water resource development projects in particular has been widely publicized during the 1970's. Consequently, there is a necessity for systematic environmentally oriented studies and assessments for each individual development scheme to identify major possible environmental impacts and plan mitigating measures. The need for a set of guidelines for conducting environmental impact assessments for the development programmes of tropical river basins has been recognized, and this book summarizes appropriately developed guidelines. The "guidelines" stress the opportunities that environmental impact assessments present in making choices regarding the proper management of development in tropical regions. Too often those concerned with the environmental impacts of development have been so anxious to impose a new discipline on the planning and implementation process that they have been unduly rigid in their approach. It is true that the consequences of neglecting environmental factors, from the early stage of project identification to the later detailed work of project design and construction, can be devastating, both for the natural environment and for the human populations it sustains. It does not follow however that there are many wrong ways to plan and implement development and that there is only one right way to do it, or that one particular set of procedures, methods and techniques of impact assessment have to be incorporated into development planning and decision-making to guarantee the right result.

Johnson, S.J., Krinitzsky, E.L. and Dixon, N.A., "Reservoirs and Induced Seismicity at Corps of Engineers Projects", Miscellaneous Paper S-77-3, Jan. 1977, U.S. Army Engineer Waterways Experiment Station, Vicksburg, Mississippi.

A review was made of major reservoirs of the Corps of Engineers with regard to their experiences with induced seismicity. The Corps has 24 large reservoirs, i.e., those with dams 60.96 m (200 ft) in height or greater, and with volumes of 1.2335×10^9 cu m (1,000,000 acre-ft) or greater. There also are 10 dams that are 91.44 m (300 ft) high or higher, but with reservoirs of less than 1.2335×10^9 cu m. Thirteen reservoirs have been instrumented for monitoring microearthquakes. They range from large to small and are located in aseismic to highly seismic areas. Only one Corps reservoir, Clark Hill in Georgia-South Carolina, has experienced a felt earthquake, but the 21-year interval between impoundment and the occurrence of the earthquake is not typical of reservoir-induced earthquakes; and a direct relationship appears improbable.

Kadlec, J.A., "Effects of a Drawdown on a Water Fowl Impoundment", Ecology, Vol. 43, No. 2, Spring 1962, pp. 267-281.

A pilot drawdown on the Backus Lake flooding project in north-central lower Michigan was evaluated during the summer of 1958 along with its effect on vegetation, waterfowl, soil, water, and bottom fauna. Plant species composition was not notably affected. Common perennials were able to survive drainage for one growing season. Many submerged and floating-leaf species were reduced in abundance. Water lilies were little affected except in severely dried areas. Water smartweed and bushy pondweed grew luxuriantly after the drawdown. Most emergents spread and increased in abundance. Sedges and woolgrass were most abundant on dry portions of the study area. Cattail, bulrush, and burreed were more abundant where soil moisture was returned throughout the drawdown. Rice cutgrass and mannagrasses were generally distributed. Waterfowl utilization of the area increased in late summer 1959. Abundant food attracted ducks, and increased cover caused increased use by breeding waterfowl in 1960. Soil and water analyses indicated a definite increase in plant nutrients, especially soil nitrates due to aerobic nitrification. Invertebrate populations were considerably reduced after the drawdown.

Kay, D. and McDonald, A., "Reduction of Coliform Bacteria in Two Upland Reservoirs: The Significance of Distance Decay Relationships", Water

Research, Vol. 14, No. 4, 1980, pp. 305-318.

This paper reported on the coliform bacterial densities observed between September 1976 and September 1977 in two British upland reservoirs having multiple-use catchment areas. The level of catchment use was defined in terms of agricultural and recreational activity, and the rate of bacterial reduction in the reservoir impoundments was investigated. The applicability of previous studies (which examined rates of coliform reduction in different situations) to the British upland reservoir was discussed, and a set of calculated purification rates, observed during different limnological conditions, were presented.

Keeney, D.R., "A Prediction of the Quality of Water in a Proposed Impoundment in Southwestern Wisconsin, USA", Environmental Geology, Vol. 2, No. 6, 1978, pp. 341-349.

Due to large estimates of nutrient loadings and resultant eutrophication and water quality in the proposed Lake La Farge impoundment on the Kickapoo River, Wisconsin, federal funding was withdrawn from the project. Phosphorus loadings to the impoundment are estimated to be 4.4 g/sq m/yr. To reach Vollenweider's permissible levels, P loadings must be 0.2 g/sq m/yr; this cannot be achieved since base flow contributes nearly one-third of total P loading. Runoff supplies a major portion of Nitrogen (39%) and P (30%) to the impoundment. Base flow is also an important nutrient source, providing 35% of N and 32% of P. Recognizable point sources (sewage treatment plants and farmyards) supply 26% and 37% of N and P, respectively. Tertiary sewage treatment for the small community is difficult to justify economically. Area farms currently practice well-above-average conservation practices, and little improvement in nonpoint nutrient removal would occur without considerable expense. However, in-lake treatment to remove weeds and algae from high-use areas could be done at minimal costs. The large multipurpose impoundment providing flood control and recreation was first proposed in the 1960's. Construction commenced in 1971-72, while about 1970-71 several environmental groups petitioned against completing the impoundment. Major contentions involved recreational benefits which assumed water quality good enough for recreational use of the impoundment.

Kelly, D.M., Underwood, J.K. and Thirumurthi, D., "Impact of Construction of a Hydroelectric Project on the Water Quality of Five Lakes in Nova Scotia", Canadian Journal of Civil Engineering, Vol. 7, No. 1, 1980, pp. 173-184.

Water quality of five highland lakes was monitored for 2.5 years during the construction of the Wreck Cove hydroelectric project, Nova Scotia, starting in February, 1975. Tunneling, dam construction, reservoir clear-cutting, slashing, burning, and building access roads had a pronounced impact on Surge Lake, and less on the other lakes--Big, Cheticamp, Gisborne, and Long. Ammonia and nitrate-nitrite concentrations in Surge Lake increased 833 and 7300%, respectively, following clear-cutting and initial construction activities; chlorophyll-a decreased by 1825%. Other water quality parameters which increased on the order of 50-450% were turbidity, suspended solids, conductivity, Ca, Mg, Na, K, and total and ortho-P. The following were more extreme in the Surge Lake watershed than in the other watersheds: slopes, intensity and length of construction, clear-cutting and burning, and a forest fire. In addition, soil was more erosive (less clay), and more vegetative cover, including roots, was removed.

Kenyon, G.F., "The Environmental Effects of Hydroelectric Projects", Canadian Water Resources Journal, Vol. 6, No. 3, 1981, pp. 309-314.

The advantages of hydroelectric power--simple, renewable, and nonpolluting--must be balanced against the environmental impacts, particularly of large projects. Flooding behind a dam takes fertile bottomland out of production; destroys prime forest land; disturbs wildlife, fish, and human habitats; increases water temperature; changes water quality; and inundates scenic rivers. Fluctuating water levels and changes in flow regimes during operation of a hydroelectric station create a dead littoral zone around the reservoir, drown forests, affect recreational use, disturb waterfowl nesting sites, cause scour and flushing downstream, and can produce nitrogen supersaturation in the spillway area. Small hydroelectric projects may retain many of the benefits without the unfavorable environmental effects produced by megaprojects.

King, D.L., "Environmental Effects of Hydraulic Structures", Journal of the Hydraulics Division, American Society of Civil Engineers, Vol. 104, No. 2, Feb. 1978, pp. 203-221.

In response to objectives outlined by the Hydraulic Structures Committee, the Task Committee has attempted to: (1) define what structure types should be selected for study in view of current research activity and potential effects on the environment; (2) provide a general survey of how hydraulic structures interrelate with their environment; and (3) make recommendations for future task or full committees to examine specific program areas. Tables showing categories of hydraulic structures and categories of environmental effects are presented. The Task Committee preference to use matrices as an assessment method is illustrated with several examples. Finally, an example is developed to examine the effects of impoundment on downstream water quality. Data for the example were obtained by a limited survey of engineers and biologists. Results are presented and discussed.

LaBounty, J.F. and Roline, R.A., "Studies of the Effects of Operating the Mt. Elbert Pumped Storage Powerplant", Proceedings of the Clemson Workshop on Environmental Impacts of Pumped Storage Hydroelectric Operations, FWS/OBS-80/28, Apr. 1980, U.S. Fish and Wildlife Service, Washington, D.C., pp. 54-66.

The Mt. Elbert Pumped Storage Powerplant is located on the northwest shore of the lower lake of Twin Lakes near Leadville, Colorado. Construction of the powerplant started in February 1972, and initial operation of the first of two units was planned for 1981. In 1971, the Water and Power Resources Service initiated studies to obtain limnological and fishery data in an attempt to better understand the environmental impacts of pumped storage. Past, ongoing, and planned research on the Twin Lakes ecosystem is reviewed. Expected impacts resulting from powerplant operation includes fish mortality, stirring of lake sediments, and changes in the physical environment of Twin Lakes. To assess and quantify the impacts of pumped storage on Twin Lakes biota, sampling facilities in the tailrace of the lower lake, and in the Mt.

Elbert forebay, are being designed and constructed. These structures and techniques will allow sampling of aquatic organisms at any time during the pumping and generation modes of the powerplant.

Lewke, R.E. and Buss, I.O., "Impacts of Impoundment to Vertebrate Animals and Their Habitats in the Snake River Canyon, Washington", Northwest Science, Vol. 51, No. 4, 1977, pp. 219-270.

Impoundment by Lower Granite Dam on the Snake River inundated 1,319 ha of wildlife habitat; 210 ha represented tree-shrub riparian habitat and 1,109 ha represented weedy-floodplain habitat. These habitat were of paramount importance to wildlife, particularly in the winter. Sixty-five of 129 species studied were significantly dependent upon tree-shrub riparian habitat. Thirty-four species were significantly dependent upon riverbank-floodplain habitat. Birds forced from these habitats by inundation will not be able to re-establish themselves in remaining above-pool habitats which were filled to capacity before impoundment occurred. Furthermore, the loss of below-pool habitats will indirectly cause a decrease in the number of birds remaining in above-pool habitats. The greater a species' dependence on tree-shrub riparian or riverbank-floodplain habitat, the greater the impact of impoundment on that species. Impoundment is estimated to cause a loss locally of nearly 14,000 birds in the summer and over 30,000 birds in the winter. Migrant birds did not utilize the lower Snake River Canyon habitats to a great degree, and thus will not be affected as much as resident birds. Since vertebrate species will be affected differently by habitat loss, population density will decline disproportionately. Some species will be forced completely from the area. All upland game birds will decrease after impoundment; ring-necked pheasants will be reduced the most because of this species' strong dependence upon the floodplain. Although chukar partridge populations declined 46 percent from 1972 to 1973, this upland game bird will very probably maintain relatively high population densities in future years. The overall impact to vertebrates will be a high loss in numbers and a decrease in diversity. Opportunities for management in the area of Lower Granite Reservoir are very limited. The most productive soils are now inundated, and only shallow rocky soils extend above the pool level. These shallow soils will not support high quality riparian and floodplain habitats, thus resulting in a permanently reduced and simplified vertebrate fauna.

Lewke, R.E., "Dams and Wildlife", Passenger Pigeon, Vol. 40, No. 3, Fall 1978, pp. 429-442.

Nearly 100 percent of the Lower Snake River floodplain between Lower Granite Dam and Clarkston, Washington, was inundated in February, 1975. Of the area inundated, 16 percent represented valuable riparian habitat and 84 percent represented weedy floodplain habitat. Evidence indicated that most displaced birds would not be able to re-establish themselves in remaining above-pool habitats, and that the loss of below-pool habitats would indirectly cause a decrease in the number of birds remaining in above-pool habitats. Of the 129 bird species observed in the study area before inundation, 50 percent were found to be significantly dependent upon the riparian physiognomic type. Impoundment was estimated to cause the loss locally of over 14,000 birds in the summer and 30,000 in the winter. Migrant species of birds, excluding waterfowl, did not utilize the Lower Snake River Canyon habitats to a great degree and thus will not be affected nearly as much as permanent, summer, and winter resident species. Since different species will not be affected equally by habitat loss, the population density of each species will decline disproportionately.

Manning, R.E., "Impacts of Recreation on Riparian Soils and Vegetation", Water Resources Bulletin, Vol. 15, No. 1, Feb. 1979, pp. 30-43.

The seemingly magnetic attraction of water resources for recreation has direct implications for proximate land resources which are needed to provide access and support facilities. This paper reviewed and synthesized the literature dealing with the impacts of recreation use on riparian soils and vegetation. Part One of the paper sets forth the major negative impacts of recreation used on soils and vegetation. A 7-step soil impact cycle was identified, beginning with the scuffing away of leaf litter and other organic material and working through the soil erosion and sedimentation process. Four major kinds of impacts of recreation use on vegetation were outlined, and the "vicious circle" relationships between impacts on soil and vegetation was demonstrated through a Soil/Vegetation Impact Diagram. Part Two identified several spatial and temporal patterns of environmental impact caused by recreation use, campground and trail expansion, ground cover response and succession, rates of soil compaction, and resource response to various intensities of recreation use. The final part of the paper dealt with measuring environmental impacts caused by recreation use. Management implications of the research findings were considered throughout the paper.

Matter, W. et al., "Movement, Transport, and Scour of Particulate Organic Matter and Aquatic Invertebrates Downstream from a Peaking Hydropower Project", Tech. Report E-83-12, May 1983, U.S. Army Engineer Waterways Experiment Station, Vicksburg, Mississippi.

The Savannah River below Lake Hartwell, Georgia-South Carolina, receives hypolimnetic water discharged from the reservoir for peak power generation. Invertebrates and particulate organic material (POM) in the water column were collected during a 24-hr release cycle at sites 1.0, 4.5, and 12.5 km downstream from the dam. Water released during generation reached a maximum discharge of 6888 m^3 sec^{-1}. River discharge was less than 10 m^3 sec^{-1} during nongeneration periods. Highest POM concentrations were associated with the initial downstream surge of water at the start of power generation; values were 200 to 400 times greater than those during nongeneration periods. POM concentrations rapidly decreased to less than one-tenth of the initial surge levels during high flow. Much of the POM originated in the tailwater, and concentrations increased at successive downstream sites. Of the drifting invertebrates, 80 to 93 percent originated in the reservoir; the rest, primarily Oligochaeta, Diptera, and Ephemeroptera, were from the tailwater. Densities of benthic invertebrates were highest during passage of the initial release surge, whereas densities of invertebrates originating in the reservoir peaked 2 to 3 hr after the initial surge at each station, during maximum release. Densities of drifting benthic organisms decreased rapidly after the initial surge and increased with increasing distance downstream.

McClellan, B.E. and Frazer, K.J., "An Environmental Study of the Origin, Distribution, and Bioaccumulation of Selenium in Kentucky and Barkley Lakes", Research Report No. 122, 1980, Water Resources Research Institute, University of Kentucky, Lexington, Kentucky.

Samples of water, bottom sediment, and fish were analyzed for toxic metal ion content. The samples were collected from several selected sites along Kentucky and Barkley Lakes as well as the Cumberland River, and several subimpoundments along these aquatic systems. Emphasis was placed on selenium, although several metal ions were determined. The results showed that there are no serious pollution problems with As, Cd, Cr, Cu, Hg, Pb, Se, Sr, Zn, or Zr at any of the sites examined. Actually, none of the trace metals examined even come close to the EPA limits on fish, with the exception of lead in the White Crappie and mercury in the Bass. There appears to be no serious problem with lead in White Crappie as only 2 fish out of a total of 19 were above the limit of 2 mg/l level. Water and sediment analyses for the above-mentioned metal ions fell well within expected "normal" limits for unpolluted fresh water systems. No point sources could be identified for any of the metal ions. This is in contrast with results obtained on the lower Tennessee River in which a large chemical complex was found to contribute significant quantities of trace metals. No significant seasonal variation of trace element content was observed in any of the sample types. Since the selenium content of all samples was so low, no laboratory bioaccumulation data were obtained. No general correlation between fish length and trace element content could be established.

Miracle, R.D. and Gardner, Jr., J.A., "Review of the Literature on the Effects of Pumped Storage Operations on Ichthyofauna", Proceedings of the Clemson Workshop on Environmental Impacts of Pumped Storage Hydroelectric Operations, FWS/OBS-80/28, Apr. 1980, U.S. Fish and Wildlife Service, Washington, D.C., pp. 40-53.

Pumped storage operations pass large volumes of water from one reservoir to another, ultimately affecting fishes, either by entrainment in withdrawn waters or by modification of the aquatic environment. Mortality caused during entrainment is primarily due to the following physical stresses: abrasion and collision; pressure changes; velocity changes; and acceleration effects. Fish passage or entrainment is influenced by: size and life stage of fish; susceptibility of the fish which directly relates to life history aspects; and physical characteristics of the pumped storage facility. Pumped storage facilities can adversely affect reservoir hydrology through changes in water level, water temperature, dissolved oxygen and water velocity. Although adverse effects of pumped storage can be severe when drastic modifications results from plant operations, most fish populations are able to adapt to the changing environments encountered at facilities in the United States.

Nelson, W. et al., "Assessment of Effects of Stored Stream Flow Characteristics on Fish and Wildlife, Part A: Rocky Mountains and Pacific Northwest (Executive Summary)", Publication No. FWS/OBS-76/28, Aug. 1976, Environmental Control, Inc., Rockville, Maryland.

The report contains the findings, conclusions and recommendations regarding changed flow regions below dams, the impact on fisheries, and methodologies used to assess flow requirements for 96 dams and diversions in the Rocky Mountain and Pacific Northwest regions.

Nix, J., "Distribution of Trace Elements in a Warm Water Release Impoundment", Oct. 1980, Water Resources Center, University of Arkansas, Fayetteville, Arkansas.

The results of a detailed water quality survey conducted from 1969-1975 on DeGray Reservoir in southwest Arkansas were used to describe physical/chemical processes affecting reservoir and river system constituent concentrations. Dissolved oxygen (DO) and turbidity profiles delineated water movement through the reservoir. Development of metalimnetic DO minima appeared as related to advective transport within the metalimnion (with metalimnetic DO depletion greatest during years when spring rains occurred after stratification onset). Dissolved oxygen data showed new reservoir "aging" (with hypolimnetic DO depletion severe during early impoundment years, but moderated after four years). Iron (Fe) and manganese (Mn) expectedly correlated well with DO depletion; other trace metals studied did not correlate with either Fe or Mn. Cobalt correlated significantly with nickel, probably due to their geochemical similarity. Storm event intervention was the most significant factor determining concentrations of calcium and other conservative constituents. Calcium dropped as storm water entered the reservoir.

Pastorok, R.A., Lorenzen, M.W. and Ginn, T.C., "Environmental Aspects of Artificial Aeration and Oxygenation of Reservoirs: A Review of Theory, Techniques, and Experiences", Tech. Report E-82-3, May 1982, U.S. Army Engineer Waterways Experiment Station, Vicksburg, Mississippi.

Artificial circulation and hypolimnetic aeration have been used successfully in management of eutrophic reservoirs to alleviate water quality problems, control algal blooms, and improve fish habitat. This report includes: (a) a comprehensive review of aeration/circulation techniques and past experiences encompassing literature from January, 1972, through December, 1980; (b) statistical analyses of artificial circulation experiences to examine the causes of alternative responses to treatment; (c) a summary of morphometric and water quality data for 107 reservoirs managed by the U.S. Army Corps of Engineers; and (d) a generic evaluation procedure for alternative management applications in reservoirs. Artificial destratification by mechanical pumping or diffused-air mixing usually elevates dissolved oxygen content of the lake by bringing anoxic bottom waters to the lake surface where aeration occurs through contact with the atmosphere. Oxygenation may cause precipitation of phosphate compounds and inhibition of nutrient release from sediments, but invasion of benthic macroinvertebrates into the profundal zone may play a role in maintenance of high phosphorus release rates from oxygenated surficial sediments. Water quality generally improves after treatment, but undersizing of water pumps or improper timing of destratification relative to occurrence of algal blooms can aggravate existing oxygen deficits.

When the mixed depth is increased, models of algal production predict a decline in the ratio of photosynthetic rate to respiration rate and a consequent decline in algal biomass per unit area of lake surface.

With sufficient mixing, a drop in pH of the upper waters is observed, followed by a shift from nuisance blue-green algae to a mixed assemblage of green algal species. Zooplankton and benthic macroinvertebrates often increase during artificial circulation as a result of habitat expansion and possible enhancement of food resources. Although short-term increases in fish growth and yield have been attributed to improvement of food and habitat, long-term observations are unavailable.

Hypolimnetic aeration improves water quality without disrupting thermal stratification. Although the potential benefits of hypolimnetic treatment in controlling algal blooms are more limited than those realized with whole lake mixing, the risk of adverse impacts appears to be lower for hypolimnetic aeration. Oxygenation of downstream reaches can be achieved by hypolimnetic aeration or oxygenation, localized mixing, aeration in the outlet works, and tailwaters aeration.

Petts, G.E., "Long-Term Consequences of Upstream Impoundment", Environmental Conservation, Vol. 7, No. 4, Winter 1980, pp. 325-332.

The environmental consequences of dam and reservoir construction on a river system are considered in a framework consisting of first-, second-, and third-order impacts. The immediate and simultaneous effects of the activity (dam building) are first-order impacts, for example, reductions in peak flow, entrapment of sediment load, reduction in suspended sediment load, induced erosion immediately below the dam, and channel changes. These induce second-order impacts, changes in channels and invertebrate populations taking place over a longer period after construction--perhaps as long as 50 years. Channel cross-sectional reduction dominates morphological changes in impounded rivers--depth increase from erosion, depth increase from sedimentation, width reduction from redistribution of the floodplain and channel boundary materials, and width reduction from sediment deposition. After initial changes in aquatic life during the second-order stage, further adjustments occur as part of a third-order impact. For example, accumulation of fine sediments in pools discourages growth of invertebrates and encourages establishment of rooted aquatic plants. These plants in turn can further affect channel morphology. Changes in fish habitats, intimately associated with changes in channel width, depth, and sediment composition and in flora and fauna, may continue to occur for many years after the initial construction. Planners should not base environmental impact assessments on short-term effects alone, but must consider the complex interactions taking place over many years before a morphological and ecological equilibrium is reached.

Petts, G.E., "Morphological Changes of River Channels Consequent Upon Headwater Impoundment", Journal of the Institution of Water Engineers and Scientists, Vol. 34, No. 4, July 1980, pp. 374-382.

Geomorphological data from 14 impounded rivers located throughout Britain have been used to assess the long-term impact of reservoirs upon river channels downstream, and to identify the potential problems for river management. Changes of channel size and shape have halved the water conveyance capability of 11 of the 14 rivers studied. In the absence of adequate discharge data, two surrogate variables, reservoir surface area and time-to-peak of the reservoir inflow hydrograph, are used to explain the gross variation of channel morphology downstream from reservoirs. Four mechanisms, channel erosion, redistribution of the boundary sediment, channel side deposition, and channel bed aggrandizement, contributed to the morphological changes. Long-term changes downstream from reservoirs can affect land drainage, coastal stability, water quality, navigation, structural stability, fisheries, and aesthetics, as well as the aquatic flora and fauna.

Pickering, J.A. and Andrews, R.A., "An Economic and Environmental Evaluation of Alternative Land Development Around Lakes", Water Resources Bulletin, Vol. 15, No. 4, Aug. 1979, pp. 1039-1049.

Reported herein is a study which made an evaluation of alternative land developments around New Hampshire lakes. Alternative development patterns, evaluated by their impacts on the lake area environment and area economy, included residential patterns, commercial patterns, and combinations of these two types. Phosphorus loading of the lake water was used as a proxy variable for changes in the lake water quality. Commercial developments yielded the highest revenues to the town and local area. It also attracted the most lake users to the area as well as contributing the largest phosphorus loading in the lake waters. Residential developments, although contributing high revenues to the businessmen in the area, yielded less net income to the town. Phosphorus loading levels from residential developments were much lower than lake phosphorus loading by commercial developments.

Ploskey, G.R., "Fluctuating Water Levels in Reservoirs: An Annotated Bibliography on Environmental Effects and Management for Fisheries", Tech. Report E-82-5, May 1982, U.S. Army Engineer Waterways Experiment Station, Vicksburg, Mississippi.

This report contains 367 annotations describing the effects of fluctuating reservoir water levels on fish. Citations on phytoplankton, zooplankton, and water quality that pertain to reservoir fisheries are also included. An index to facilitate location of references dealing with specific topic areas is included as an appendix.

Ploskey, G.R., "A Review of the Effects of Water-level Changes on Reservoir Fisheries and Recommendations for Improved Management", Tech. Report E-83-3, Feb. 1983, U.S. Army Engineer Waterways Experiment Station, Vicksburg, Mississippi.

This report synthesizes and summarizes information gathered from available sources about the physicochemical and biological effects of water-level changes on reservoir ecosystems. It describes how variations in both the physical environment (i.e., basin morphometry, bottom substrates and structures, erosion, turbidity, temperature, and water-retention time) and the chemical environment (i.e., nutrients and dissolved oxygen) caused by water-level changes can directly influence a reservoir's production of fish. It also describes the complex ways in which water-level changes affect aquatic plants, zooplankton, and the benthos and how these trophic variations can eventually affect the growth, reproduction, and harvest of fish. The final part of the report summarizes the effects of drawdown and flooding on reservoir fish populations and recommends ways to manage reservoir fluctuation zones by making controllable variables as favorable as possible for fish survival, spawning, and feeding.

Raymond, H.L., "Effects of Dams and Impoundments on Migrations of Juvenile Chinook Salmon and Steelhead from the Snake River, 1966 to 1975", Transactions of the American Fisheries Society, Vol. 108, 1979, pp. 505-529.

Migrations of juvenile chinook salmon, Oncorhynchus tshawytscha, and steelhead, Salmo gairdneri, from tributaries of the Snake River were monitored as far downstream as The Dalles Dam on the Columbia River in most years during the period 1966 to 1975. New dams constructed on the Snake River adversely affected survival and delayed migrations of juveniles. Significant losses of juveniles in 1972 and 1973 were directly responsible for record-low returns of adults to the Snake River in 1974 and 1975. Major causes of mortality were passage through turbines at dams, predation, and delays in migration through reservoirs in low-flow years, and prolonged exposure to lethal concentrations of dissolved gases caused by spilling at dams during high-flow years. Migrations of juvenile steelhead were generally later than those of chinook salmon and generally coincided with maximum river discharge. Lack of river runoff in 1973 caused a significant number of steelhead to stop migrating and to hold over in reservoirs. Mortality of chinook salmon and steelhead resulting from new dams has differed with respect to area and cause. Magnitude and composition of seaward migration has changed from 3 to 5 million fish of both wild and hatchery origin in the 1970's.

Reynolds, P.J. and Ujjainwalla, S.H., "Environmental Implications and Assessments of Hydroelectric Projects", Canadian Water Resources Journal, Vol. 6, No. 3, 1981, pp. 5-19.

Hydroelectric power development can be compatible with the environment if projects are planned properly with input from the public and government agencies. Possible benefits of power development are construction of roads, communications, and services; new jobs and industry; possible opportunities for hunting and fishing; reforestation; increases in tourist activities; and building of new communities. The many "disbenefits" include damage of fishing, wildlife, forests, shorelines, and water quality; flooding; recreational losses; dislocation of people; and disturbance by high voltage transmission lines. Even after 10 years of formal environmental impact assessments in Canada problem areas remain: (1) inadequate analytical techniques, (2) lack of assessment review procedures and manuals, (3) hurried environmental assessment reviews, and (4) absence of a detailed monitoring program supervised by a regulatory agency. It is suggested that cost-benefit analyses of environmental investigations be conducted.

Roseboom, D.P. et al., "Effect of Bottom Conditions on Eutrophy of Impoundments", Illinois State Water Survey Circular 139, 1979, Illinois State Water Survey, Urbana, Illinois.

Two man-made impoundments, Lakes Eureka and Canton in central Illinois, were studied to develop procedures for assessing behavior characteristics and thereby relative eutrophy of waters in man-made lakes. Likenesses and differences for the two lakes in productivity, water chemistry, and sediment oxygen demand were identified. Both lakes thermally stratify and maintain an anaerobic zone in bottom strata for a 5-month period in the summer; stratification is weak in the shallower Lake Eureka, but well defined in Lake Canton. The rate of anaerobic zone formation is greater in Lake Eureka, but the zone's extent from the bottom is less than in Lake Canton; overall, oxygen depletion conditions in Lake Canton are the most severe. The thermal stratification of the lakes is not a barrier to the extent the anaerobic zone penetrates the overlying waters. Both lakes support blue-green algae blooms. The density of algae at Lake Canton exceeds that at Lake Eureka; however, alkalinity reductions at the surface indicate algal productivity of Lake Eureka exceeds that of Lake Canton by 30 percent, and alkalinity increases in the bottom waters indicate microbial activity in bottom muds in Lake Eureka exceeds that in Lake Canton by 44 percent. Differences in alkalinity between the surface and bottom waters in both lakes were significant. Nonetheless, higher total and dissolved phosphorus contents in Lake Canton bottom waters give that lake a greater potential for productivity. Significant quantities of iron and manganese (mainly Fe(++) forms) are released from the bottom muds of both lakes during summer stagnation. Chlorine demand of bottom waters is greater than that in upper water layers in both lakes.

Sargent, F.O. and Berke, P.R., "Planning Undeveloped Lakeshore: A Case Study on Lake Champlain, Ferrisburg, Vermont", Water Resources Bulletin, Vol. 15, No. 3, June 1979, pp. 826-837.

Development of private lakeshores usually precedes establishment of public access, resulting in the best access areas being occupied before public access is provided. The authors present an analytical procedure developed to enable planners to classify undeveloped lakeshore areas according to their suitability for public and private uses. The undeveloped lakeshore evaluation system rated the physical characteristics of the lakeshore for five of the most common uses: (1) public beaches; (2) camping and picnicking areas; (3) boat access areas; (4) marinas; and (5) development-vacation homes, cottages, and motels. Standards for rating the lakeshore potential have been developed for the following site requirements: size, slope, soil suitability, shoreline type (sand, rock, muck, or gravel), water quality, site location, scenery, and road access. Each of these items are measured and rated on a numerical scale from 1 to 5; the basis of the rating is a comparison with other sites for a similar use in the town or region. The system is applied to Ferrisburg, Vermont, located on the southern reach of Lake Champlain, with a shoreline of 14 miles, 7 of which are undeveloped. Data sources for the ratings were air photos, USGS topographical maps, Soil Conservation Service maps, and other previous studies, plus a land and water visit to each site. The evaluation system was used to develop a lakeshore land use plan, the goals of which are: (1) protection of the natural lakeshore environment, including using all wetlands as conservation areas; (2) increasing the town tax base; and (3) providing more public access to the lake through a public marina and designated swimming areas.

Schreiber, J.D. and Rausch, D.L., "Suspended Sediment-Phosphorus Relationships for the Inflow and Outflow of a Flood Detention Reservoir", Journal of Environmental Quality, Vol. 8, No. 4, Oct.-Dec. 1979, pp. 510-514.

Callahan reservoir, located in an agricultural area near Columbia, Missouri, was studied for 3 years to determine the inflow and outflow suspended sediment phosphorus (P) and the solution (P) relationships. During the study, the mean inflow solution ortho-P was 0.085 mg/l as compared with 0.041 mg/l for the outflow. Concentrations of ortho-P were highest in both the inflow and outflow during the fall and winter. As yearly suspended sediment concentrations decreased in both the inflow and outflow due to lower runoff and sediment yields, solution ortho-P, as well as sediment total, inorganic, organic, and exchangeable P concentrations, increased. Similarly, as a result of coarse sediment deposition within the reservoir during individual storm events, outflow sediments were enriched in clay and had higher concentrations of total, inorganic, organic and exchangeable P than inflow sediments. However, because of sediment deposition within the reservoir, outflow volume concentrations of sediment total P decreased four-fold as compared to inflow sediments.

Swanson, G.A. and Meyer, M.I., "Impact of Fluctuating Water Levels on Feeding Ecology of Breeding Blue-Winged Teal", Journal of Wildlife Management, Vol. 41, No. 3, July 1977, pp. 426-433.

As a result of low-water levels in 1973, seasonal wetlands dried early in the spring, and blue-winged teal used semi-permanent lakes that were entering a drawdown phase. When water levels fall on semi-permanent lakes, a short-term increase in invertebrate availability to waterfowl may result due to shallow water conditions and the concentration of invertebrates by a reduced water volume. Blue-winged teal shifted from a diet high in snails when seasonal wetlands were relatively abundant to one dominated by midge larvae when semi-permanent lakes were the main wetlands used.

Teskey, R.O. and Hinckley, T.M., "Impact of Water Level Changes on Woody Riparian and Wetland Communities. Volume I: Plant and Soil Responses to Flooding", Report No. 77/58, Dec. 1977, Office of Biological Services, U.S. Fish and Wildlife Service, Washington, D.C.

A comprehensive literature review of general plant physiological responses to managed or natural changes in water levels, (i.e., submersion, flooding, soil saturation) and on plant tolerance mechanisms involved in water level changes is presented. The major effect of flooding is the creation of an anaerobic environment surrounding the root system, and the maintenance of proper root functioning is the factor which determines tolerance to flooding. Physical tolerance mechanisms involve processes designed to increase oxygen content in the roots either by transport of oxygen from the stem or from parts of the root system where oxygen is more available. Metabolic mechanisms enable the plant to utilize less toxic endproducts. Tolerant species are able to maintain root systems with a minimum of stress by incorporating a variety of tolerance mechanisms. Soil factors may ameliorate or accentuate the problem caused by flooding on physiological changes. Soils which are flooded show a decrease in oxygen concentration which leads to changes in soil chemistry (pH and redox potential) and nutrient availability. Five water level factors--time of year, flood frequency, duration, water depth, and siltation--are considered critical in determining a plant's physiological responses.

Teskey, R.O. and Hinckley, T.M., "Impact of Water Level Changes on Woody Riparian and Wetland Communities. Volume II: The Southern Forest Region", Report No. 77/59, Dec. 1977, Office of Biological Services, U.S. Fish and Wildlife Service, Washington, D.C.

A description and documentation of the natural plant ecoregions (communities) occurring in the Southern Forest region as affected by flood inundation is presented. The four ecoregions described are: Southern Mixed Forest, Southern Floodplain Forest, Beech-Sweetgum-Magnolia-Pine-Oak Forest, and Bluestem Prairie. Within each ecoregion, site characteristics for dominant species are followed by lists of associated species arranged in order of increasing site-soil moisture. In addition, the climate, soils, general physiography and bottomland successional pattern for each ecoregion is developed. A table summarizes information from existing literature regarding mature tree and seedling survival in 66 species under three water conditions: total submersion, partial submersion, and soil saturated. The information is also divided into flood periods during the growing season, the dormant season, and year-round. Associated with this table is a ranking of relative tolerance to flooding of these species.

Teskey, R.O. and Hinckley, T.M., "Impact of Water Level Changes on Woody Riparian and Wetland Communities. Volume III: The Central Forest Region", Report No. 77/60, Dec. 1977, Office of Biological Services, U.S. Fish and Wildlife Service, Washington, D.C.

A description and documentation of the natural plant ecoregions (communities) occurring in the Central Forest Region as affected by flood inundation is presented. The three ecoregions described are: Oak-Hickory Forest, Oak-Hickory Bluestem Parkland, and Oak-Bluestem Parkland. Within each ecoregion, site characteristics for dominant species are followed by lists of the associated species arranged in order of increasing site-soil moisture. In addition, the climate, soils, general physiography and bottomland succession pattern for each ecoregion is developed. A table summarizes information from existing literature regarding mature tree and seedling survival in 53 species under three water conditions: total submersion, partial submersion, and soil saturated. The information is also divided into flood periods during the growing season, the dormant season, and year-round. Associated with this table is a ranking of relative tolerance to flooding of these species.

United Nations Environment Program, "Environmental Issues in River Basin Development", Proceedings of the United Nations Water Conference on Water Management and Development, 1978, Vol. 1, Part 3, Pergamon Press, New York, New York, pp. 1163-1172.

Environmental effects of dam construction in river basins and methods of minimizing such effects are described. Large dams tend to have particularly significant and complex effects on aquatic ecosystems which must be carefully evaluated prior to construction. Rational development involves: (1) a comprehensive basinwide assessment of social, economic, and ecological characteristics and of the effects of development; and (2) evaluation of development alternatives reflecting social, economic, and environmental factors to provide the basis for environmental management. Dam construction can provide the water supply, hydroelectric power, and flood control, and can greatly improve agriculture, forestry, and livestock management. Excessive use of water and agricultural chemicals can however, cause waterlogging and salinization, and can affect water quality and quantity. Dams produce a permanent physical transformation, inundating settled areas and destroying habitats, affecting the ground water regime and water table, possibly increasing seismic tendencies, and often leading to explosive aquatic weed growth and the spread of schistosomiasis and other communicable diseases. Dams in tropical areas tend to favor weed propagation and vectors of parasitic diseases, while temperate-zone dams often interfere with fish migration. Resettlement of population displaced by dams often leads to housing, disease, and social problems. Loss of wetlands endangers many plant and animal species.

Weiner, R.M. et al., "Microbial Impact of Canada Geese (Branta canadensis) and Whistling Swans (Cygnus columbianus columbianus) on Aquatic Ecosystems", Applied and Environmental Microbiology, Vol. 37, No. 1, Jan. 1979, pp. 14-20.

Quantitative and qualitative analyses of the intestinal bacterial flora of Canada geese and whistling swans indicated that wild birds harbor significantly more fecal coliforms than fecal streptococci. The reverse was typical of captive and fasting birds. Neither Salmonella spp. nor Shigella spp. were isolated from 44 migratory waterfowl wintering in the Chesapeake Bay region. Enteropathogenic Escherichia coli were detected in 7 birds. Geese eliminated 10^7 and swans 10^9 fecal coliforms/d. In situ studies showed that large flocks of waterfowl can elevate fecal coliform densities in the water column. From the data the microbial impact of migratory waterfowl upon aquatic roosting sites can be predicted.

Williams, D.T., "Effects of Dam Removal: An Approach to Sedimentation", Technical Paper No. 50, Oct. 1977, U.S. Army Corps of Engineers, Hydrologic Engineering Center, Davis, California.

In recent years hydraulic structures such as dams have been removed due to deterioration, increased maintenance cost, or obsolescence. Investigation of the hydraulic, hydrologic, and sediment transport consequences of the removal of these structures have been very limited, thus necessitating the establishment of analytical techniques and procedures to predict adequately these effects. A mathematical model (HEC-6) was selected because of its success in the prediction of sediment transport when applied to a wide variety of cases. The removal of the Washington Water Power Dam (WWPD) on the Clearwater River near Lewiston, Idaho was selected for study. Procedures and techniques of calibration and verification were developed, comparisons of actual and predicted volumes of sediment transported were made, where the sediment scoured or deposited was predicted, and their rates were presented. There was discussion of the applicability of the model to this type of problem, limitations of a one-dimensional model, and interpretation of the results. The comparison of measured and computed final bed elevations, with the dam removed, was very satisfactory. Overall long-range trends for each operating condition was as expected. The calculated rate of scour was accurate at the WWPD site (River Mile 4.62), but lagged by approximately ten months at other upstream sections. This difference can be attributed to localized scour and "layering" of the bed particle distribution. Neither can be modeled by HEC-6. Some of the variations in the rate of scour and deposition can be attributed to the limitations of a one-dimensional model.

World Health Organization, "Environmental Health Impact Assessment", EURO Reports and Studies No. 7, 1979, Copenhagen, Denmark.

This document briefly summarizes a WHO seminar held in Greece in October, 1978. The purpose of the seminar was to review, on the basis of a number of case studies, the experience gained by various countries in assessing the environmental health impact of economic development, and to place this experience at the disposal of the Greek government and other governments; further, to outline a plan of action leading to the development of a model code of practice with regard to the health component of the environmental impact analysis process. This report contains the main deliberations, conclusions and recommendations of the seminar.

Yousef, Y.A., "Assessing Effects on Water Quality by Boating Activity", EPA/670/2-74-072, Oct. 1974, U.S. Environmental Protection Agency, Cincinnati, Ohio.

This research study was directed towards an assessment of effects on water quality in shallow water bodies (less than 30 feet deep) due to mixing by boating activity. Definition of the problem, isolation of effects and conditions, and determination of areas for further research were stressed. Four shallow lakes in Orange County, Florida, namely Lake Mizell, Lake Osceola, Lake Maitland, and Lake Claire were studied. Changes in several water quality parameters before and after limited boating activity were monitored. Agitation and mixing by boating activity destratified the lake, and in some cases, increased the oxygen concentration and the rate of oxygen uptake by suspended matter. An increase in turbidity was observed and was generally dependent on water depth, motor power, and nature of bottom deposits.

Yousef, Y.A. et al., "Mixing Effects Due to Boating Activities in Shallow Lakes", Technical Report No. 78-10, June 1978, Environmental Systems Engineering Institute, Florida Technological University, Orlando, Florida.

Recreational boats equipped with engines varying from 28 to 165 horsepower were used for agitation on Lake Claire, Mizell, and Jessup in central Florida. Primary and secondary waves generated by motorboats at the water-sediment interface were recorded. When boat velocities were greater than the square root of the acceleration of gravity and water depth, the average primary wave amplitudes decreased exponentially with depth. An empirical relationship between average wave amplitude, engine horsepower and the water depth was developed. Scour velocities for sediment particles from known wave amplitudes were calculated and verified. Mixing by motorboats in the lakes resuspended bottom sediments and increased turbidity. The increase in turbidity was accompanied by an increase in the phosphorus content, chlorophylla, and respiration rates within the waterbody. Also, a reduction in the dissolved oxygen content may result.

Yousef, Y.A., McLellon, W.M. and Zebuth, H.H., "Changes in Phosphorus Concentrations Due to Mixing by Motor Boats in Shallow Lakes", Water

Research, Vol. 14, No. 7, July 1980, pp. 841-852.

Changes in water quality due to mixing by motorboats were studied in shallow lakes of central Florida. The three lakes, Lakes Claire, Mizell and Jessup, differed in average water depth, sediment characteristics and trophic state. Significant increases in turbidity and ortho- and total phosphorus concentrations were demonstrated in water samples collected after mixing by motorboats in Lakes Claire and Jessup. The results obtained from Lake Mizell were not as conclusive. The increase in the orthophosphorus content for Lakes Claire, Mizell and Jessup averaged 43, 16, and 73 percent; the increase in total phosphorus content for the same three lakes was 39, 28, and 55 percent. Positive correlations existed between turbidities and the phosphorus content in the water column. The authors also demonstrated that the rate of increase in the phosphorus content with mixing time is much higher than the rate of decline after cessation of mixing. Data indicated substantial water quality effects are possible due to recreational boating on shallow lakes.

Zimmerman, E.G., Anderson, K.A. and Calhoun, S.W., "Impact of Discharge from Possum Kingdom Reservoir (Texas) on Genic Adaptation in Aquatic Organisms", Aug. 1980, Department of Biological Sciences, North Texas State University, Denton, Texas.

Extensive river and stream regulation has occurred in the United States over the last several decades. A recent international symposium reviewed the overall pattern of altered species composition and productivity caused by stream regulation. Construction of dams basically results in alteration of aquatic environments which influences biota for a considerable distance downstream from the impoundment. The purpose of this study was to investigate the response of one component of the aquatic community, the forage fishes, to these conditions. Data concerning fish physiology and genetics above and below Possum Kingdom Reservoir and its relationship to the ecological modifications which have resulted from the impoundment have not been gathered prior to this study. The general focus of this study was to characterize differences in the forage fish communities, physiological response, and genetic alteration above and at several sites below the reservoir, and attempt to relate the observed differences to physiological/behavioral responses of nonimpacted populations. Specific objectives of this study were to: (1) quantify changes in water quality which occur as a result of impoundment of the Brazos River; (2) characterize the forage fish fauna above and below the reservoir and relate any differences to alterations in habitat and/or water quality differences which have been manifested since completion of the dam; (3) determine physiological and behavioral responses of selected key species to variations in environmental salinity in an attempt to explain observed distributional patterns; and (4) examine physiological and genetic correlates of observed patterns among populations of N. lutrensis below Possum Kingdom Reservoir.

APPENDIX B

IMPACTS OF CHANNELIZATION PROJECTS

Bastian, D.F., "The Salinity Effects of Deepening the Dredged Channels in the Chesapeake Bay", Report NWS-81-S1, Dec. 1980, U.S. Army Institute for Water Resources, Fort Belvoir, Virginia.

Tests on the Chesapeake Bay Model, the world's largest estuarine model, were used to assess the effects of increasing the approach channels to Baltimore, Maryland from 13 to 15 m. There are four sections of dredged channels comprising 55 km of the 277 km distance from the bay mouth to Baltimore. The increased depth of channel would extend the length of dredged channels to 79 km. Base tests using the existing 13-m channels were conducted to determine the synoptic velocity, salinity, and tidal conditions at a number of locations throughout the bay, but primarily in the dredged channels. A 2½ year hydrographic period was simulated in the model to enhance the evaluation by adding a variable discharge as a parameter. A 12-constituent harmonic tide was used, giving a 28 lunar day tidal sequence which simulates a lunar month. The entire test was repeated, but with the 15-m channel installed. The tide is far more significant than discharge in affecting the velocities for the stations observed. No shift in flow predominance was identified between the base and the plan which could be used for conclusions about the effects of depth change on sediment transport. The deepened plan depths appear to induce higher salinities in the Patapsco and surrounding bay area. Salinity differences, base to plan, in this area attenuate with distance from the deepened channels. The model appears to be more saline during the plan test than the base test.

Benke, A.C., Gillespie, D.M. and Parrish, F.K., "Biological Basis for Assessing Impacts of Channel Modification: Invertebrate Production, Drift, and Fish Feeding in a Southeastern Blackwater River", Report No. ERC 06-79, 1979, Environmental Resources Center, Georgia Institute of Technology, Atlanta, Georgia.

Invertebrate production dynamics in the Satilla River in Georgia were suited to determine the effect of channelization or other river alterations on animal diversity, productivity, and general river ecology. The 362 km river has a drainage basin of 9,143 sq km with good to excellent water quality. Invertebrates were studied in three major habitats: snags, submerged wooden substrates; the main channel, sandy benthic conditions; and the muddy benthic habitat of backwater areas. Sampling was conducted at two sites on the river, one near Waycross 290 km from the Atlantic Ocean, and one at Atkinson 129 km from the ocean. Samples taken to determine the relative importance of the various invertebrate habitats included: (1) quantitative sampling to estimate invertebrate production; (2) invertebrate drift sampling with notation of habitat of origin for drift organisms; and (3) sampling of the feeding habitats of major fish species to determine trophic pathways and the habitats of origin of the fish. Results show that the snag habitat had the greatest species diversity, standing stock biomass, and total production with weights of 57 to 72 g dry wt/sq m. Drift samples revealed that approximately 80 percent of the animals found in the drift originated from the snags. Drift densities were high compared to other rivers with roughly 3 invertebrates/cu m. The major insectivorous fish species, redbreast, bluegills, and large-mouth bass, are largely dependent upon snags as sources of food. Due to this importance to invertebrates, the removal of snag habitats and/or channelization of the river would cause a significant decline in animal diversity and productivity.

Duvel, W.A. et al., "Environmental Impact of Stream Channelization", Water Resources Bulletin, Vol. 12, No. 4, Aug. 1979, pp. 799-812.

Geologic, engineering, and biological investigations of six Pennsylvania coldwater streams were undertaken to determine the impact of channel modifications instituted both prior to and following Hurricane Agnes. The primary focus of the study was on the ecological changes brought about by stream channelization. No long-term deleterious effects on water quality, attached algae, benthic fauna, or forage fish populations were found. Trout, however, were found to be greater in number and weight in natural than in channelized stream reaches. Lack of suitable physical habitat appeared to be the primary cause of reduced trout populations in stream reaches which have been channelized.

Erickson, R.E., Linder, R.L. and Harmon, K.W., "Stream Channelization (PL. 83-566) Increased Wetland Losses in the Dakotas", Wildlife Society Bulletin, Vol. 7, No. 2, Summer 1979, pp. 71-78.

Stream channelization under the Small Watershed Program (PL. 83-566) was the major influence on wetland drainage in the Wild Rice Creek Watershed in North and South Dakota. Drainage rates were 2.6 times higher in the channeled area than in the unchanneled area during project planning, and 5.3 times higher during and following construction. Although the channel's claimed benefits were watershed protection and flood control, the channel permitted and stimulated wetland drainage deleterious to wildlife. Substantial legal, policy, and administrative changes are necessary to prevent such habitat loss. Basic in any change is an overriding consideration of public values, such as wildlife, when public funds are involved, with cost-sharing rates that reflect and promote this concern. Especially when wetlands are present, base acreages need to be established early in project planning, and the sponsor should be required to show how wetlands will be protected as a condition for receiving public funds. Recent executive orders and policy guidelines on wetlands and channelization express a national concern for wetlands and may be helpful, but do not address fundamental cost-share issues and therefore are not believed to be adequate to protect wetlands in these projects.

Frederickson, L.H., "Floral and Faunal Changes in Low Land Hardwood Forests in Missouri Resulting from Channelization, Drainage, and Impoundment", FWS/OBS-78-91, Jan. 1979, U.S. Fish and Wildlife Service, Washington, D.C.

This study was designed to gather data on the effects of decreasing forest cover in the lower Mississippi Valley. The objectives were to identify the effects of channelization, impoundments, and drainage on plant and animal communities, and to develop reliable techniques to monitor and predict changes in these communities as a result of such actions. The Mingo Swamp, along a portion of the St. Francis River in Southeastern Missouri, was the study area. Photo interpretation provided overall information on the nature of tree cover with line transects gathered; data on species composition and density were gathered along line transects selected randomly along the river. Among the conclusions reached were that stream channelization reduced or changed riparian habitat, decreasing forest area by as much as 78 percent as compared to no more than 7 percent in unchannelized areas; sites inundated the longest had plants most tolerant to flooding; bird populations tended to avoid channelized streams; and that channelization does reduce flooding and benefit agriculture. Recommendations made indicated that other developmental alternatives, such as construction of levees and government purchase of land to protect riparian habitat, are preferred over the adverse effects of channelization, and farmers should be made aware of the changes effected on forest composition from the drainage of their land.

Headrick, M.R., "Effects of Stream Channelization on Fish Populations in the Buena Vista Marsh, Portage County, Wisconsin", Sept. 1976, U.S. Fish and Wildlife Service, Stevens Point, Wisconsin.

Fish populations from ditches 6 to 8 years old and 52 to 62 years old were compared with populations in adjacent portions of natural streams. Two study areas were selected: an upstream zone of good brook trout habitat and a downstream zone of marginal trout habitat. Loss of year-round instream cover through channelization limited brook trout density, which reduced annual brook trout production to 28.8 kg/stream km in the upstream new ditch, compared to 72.2 kg/km and 65.5 kg/km in the upstream old ditch and the upstream natural stream, respectively. Angler success was also reduced in the upstream new ditch. Midsummer water temperatures reached upper lethal levels for brook trout in the downstream ditches where current velocity was reduced and white sucker were abundant. Mottled sculpin were consistently absent from the upstream new ditch and scarce in the downstream new ditch. The natural stream had the greatest number of fish species in both study areas; the new ditch had the fewest species. Recovery was more rapid in ditches where spoil was spread on adjacent fields and bank vegetation left in place than where spoil was left on the banks.

Huang, Y.H. and Gaynor, R.K., "Effects of Stream Channel Improvements on Downstream Floods", Research Report No. 102, Jan. 1977, Kentucky Water Resources Research Institute, University of Kentucky, Lexington, Kentucky.

A self-calibrating watershed model is presented for predicting the effect of channel improvements on downstream floods. The model is called MOPSET because it is a modified version of OPSET developed several years ago at the University of Kentucky. OPSET is a computerized procedure for determining an optimum set of parameter values by matching synthesized flows with recorded flows. Major modifications include the replacement of the modified Muskingum method of channel routing by a kinematic finite difference method, the division of the watershed into a number of segments, and the inclusion of a storage routing procedure to take care of any reservoirs or flood control structures located in the watershed. The computer program is well documented and can be used not only as a flood predicting model but also as a general model for hydrologic simulations. The model was applied to three different watersheds in Kentucky. It was found that an optimum set of parameter values obtained automatically by the model was not unique and might not yield the most desirable solution. For this reason, new features were added so that the user can exercise his judgment in selecting the most desirable parameter values. The synthesized flows obtained from these watersheds are presented and compared with the recorded flows. The effects of channel improvements, flood control structures, and routing procedures are discussed.

Klimas, C.V., "Effects of Permanently Raised Water Tables on Forest Overstory Vegetation in the Vicinity of the Tennessee-Tombigbee Waterway", Misc. Paper E-82-5, Aug. 1982, U.S. Army Engineer Waterways Experiment Station, Vicksburg, Mississippi.

Water tables are expected to rise somewhat in the vicinity of planned impoundments along the Tennessee-Tombigbee Waterway. Where permanent flooding of timber is anticipated, tree mortality is expected and has been accounted for in the project planning process. However, where water table levels will be raised significantly but permanent flooding does not occur, the effects on forests are very unclear. This report identifies factors that may be pertinent in assessing this type of impact. The information presented has been drawn from the scientific literature on species flood tolerance and related topics that provide insight on potential water table effects. This review concentrates on induced mortality and changes in growth rates of mature trees. Raised water tables can detrimentally affect trees primarily by causing oxygen depletion in the root zone. The ability of an individual tree to withstand this type of stress may be related to a variety of factors, including the tree species, rooting depth, soil conditions, and the timing, duration, and frequency of encroachment of the water table into the root zone. In some situations, trees may show improved growth as a result of raised water tables, although this response may be short-lived. The forest communities likely to be most sensitive to partial root-zone saturation are those typically found on well-drained upland sites. Where such communities are subjected to drastic changes in the root environment, mortality may be widespread and rapid. In contrast, swamp forest communities (cypress-tupelo) are generally much less sensitive to increases in site moisture status, and may be affected primarily with respect to long-term growth and reproduction, or not at all. The complex floodplain forest (bottomland hardwood) communities present the greatest difficulty in predicting potential impacts resulting from raised water tables. Possible effects range from near-complete mortality in areas where permanent surface saturation occurs to no effect or growth rate increases in response to very minor, subsurface water table rises. The complexity of this issue suggests that only very general impact predictions may be possible except in areas where extreme stresses are anticipated.

Maki, T.E., Hazel, W. and Weber, A.J., "Effects of Stream Channelization on Bottomland and Swamp Forest Ecosystems", Completion Report, Oct. 1975, School of Forest Resources, North Carolina State University, Raleigh, North Carolina.

The general objective of this study was to develop a better understanding of the swamp forest ecosystem, including its dominant and subsidiary vegetation communities, its associated fauna, soil properties and surface stratigraphy, and the hydrology in relation to alterations of the "natural" drainage regime. Sixteen well lines were established, representing 8 channelized situations and 8 nonchannelized situations in 10 stream bottoms from Craven County northward to Gates County, North Carolina, covering over a hundred miles of Coastal Plain. Within each of the 8 well-line categories, 4 lines are in recently cut-over tracts and 4 are within tracts that have not seen logging activities for several decades. Some tentative conclusions from the first year's study are: (1) the swamp forest ecosystems of eastern North Carolina are largely dominated by quantities of cull trees which are the residual elements from high-grading logging operations during the past 150 years or more; (2) surface soils to depths of 40 cm or so are quite uniform in texture and composition, with texture averaging 70 to 80 percent clay; (3) channelized streams maintain perennial and clear flow during drought periods in sharp contrast to the murky water in natural streams during low flow; and (4) clear-cutting substantially affects the soil moisture regime because of the temporary reduction in evapotranspiration draft.

Parrish, J.D. et al., "Stream Channelization Modification in Hawaii, Part D: Summary Report", Report No. FWS/OBS-78/19, Oct. 1978, U.S. Fish and Wildlife Service, Hawaii Cooperative Fishery Research Unit, Honolulu, Hawaii.

A three-year statewide study was made of the occurrence and consequences of channelization in Hawaiian streams. The 366 perennial streams of the state were inventoried for the first time, and some basic information was catalogued on their physical characteristics, complete status of channel alteration, and macrofaunal communities. Fifteen percent of the state's streams have channels altered in at least 1 of 6 forms. Forty percent of the modified channel length is concrete lined--the form of alteration found to be most ecologically damaging. Field measurements showed that channel alterations commonly caused large changes in pH values, conductivity, dissolved oxygen and daily temperatures. The native species tested in the laboratory had less tolerance of high temperatures than exotics, and some natives had upper lethal temperature limits within the temperature range measured in the channelized stream. Twenty-five species of fish and decapod crustaceans were collected statewide, of which only 9 percent were native. Native species were not abundant in most heavily channelized sections, whereas exotic species were almost entirely absent. Recommendations for mitigating the impact of channel modification are included.

Pennington, C.H. and Baker, J.A., "Environmental Effects of Tennessee-Tombigbee Project Cutoff Bendways", Misc. Paper E-82-4, Aug. 1982, U.S. Army Engineer Waterways Experiment Station, Vicksburg, Mississippi.

Biological and physical data were collected from four bendways within the river portion of the Tennessee-Tombigbee Waterway (TTW) from Columbus, Mississippi, to Demopolis, Alabama: Rattlesnake Bend, Cooks Bend, Big Creek Bendway, and Hairston Bend. During this study, the four bendways had not all been cut off and had been impounded for various lengths of time. At the completion of the TTW project, all four of the bendways will be severed from the main navigation channel. Four distinct areas within each bendway were compared: above the bendway, within the bendway, below the bendway, and within the cut. Sampling was conducted from January 1979 to September 1980 to coincide with four different river stage/water temperature regimes.

Sediment analysis and bottom profiles indicated that the substrate composition of some of the bendways is changing. Overall, the substrate of the study area is changing from a sand-gravel-fines mixture to one of predominantly sand and fines. Areas of some bendways, in particular the upper areas, were accumulating sediments. At Big Creek Bendway, this accumulation completely blocked water exchange between the river and the within-bendway areas. Few significant differences in water quality were documented for either within individual bendways or among the four bendways. Only at Big Creek Bendway were consistent differences found between within-bendway samples and river samples. Phytoplankton composition and chlorophyll concentrations showed only small differences among bendways. Aquatic macrophytes were scatted and uncommon in the four bendways. Water-willow (Justicia sp.) was most commonly encountered, particularly in Rattlesnake Bend where numerous small beds were found. Based upon total collections, a consistent family assemblage of macroinvertebrates characterized the four bendways. Although 60 family-level taxa were collected, nine families of macroinvertebrates accounted for between 93.5 and 97.2 percent of the benthos. The importance of these families varied among bendways and appeared to reflect differences in physical bendway conditions, particularly substrate type and current velocities. Eighteen species of Unionid mollusks, plus the Asian clam Corbicula, were collected during the surveys. Nearly all the specimens were found at Big Creek Bendway; none were collected at Hairston Bend. With the exception of three species of Pleurobema, no unusual or uncommon mollusk species were found.

Based on overall ichthyofaunas, two groups of bendways were delineated that corresponded to impoundment and riverine habitats. Rattlesnake Bend and Cooks Bend were located in lower pool sections, where impoundment conditions prevailed, and their ichthyofaunas were dominated by clupeids (shad) and centrarchids (sunfishes, crappies, and basses). Hairston Bend, essentially a riverine reach during this study, was dominated by cyprinids (minnows), ictalurids (catfishes), and catostomids (suckers). Big Creek Bendway, unique in having both riverine and lacustrine habitats, was faunistically most similar to Hairston Bend, but also showed moderate similarities to the other bendways.

Possardt, E.E. and Dodge, W.E., "Stream Channelization Impacts on Songbirds and Small Mammals in Vermont", Wildlife Society Bulletin, Vol. 6, No. 1, Spring 1978, pp. 18-24.

Stream channelization effects were documented in the White River watershed during the first and second years after channelization. Birds were mist-netted during four sampling periods. Percentages of birds collected from channelized areas were 33%, 27%, 38%, and 46% for fall 1974, spring 1975, summer 1975, and late summer 1975, respectively. Species diversity was significantly less in channelized areas for fall 1974 and early summer 1975; highly significantly less for spring 1975; no significant difference existed for later summer 1975. Swallows and spotted sandpipers (Actitis macularia) were more abundant in channelized areas; thrushes, vireos, and particularly warblers were more abundant in nonchannelized areas. Small mammals were live-trapped during three sampling periods. Percentages of mammals collected from channelized areas were 28%, 39%, and 39% for fall 1974, early summer 1975, and late summer 1975, respectively. Shrews and jumping mice were most adversely affected; the white-footed mouse (Peromyscus leucopus), the most abundant small mammal collected, recovered rapidly in channelized areas. Impacts on small mammals and songbird populations was most dramatic where streamside vegetation had been extensively destroyed.

Prellwitz, D.M., "Effects of Stream Channelization on Terrestrial Wildlife and Their Habitats in the Buena Vista Marsh, Wisconsin", Report FWS/OBS-76-25, Dec. 1976, Wisconsin Cooperative Fishery Research Unit, Stevens Point, Wisconsin.

Stream channelization affected wildlife in Buena Vista Marsh by draining wetlands, setting back plant succession, and decreasing habitat diversity along streambanks by removing or burying plants. Plant and animal species composition and abundance were studied in a continuum of plant successional stages from grassland to mature woods on streambanks adjacent to recently dredged (6 years), old dredged (50 years), and natural streams. Sheet-water area and longevity and wildlife use of three sheet-water areas with various degrees of drainage were compared. Bird and mammal species diversity and bird abundance increased as streambank plant succession advanced, until a mature wooded stage was reached. Abundance of small mammals was related to the amount of ground cover and diversity of habitats along the streambanks. Sheet-water area and longevity were greatest on undrained wetlands, and least near recently dredged channels. Waterfowl use, bird nesting, and reptile and amphibian abundance were also greatest on undrained areas. Invertebrates and various seeds made up 98.4 and 1.6 percent, respectively, of the diet of breeding blue-winged teal using sheet-water areas.

Stone, J.H. and McHugh, G.F., "Simulated Hydrologic Effects of Canals in Barataria Basin: A Preliminary Study of Cumulative Impacts", Final Report, June 1977, Louisiana State University Center for Wetland Resources, Baton Rouge, Louisiana.

Computer simulation of the hydrography of Barataria Basin indicates significant hydrologic changes due to navigation and transportation canals. The simulations compared hydrologic parameters in the Basin before and after the construction of the Barataria and Intracoastal Waterways, and the canals associated with eight oil and gas fields. The waterways accounted for about 90% of the simulated changes; the remaining 10% was due to the canals of the eight oil and gas fields. The effects of oil and gas canals are individually small, but they are cumulative and, overall, probably are just as significant as those due to waterways, since there are approximately 90 fields in the Basin. Water heights increased on the average by about 2 inches, or more than 9% above normal conditions. Total water flow over a 25-hour tidal cycle increased between 7 and 16% in the intermediate marshes. Water flow at individual exchange points in the freshwater marshes was reduced to between 20 and 70% of normal. Water flow through tidal passes was not changed. Systematic studies of these effects, how to mitigate them, and how they are related to biological production are badly needed. Large scale development projects, which require waterways or canals, should be carefully evaluated and monitored to determine their full environmental effects.

Stone, J.H., Bahr, Jr., L.M. and Way, Jr., J.W., "Effects of Canals on Freshwater Marshes in Coastal Louisiana and Implications for Management",

Freshwater Wetlands: Ecological Processes and Management Potential, 1978, Academic Press, New York, New York, pp. 299-320.

Water flow and quality determine and control species composition and function in the freshwater marshes of coastal Louisiana. Man's activities alter this when he disrupts or removes the marsh. For example, man-made canals can change the hydrologic regime, depending on its alignment and local elevations, from -1% to -35% of normal flow. This in turn likely accelerates land loss from increased wave action. It is estimated that perhaps 172 hectares per year of freshwater marsh in coastal Louisiana is being lost due to man's activities. Canals also tend to divert runoff water away from the marsh (where it would be purged of pollutants) to open water bodies, thereby probably causing eutrophication. For example, the P loading rate of several freshwater lakes in the upper Barataria Basin is estimated between 1.5 and 4.3 g/sq m/yr, which may be from 4 to 11 times greater than the loading limit for eutrophication.

Thackston, E.L. and Sneed, R.B., "Review of Environmental Consequences of Waterway Design and Construction Practices as of 1979", Tech. Report E-82-4, Apr. 1982, U.S. Army Engineer Waterways Experiment Station, Vicksburg, Mississippi.

Water projects designed and constructed by the Corps of Engineers (CE) include dikes, revetments, levees, and channel modifications for flood control and navigation purposes. A review of existing CE design guidance in the form of Engineer Manuals (EMs) and personal contacts with CE field offices was performed to appraise current practices regarding consideration of environmental effects in the design and construction of waterway projects. A computer-assisted review of the technical literature was also performed in order to determine observed adverse environmental effects of waterway projects. Possible adverse effects presented in the literature include wetlands drainage, loss of native vegetation, cutoff of oxbows and meanders, water table drawdown, increased erosion and sedimentation, and change of aesthetics. Other possible effects on the aquatic system may include the loss of aquatic habitat, productivity, and species diversity, and the degradation of water quality.

Alternatives to traditional channel modification were identified through the literature review. Structural alternatives to channel modification include levees, floodways, reservoirs, and land treatment measures. Additional alternatives include various forms of flood plain management; flood plain zoning; construction of bypass channels around sensitive wetland areas; construction of numerous, very small, water-retention structures; and substitution of clearing and snagging, or only snagging, for complete channelization. Current efforts to minimize adverse environmental effects of dikes were investigated through personal contact with CE field offices. Programs of notching or gapping dikes are in progress on the Missouri and Mississippi Rivers. Studies are under way to quantify the effects of such notches on the riverine ecosystem. Case studies of the planning process are presented for two recently designed waterway projects.

White, T.R. and Fox, R.C., "Recolonization of Streams by Aquatic Insects Following Channelization", Technical Report 87, Vol. I, May 1980, Water

Resources Research Institute, Clemson University, Clemson, South Carolina.

Recovery, defined as development of an aquatic insect fauna similar to that of the control streams, does not occur in South Carolina streams following channelization. Five channelized streams and two natural streams were sampled over the 1975 to 1979 period to determine channelization effects on species composition and diversity of aquatic insects in the Piedmont/Coastal Plain regions. Stream samples revealed that channelization yields a fauna composed principally of very tolerant or normally pond-inhabitating species. Taxa preferring fast currents, vegetation, and low turbidity were rarely found in the channelized streams. Surber square foot and core samplings afforded the greatest utility for characterizing stream aquatic insect diversity; kick-screen sampling was found preferable in determining population composition. The results supported the contention that, if a creek must be altered, it should be by means other than classical channelization for best running water insect faunal preservation. Recommendations to minimize impacts were developed for cases in which a stream must be altered and no alternative exists.

Wright, T.D., "Potential Biological Impacts of Navigation Traffic", Misc. Paper E-82-2, June 1982, U.S. Army Engineer Waterways Experiment Station, Vicksburg, Mississippi.

A literature search was conducted to identify research relating physical and chemical changes associated with navigation traffic to potential biological impacts. It was found that, although some information on physical and chemical changes was available, documentation to demonstrate biological impacts was generally lacking. Where possible impacts were identified, they were observed at the organism level, rather than at the population, community, or ecosystem level. A particular problem was encountered in attempting to separate impacts of navigation from those caused by natural and/or anthropogenic perturbations. It was concluded that biological impact analysis beyond the organism level may not be within the existing state-of-the-art technology.

Zimmer, D.W. and Bachmann, R.W., "Channelization and Invertebrates in Some Low Streams", Water Resources Bulletin, Vol. 14, No. 4, Aug. 1978, pp. 868-883.

Habitat diversity and invertebrate drift were studied in a group of natural and channelized tributaries of the upper Des Moines River during 1974 and 1975. Channelized streams in this region had lower sinuosity index values than natural channel segments. There were significant (P=0.05) positive correlations between channel sinuosity and the variability of water depth and current velocity. Invertebrate drift density, expressed as biomass and total numbers, also was correlated with channel sinuosity. Channelization has decreased habitat variability and invertebrate drift density in streams of the upper Des Moines River basin, and probably has reduced the quantity of water stored in streams during periods of low flow.

APPENDIX C

IMPACTS OF DREDGING PROJECTS

Allen, K.O. and Hardy, J.W., "Impacts of Navigational Dredging on Fish and Wildlife: A Literature Review", FWS/OBS-80/07, Sept. 1980, U.S. Fish and Wildlife Service, Washington, D.C.

This review covers impacts to fish, other aquatic organisms, and wildlife, as well as habitat enhancement opportunities, resulting from construction of new navigational channels and maintenance dredging of existing channels. The type of dredging equipment used determines, to a great extent, the viable disposal alternatives, the type and magnitude of potential impacts, and the potential for habitat development. About 99 percent of the dredging volume in the United States is accomplished by hydraulic dredges. The composition of dredged material from a particular site affects its pollution potential as well as the potential for habitat development or other beneficial uses. Dredged material from new channels or new works projects often has chemical and engineering properties which create fewer environmental problems than material from maintenance projects. Material removed during maintenance dredging of navigational channels is an accumulation of detached soil particles which have been transported by wind and water, and may contain a variety of contaminants.

Bohlen, W.F., Cundy, D.F. and Tramontano, J.M., "Suspended Material Distributions in the Wake of Estuarine Channel Dredging Operations", Estuarine and Coastal Marine Science, Vol. 9, No. 6, Dec. 1979, pp. 699-711.

Field sampling of the suspended material field downstream of a large volume bucket dredge operating in the Lower Thames River estuary near New London, Connecticut, was conducted in order to examine the magnitude and character of the dredge-induced resuspension and to evaluate typical operational efficiency. These data indicated that approximately 1.5 to 3 percent of the sediment volume in each bucket-load is introduced into the water column producing suspended material concentrations adjacent to the dredge of 200 mg/l to 400 mg/l. These values exceeded background levels by two orders of magnitude. Analysis of particulate organic carbon and grain size characteristics indicated that resuspension also alters suspended load composition increasing the percentage of inorganic materials and median grain size. Proceeding downstream, material concentrations along the centerline of the dredge-induced plume decrease rapidly, approaching background within approximately 700 m. Compositional variations display similar trends with the major perturbations confined to the area within 300 m of the dredge. The observed spatial distributions indicated that the dredge-induced resuspension is primarily a near-field phenomenon producing relatively minor variations as compared to those caused by naturally occurring storm events. Previous work has shown that these latter systems can produce estuary-wide variations in suspended material concentrations, increasing the mass of material in suspension by at least a factor of two. This increase in total suspended load is nearly an order of magnitude larger than that produced by the dredge.

Brannon, J.M., "Evaluation of Dredged Material Pollution Potential", Technical Report No. DS-78-6, Aug. 1978, U.S. Army Engineer Waterways Experiment Station, Vicksburg, Mississippi.

This report synthesizes data from seven research projects that investigated pollution properties of dredged material, and procedures for determining their potential for effects on water quality and aquatic organisms. Short-term impacts of dredged material on water quality and aquatic organisms are related to the concentration of chemically mobile, readily available contaminants rather than total concentration. The Elutriate Test, which measures concentrations of contaminants released from dredged material, can be used to evaluate short-term impacts on water quality. The only constituents generally released from dredged material are manganese and ammonium -N. Elevated concentrations of these constituents, however, are of short duration because of rapid mixing and are of low frequency due to the intermittent nature of most disposal operations. Short-term chemical and biological impacts of dredging and disposal have generally been minimal. Longer term impacts of dredged material on water quality have generally been slight and can be evaluated by means of the Elutriate Test and analysis of mobile forms of sediment contaminants. The greatest hazard of dredged material disposal is the potential effect of material on benthic organisms. Most dredged material has not proven particularly toxic. Some dredged material, however, can be extremely toxic or of unknown toxicological character.

Chen, K.Y. et al., "Confined Disposal Area Effluent and Leachate Control Laboratory and Field Investigations", Technical Report No. DS-78-7, Oct. 1978, U.S. Army Engineer Waterways Experiment Station, Vicksburg, Mississippi.

This report summarizes the findings of five work units concerned with the impacts of dredged material disposal in confined land disposal areas. Three work units dealt with active disposal operations at 11 sites; impact was assessed by comparing the quality of influents and effluents at each site with background surface receiving water. Two work units evaluated the impact of confined disposal area leachates on ground waters. Leachate studies included laboratory column elutions (3-9 month period) of each of five types of dredged material overlying one of two different soils; four field sites were also monitored for changes in leachate and ground water quality in a 9-month study. Field sites included fresh and brackish water dredging environments in geographical areas where contamination problems were anticipated. Dredged material and environmental features varied greatly at different sites. In most cases soluble concentrations of most chemical constituents were very low. Only soluble manganese and ammonia nitrogen levels failed to meet most criteria. Leachate studies suggest the disposal of brackish water dredged material in upland disposal areas may render subsurface water unsuitable for public water supply or irrigation purposes. Guidelines for evaluation of potential disposal sites should be developed in steps not requiring complete execution of the total program to determine site suitability.

Conner, W.G. and Simon, J.L., "The Effects of Oyster Shell Dredging on an Estuarine Benthic Community", Estuarine and Coastal Marine Science, Vol. 9, No. 6, Dec. 1979, pp. 749-758.

This paper describes the extent and nature of the effects on the benthos of physical disruptions associated with dredging fossil oyster shell. Two dredged areas and one undisturbed control area in Tampa Bay, Florida, were quantitatively sampled before dredging, and for one year after dredging. The immediate effects of dredging on the soft-bottom community were reductions in numbers of species (40% loss), densities of macroinfauna (65% loss), and total biomass of invertebrates (90% loss). During months 6 to 12 after dredging, the analysis used (Mann-Whitney U Test, alpha = 0.05) showed no difference between dredged and control areas in number of species, densities, or biomass (except the E sub 1 area). Community overlap (Czeckanowski's coefficient) between dredged and control areas was reduced directly after dredging, but after 6 months the predredging level of similarity was regained.

Conrad, E.T. and Pack, A.J., "A Methodology for Determining Land Value and Associated Benefits Created from Dredged Material Containment", Technical Report No. D-78-19, June 1978, U.S. Army Engineer Waterways Experiment Station, Vicksburg, Mississippi.

This report presents a methodology for determining land values and associated benefits from the productive use of dredged material containment sites. A discussion of productive uses of dredged material sites, their physical characteristics, institutional and legal constraints, and local land demand is included, as well as an overview of property valuation. Site description, establishment of use potential, value estimation, and associated benefits and impacts are discussed. Working tables are presented. The resulting land value and the associated benefits and impacts created by dredged material containment should be explicit inputs to the formulation of plans in accordance with Principles and Standards for Water and Related Resource Planning, and Corps of Engineers regulations. Fifteen case studies of productively used dredged material containment sites were conducted to validate and refine the methodology. One of the case studies was used in this report as a site-specific example of how the methodology can be applied.

Eichenberger, B.A. and Chen, K.Y., "Methodology for Effluent Water Quality Prediction", Journal of the Environmental Engineering Division, American Society of Civil Engineers, Vol. 106, No. EE1, Feb. 1980, pp. 197-209.

A methodology for the prediction of effluent water quality from the disposal of dredged material in confined areas was presented. The computation method was based on the use of laboratory determination of contaminants in different particle size fractions together with the application of Hazen's theory of sedimentation for longitudinal basins. Hazen's equation was applied to each particle size fraction. The removal efficiency of each parameter can be obtained from the data of suspended solids removal and its concentration in that size fraction. The validity of this methodology has been verified in three active disposal sites.

Engler, R.M., "Impacts Associated with the Discharge of Dredged Material Into Open Water", Proceedings of the Third U.S.-Japan Expert's Meeting on Management of Bottom Sediments Containing Toxic Substances, EPA-600/3-78-084, 1978, U.S. Environmental Protection Agency, Washington, D.C., pp. 213-223.

With few exceptions, the impacts of aquatic disposal are mainly associated with the physical effects. These possible effects are persistent, often irreversible, and compounding. Geochemically, releases are limited to nutrients, with negligible release of toxic metals and hydrocarbons. Biochemical interactions are infrequent, with no clear trends; elevated uptake of toxic metals and hydrocarbons are negligible to nonexistent.

Flint, R.W., "Responses of Freshwater Benthos to Open-Lake Dredged Spoils Disposal in Lake Erie", Journal of Great Lakes Research, Vol. 5, No. 3-4, 1979, pp. 264-275.

Temporal variations in the natural benthic macroinvertebrate community of areas subjected to open-lake dredged spoil disposal were examined in Lake Erie near Ashtabula. The two types of dredged spoil used were sandy-silt sediment from the Ashtabula harbor and coarse sand sediments from the Ashtabula River. Two untreated reference sites were also sampled for comparison. Grab samples were used to characterize the communities at the sources of dredging to determine the types of material that would be deposited at the spoil sites. Samples were taken at the disposal and control sites at intervals of 5 days, 90 days, 256 days, and 340 days after disposal. For each sample taken, a modified Spade box corer was used to collect four replicate bottom samples. Macroinvertebrates were identified to the lowest taxa possible. Results show that for the sandy silt site the density of macroinvertebrates increased while diversity decreased. Species brought in with the river and harbor sediments contributed to the increased density. Also, while the untreated sites had communities of 60 to 80 percent oligochaetes, both disposal sites had communities of 98 percent oligochaetes. The river spoils disposal sites showed a more rapid recovery to predisposal conditions. The opportunistic nature of the Oligochaeta life history pattern allows them to take advantage of and thrive in the environment existing after disposal. This dominance of Oligochaeta along with the increased population density form unstable communities at the disposal sites.

Grimwood, C. and McGhee, T.J., "Prediction of Pollutant Release Resulting from Dredging", Journal of the Water Pollution Control Federation, Vol. 51, No. 7, July 1979, pp. 1811-1815.

The standard elutriate test was used to simulate the quality of the effluent during dredging operations, and is generally reasonably accurate. In this paper, data collected at five southern Louisiana locations were analyzed to assess the predictive value of the test as a tool for estimating pollutant release. The Kolmogorov-Smirnov two-sample test was used to test the correspondence of pollutant release in the dredge effluent and the standard elutriate test for total Kjeldahl nitrogen, chemical oxygen demand, cyanide, arsenic, cadmium, copper, chromium, lead, mercury, nickel, zinc and phenol. It was found that for chemical oxygen demand, chromium, cadmium, and phenol--and to a lesser degree for total Kjeldahl nitrogen, arsenic, nickel, and zinc--the test afforded a useful prediction of release. The relative failure of the test with respect to lead, mercury, and copper, was a result, in part, of adsorption of these materials by the suspended sediment.

Gunnison, D., "Mineral Cycling in Salt Marsh-Estuarine Ecosystems; Ecosystem Structure, Function, and General Compartmental Model Describing Mineral Cycles", Technical Report No. D-78-3, Jan. 1978, U.S. Army Engineer Waterways Experiment Station, Vicksburg, Mississippi.

A nutrient and heavy metal cycling study of marsh-estuarine ecosystems was undertaken for the Dredged Material Research Program. The objective was to gather as much information as possible on mineral cycling in marsh-estuarine ecosystems. A compartmental model outlining pathways of mineral cycling within the marsh-estuarine ecosystem was developed. Approaches used in the study included literature surveys and discussions with authorities in marsh-estuarine ecology. Information from allied fields of research was used to supplement direct sources of information. Chemical release of contaminants from dredged materials may cause problems. Oxygen demands exerted by reduced materials placed in aerobic waters can deplete oxygen. Several heavy metals have increased mobility once their binding sulfides have been oxidized. Oxidation of heavily contaminated marsh soils can cause metal toxicity for plants. Eutrophication resulting from use of contaminated materials for marsh creation is not a problem, except in unproductive estuaries. Present knowledge is inadequate to permit accurate predictions of environmental impacts resulting from use of heavily contaminated materials in marsh creation. Only those dredged materials containing nutrients and heavy metals in quantities no greater than those in marsh soils at the creation site should be used for marsh development.

Gushue, J.J. and Kreutziger, K.M., "Case Studies and Comparative Analyses of Issues Associated with Productive Land Use at Dredged Material Disposal Sites", Technical Report No. D-77-43, Dec. 1977, Two Volumes, U.S. Army Engineer Waterways Experiment Station, Vicksburg, Mississippi.

This report evaluates 12 selected cases where dredged material from navigation projects was used to create productive land. Results are directly applicable as a management aid for Corps disposal planners. The principal output was the development of an overall set of "implementation factors" for disposal-productive use projects. Thirty-seven factors were identified and categorized as environmental, technical, economic/financial, legal, institutional, or planning/implementation. These factors provide a framework for ensuring that project planners address the full range of substantive and procedural considerations that are important to successful project implementation. The case studies provide documented proof that disposal-productive use project success is as much affected by procedural factors as by substantive factors. Procedural aspects of each case study are fully delineated in individual case study synopses contained in Volume II of the study report. The detailed comparative analyses of the 12 cases, which led to the identification of the important implementation factors, are also provided in Volume II. The matrix approach enables the site-specific nature of disposal planning to be retained in the analysis while providing a common basis for comparison. As a result, the set of implementation factors is applicable to all disposal situations.

Hoeppel, R.E., "Contaminant Mobility in Diked Containment Areas", Proceedings of the 5th United States-Japan Experts Meeting on Management of Bottom Sediments Containing Toxic Substances, EPA-600/9-80-044, Sept. 1980, U.S. Environmental Protection Agency, Washington, D.C., pp. 175-207.

Nine dredged material land containment areas, located at upland, lowland and island sites, were monitored during hydraulic dredging operations in fresh and brackish-water riverine, lake, and estuarine environments. Influent-effluent sampling at the diked disposal areas showed that, with proper retention of suspended solids, most chemical constituents could be removed to near background water levels. Most heavy metals, oil and grease, chlorinated pesticides, and PCB's were almost totally associated with solids in influent and effluent samples. Ammonia, manganese, total mercury and possibly iron, copper and zinc are the only contaminants which may occasionally exceed background water levels and present water quality standards after suspended solids removal. Actively growing vegetation appeared to be efficient in reducing ammonium nitrogen to low levels and for filtering out suspended solids. Confined land disposal of dredged material seems to be a viable method for the containment and treatment of most contaminated sediments, but effluent monitoring and predictive testing methodologies should include suspended solids.

Holliday, B.W., "Processes Affecting the Fate of Dredged Material", Technical Report No. DS-78-2, Aug. 1978, U.S. Army Engineer Waterways Experiment Station, Vicksburg, Mississippi.

Determination of the fate of dredged material placed on the bottom of an ocean, lake, estuary, or river is an environmental concern that requires consideration and adequate prediction in the planning of a dredging project, since various natural processes can alter the initial configuration of the deposit and subject the surrounding bottom to some level of environmental impact. In the selection process for a disposal site, consideration must be given to the eventual disposition of dredged material in order that adequate determination of the site capacity can be made. The four primary environments that may contain subaqueous dredged material deposits are oceans, estuaries, rivers, and lakes, with various energy-related zones within each environmental system. Each zone has a unique set of physical factors and sedimentological properties that will determine the potential fate of a dredged material deposit. Prediction of the fate of dredged material at a disposal site requires knowledge of: (1) currents, (2) waves, (3) tides, (4) suspended sediment concentrations, (5) seasonal energy fluctuations, (6) storms, (7) dredging/disposal operations, (8) shipping traffic, (9) fisheries activities, (10) bathymetry, (11) sedimentology, and (12) biological activity.

Holliday, B.W., Johnson, B.H. and Thomas, W.A., "Predicting and Monitoring Dredged Material Movement", Technical Report No. DS-78-3, Dec. 1978, U.S. Army Engineer Waterways Experiment Station, Vicksburg, Mississippi.

This report summarizes the results from three work units (1B06, 1B07, and 1B09) of the Dredged Material Research Program concerned with predicting and monitoring dredged material movement. Work Units 1B06 and 1B09 concerned predicting the short-term fate of dredged material discharged in open water. Work Unit 1B06 was an evaluation and calibration of the Tetra Tech disposal models (developed under Work Unit 1B02 of the DMRP by Tetra Tech, Inc.) using field data collected at several disposal sites, including the Duwamish, New York Bight, and Lake Ontario sites. This field data was collected under Work Unit 1B09 by Yale University. Work Unit 1B07 evaluated two two-dimensional finite element models (developed under Work Unit 1B05 of the DMRP by the University of California at Davis) for the long-term prediction of sediment transport in estuaries. Part II of the report discusses the modifications of the Tetra Tech models made by the U.S. Army Engineer Waterways Experiment Station and presents calibration results using field data from the Duwamish, New York Bight, and Lake Ontario disposal sites. Part III of the report discusses the factors in the long-term transport of sediment in estuaries and how they are handled by finite element models.

JBF Scientific Corporation, "Dredge Disposal Study, San Francisco Bay and Estuary. Appendix M--Dredging Technology", Sept. 1975, U.S. Army Corps of Engineers, San Francisco, California.

A study was conducted to investigate dredging technology and to advance the state of knowledge regarding the short-term fate of dredged materials dumped from barges or hopper dredges. Particular attention was given to application of the study findings to protection of the aquatic environment in the San Francisco Bay area. Field tests evaluated physical and chemical properties of dredged materials under many conditions. Parameters observed in characterizing the data included descent velocity, cloud size, impact velocity, horizontal velocity following impact, and settling patterns. Moisture content appeared to be the primary variable determining behavior of dumped materials. Engineering aspects of intertidal disposal for the purpose of marsh creation were evaluated. Other considerations in marsh building were discussed, including tidal inlet design, means of excavating in marsh areas, and movement of salt through dredged material. Certain aspects of land disposal were investigated with reference to conditions in the San Francisco Bay area. Treatment processes for removing contaminants and for hastening drying were discussed and evaluated. Productive uses of dredged material were also considered.

Johanson, E.E., Bowen, S.P. and Henry, G., "State-of-the-Art Survey and Evaluation of Open-Water Dredged Material Placement Methodology", Contract Report No. D-76-3, Apr. 1976, U.S. Army Engineer Waterways Experiment Station, Vicksburg, Mississippi.

The dumping of dredged material into a borrow pit is feasible using hopper dredges provided that an auxiliary navigation capability such as an instrumented buoy at the center of the borrow pit, or an electronic precision navigation system, is available. Borrow pit dumping is undertaken less easily with barges and tug boats because of the inability to control accurately the position of the barges. Borrow pits have to have several acres in size and have dimensions sufficient to allow radial spreading of several hundred feet. No estimate of the long-term fate of dredged material placed in a borrow pit can be made. The concept of covering polluted dredged material was not considered efficient unless undertaken immediately after disposal. It was concluded that field and laboratory studies should be initiated to identify and to quantify the physical mechanisms associated with open-water disposal operations, and that field demonstration and an evaluation of the horizontal pump-down concept should be developed.

Johnston, Jr., S.A., "Estuarine Dredge and Fill Activities: A Review of Impacts", Environmental Management, Vol. 5, No. 5, Sept. 1981, pp. 427-440.

The effects and impact of dredge and fill activities on estuaries are discussed with emphasis on biological and water quality impacts. Recommendations are made for improving and mitigating deleterious effects. The biological effects of turbidity may cause reduced visibility and reductions in the availability of food for fish. High levels of suspended solids can reduce oyster growth and may have toxic effects on various larvae. Dissolved oxygen concentrations are lower near some dredging and filling sites, and pH can be reduced. Siltation can have drastic effects, including the immediate removal of organisms through suffocation and the long-term elimination of many desirable species of flora and fauna. Resuspension of bottom materials can result in the release of nutrients as well as the possible release of toxicants. Where coastal wetlands cannot be circumvented during construction projects, bridging should be used rather than filling and embankment. The turbidity caused by dredging and filling can be reduced through the use of a turbidity diaper.

Landin, M.C., "A Selected Bibliography of the Life Requirements of Colonial Nesting Waterbirds and Their Relationship to Dredged Material Islands", Misc. Paper No. D-78-5, 1978, U.S. Army Engineer Waterways Experiment Station, Vicksburg, Mississippi.

An extensive bibliography is presented on the life requirements of colonial nesting waterbirds in the United States and their relationship to dredged material. An additional bibliography of 181 references pertaining to the vegetation and soils on dredged material islands and environmental impacts of dredged material deposition on waterbird habitats is also presented. Selected references from Canada, Europe, and Africa that pertain to related waterbirds, or those introduced to the United States, are also included. This report provides the reader access to little-known sources and gives localized data on waterbird species that are not otherwise readily available. References from the years 1844 to 1978 are included.

Laskowski-Hoke, R.A. and Prater, B.L., "Dredged Material Evaluations: Correlations Between Chemical and Biological Evaluation Procedures", Journal of the Water Pollution Control Federation, Vol. 53, No. 7, July 1981, pp. 1260-1262.

A bulk sediment-chemistry evaluation procedure (EPA Regions V and IX) and an elutriate test (U.S. Army Corps of Engineers) were compared with a bioassay technique as a means of evaluating disposal methods for dredged material. Forty sediment samples collected in 1977 from Lake Michigan were analyzed. The elutriate process required analysis of ammonia, COD, total P, total Kjeldahl N, nitrate, nitrite, chloride, sulfate, and 8 heavy metals. The sediment-chemistry technique relied on analysis of total Kjeldahl N, total P, COD, sieve size, oil, grease, cyanide, and 11 metals. Test organisms for the bioassay were Pimephales promelas Rafinesque, Hexagenia limbata Walsh, Lirceus fontinalis Rafinesque, and Daphnia magna Straus. Statistical analysis of results indicated that a combination of bulk sediment-chemistry and sediment bioassays would be the most efficient and ecologically sound approach to evaluating the effects of dredged materials on benthic communities.

Lehmann, E.J., "Dredging: Environmental and Biological Effects (Citations from the Engineering Index Data Base)", Dec. 1979, National Technical Information Service, U.S. Department of Commerce, Springfield, Virginia.

Worldwide research on dredging related to biology, nutrient composition, sedimentation, and water pollution is cited. Studies include dredging of harbors, estuaries, waterways, and for mineral recovery. The effects of the disposal of dredge spoil in containment areas, landfills, or oceans are cited, as is dredge spoil reuse in land reclamation. This bibliography contains 174 abstracts.

Maurer, D. et al., "Vertical Migration and Mortality of Benthos in Dredged Material, Part I: Mollusca", Marine Environmental Research, Vol. 4, No. 4, 1981, pp. 299-319.

Benthic invertebrates have many characteristics which make them prime candidates for burial studies in dredged material. A major concern in dredging and disposal projects is the effect of burial on the survival of benthic invertebrates. The purpose of this study was to determine the ability of estuarine benthos, and in particular three species of mollusks (Mercenaria mercenaria, Nucula proxima, and Illyanassa obsoleta), to migrate vertically in natural and exotic sediments and to determine the survival of benthos when exposed to particular amounts of simulated dredged material. Mortalities generally increased with increased sediment depth, with increased burial time, and with overlying sediments whose particle size distribution differed from that of the species' native sediment. Temperature affected mortalities and vertical migration, with both being greater in summer temperatures than in winter temperatures. It was concluded that vertical migration is a viable process which can significantly affect rehabilitation of a dredged disposal area. Under certain conditions, vertical migration should be considered, together with larvel settling and immigration from outside impacted areas, as a mechanism of recruiting a dredge-dump site.

Massoglia, M.F., "Dredging in Estuaries--A Guide for Review of Environmental Impact Statements", Report No. NSF/RA-770284, 1977, Oregon State University, Corvallis, Oregon.

The purpose of this symposium/workshop was to familiarize various user groups with the "Guidelines for Impact Assessment of Dredging in Estuaries" prepared by Oregon State University. The guidelines provide a methodology that is intended to ensure that the quality of the data contained in an Environmental Impact Statement (EIS) is adequate to comply with the intent of the National Environmental Policy Act of 1969 (NEPA). The guidelines, used in conjunction with this symposium/workshop proceedings report, are intended to assist the Corps of Engineers, the Environmental Protection Agency, the Maritime Administration, estuary managers, and individuals in the extremely important process of evaluating and preparing environmental impact statements for dredging in estuaries. The symposium/workshop consists of a series of presentations by the Oregon State University research team; four concurrent workshop sessions; presentations by individuals from public and private organizations involved in planning, decision-making, regulation, and evaluation of estuarine dredging; and a public session.

Morrison, R.D. and Yu, K.Y., "Impact of Dredged Material Disposal Upon Ground Water Quality", Ground Water, Vol. 19, No. 3, May/June 1981, pp. 265-270.

Ground water was monitored over a 2-yr period at three dredge disposal sites: Grandhaven, Michigan (fresh water); Mobile, Alabama; and Sayreville, New Jersey (both estuarine). Leachate samples collected from monitoring wells and vacuum/pressure lysimeters were analyzed for 27 water quality parameters. The Na, Cl, and K concentrations were high (maximum 4310, 8330, and 250 mg/l, respectively) in ground water sites where saline dredge material was deposited. Dilution is the major mechanism regulating transport of these ions. The Mg and Ca maxima were 720 and 380 mg/l, respectively, representing significant increases in water hardness over background levels. Mechanisms for transport of Mg and Ca are dilution, precipitation/dissolution, and ion exchange. Alkalinity ranged from undetectable to 1000 mg/l, and was controlled by dissolution of carbonates, weathering, and oxidation-reduction reactions. Total organic carbon was highly correlated with alkalinity. Maximum iron and manganese levels were 538 and 12 mg/l, respectively, both above the recommended EPA drinking water quality standards. Sulfides were generally below 20 mg/kg. Phosphate, Cd, Cu, Pb, Ni, Hg, and Zn occurred in concentrations too small to have significant effects on ground water quality. Major controlling factors for these ions are complexation, adsorption, and precipitation/dissolution.

National Marine Fisheries Service, "Physical, Chemical and Biological Effects of Dredging in the Thames River (CT) and Spoil Disposal at the New London (CT) Dumping Ground", Final Report, Apr. 1977, Division of Environmental Assessment, Highlands, New Jersey.

The impacts of dredging operations on suspended material transport in the lower Thames River estuary were confined to an area within 300 to 500 yards of the operating dredge and barge, produced an increase in total suspended load within the estuary that was small in comparison to that produced by typical aperiodic storm events, and caused no major alterations in mass transport within the estuary. Field surveys of the Thames River hydrography, phytoplankton, and trace metal concentrations in water, sediment, and shellfish suggested that effects of dredging on primary production were spatially and temporally limited. The highest concentrations of nickel, lead, cadmium and mercury in water samples were observed before or during dredging, while copper was highest after dredging but were generally higher up-river. Sediment levels of these five metals, plus zinc and organic carbon, increased in an up-river direction. Dredging-related changes in trace metal body burdens in shellfish were difficult to separate from normal seasonal variations. No gross pathology was detected in the shellfish. The physical oceanography of the disposal area showed that turbidity was higher in bottom waters than near the surface, and was not restricted to the vicinity of the spoil pile. Maximum transport was in the east/west direction with highest values occurring during the ebb tide.

Ocean Data Systems, Inc., "Handbook for Terrestrial Wildlife Habitat Development of Dredged Material", Technical Report No. D-78-37, July 1978, U.S. Army Engineer Waterways Experiment Station, Vicksurg, Mississippi.

This study of terrestrial wildlife habitat development on dredged material within the contiguous United States compiles existing published and unpublished data into a user-oriented handbook. A general list of 250 plant species with food and cover value for wildlife is indexed by life form and state; a synopsis is given for each of 100 plant species chosen from the general list on the basis of their importance to wildlife, ease of establishment, and geographic distribution. Each synopsis includes a description and discussion of habitat, soil requirements, establishment and maintenance, disease and insect problems, and wildlife value. A range map and illustrations are given along with appropriate miscellaneous comments. The handbook also outlines a suggested approach for developing terrestrial wildlife habitat on dredged material; discusses wildlife species inhabitating dredged material areas; and recommends techniques for propagation, establishment, and maintenance of plantings.

Pavlou, S.P. et al., "Release, Distribution, and Impacts of Polychlorinated Biphenyls (PCB) Induced by Dredged Material Disposal Activities at a Deep Water Estuarine Site", Proceedings of the 5th United States-Japan Experts Meeting on Management of Bottom Sediments Containing Toxic Substances, EPA-600/9-80-044, Sept. 1980, U.S. Environmental Protection Agency, Washington, D.C., pp. 129-174.

Short-term and long-term ecological impact studies were initiated at the time of open-water disposal of contaminated dredged material from the Duwamish River in Elliott Bay, Puget Sound, Washington. The experimental disposal site was located at a depth of 60 m in a marine estuary with generally weak circulation. Approximately 114,000 cubic meters of dredged material contaminated with PCB's were dumped at the site from split-hull barges. Replicate water, suspended particulate matter, and sediment samples were taken at intervals up to 9 months after disposal and investigated for PCB content. The PCB concentration in water and suspended matter generally declined over time. The sediments at the disposal site were slumping from the center of the site to the periphery during the monitoring period, and there was a slight but not significant reduction in PCB concentrations. Long-term (3-years post-disposal) PCB concentrations showed further declines, with the highest concentrations in the vicinity of the disposal site. Macrofauna, including marine worms, clams and crustaceans, was collected by grab sampler. Statistical analysis of the abundance figures suggest that some taxa may be more abundant within and close to the disposal site, and that stations in close proximity to the grid are more similar to each other than to more distant stations. Nonparametric Wilcoxon two-sample tests revealed significant differences in abundances for some taxa that were grouped as stations within the grid site versus more distant stations. All but one of the taxa exhibited greater abundances at the disposal site.

Peterson, S.A., "Dredging and Lake Restoration", EPA 440/5-79-001, 1979, U.S. Environmental Protection Agency, Corvallis Environmental Research Laboratory, Corvallis, Oregon, pp. 105-144.

Of the techniques being used for lake restoration, dredging is the only one that directly removes the products of degradation (sediment) from the lake. It deepens the lake, removes recyclable nutrients, and returns the sediment to the watershed. Environmental concerns associated with dredging include resuspension of bottom sediments, toxic substances, oxygen depletion, reduced primary production, temperature alteration, increased nutrient levels, benthic community alteration, and sediment disposal sites. Negative and positive aspects are discussed for various types of dredging equipment, including grab, bucket, clamshell, cutterhead, and specialized dredges such as mud cat, bucket wheel, and Japanese special purpose dredges. Results from dredging projects in several lakes are also discussed. Dredging costs are difficult to accurately determine because they depend on: (1) the types and quantity of sediment to be removed; (2) type of dredges used; (3) the nature of the operational environment; and (4) the geographic location and method of material disposal. Corps of Engineer average dredging costs are presented for the various U.S. geographical areas.

Raster, T.E. et al., "Development of Procedures for Selecting and Designing Reusable Dredged Material Disposal Sites", Technical Report No. D-78-22, June 1978, U.S. Army Engineer Waterways Experiment Station, Vicksburg, Mississippi.

This report presents a logical, step-by-step methodology for site selection and design of reusable dredged material disposal sites. The methodology is capable of handling anything from a single disposal site serving a single dredging location to an entire dredging program involving several dredging locations and disposal sites. Pertinent factors--legal, environmental, and technological--which influence selection of candidate disposal sites and determine their suitability as reusable and nonreusable sites are identified. Site design and operating recommendations are presented, along with a preliminary costing procedure to enable evaluation of alternative disposal options for each site and cost modifications of the entire dredging program. Numerical examples are provided to assist the reader in applying the procedures to his particular case. Although this report promotes reusable disposal sites, nonreusable sites of a nonconventional nature are also discussed for situations where reusable sites are inappropriate or economically unsound.

Slotta, L.S. et al., "An Examination of Some Physical and Biological Impacts of Dredging in Estuaries", Interim Progress Report to the National Science Foundation, Dec. 1974, Oregon State University, School of Oceanography, Corvallis, Oregon.

The major research achievements of the NSF-RANN dredging research group at Oregon State University for the first 10 months of its activity is reported. The research focuses on several major topics, including: (1) the effects of dredging on estuarine research can be used effectively by user groups, and (2) the development of concepts and techniques for monitoring impacts of dredging and other alterations to estuaries. The various kinds of impacts discussed in the study include: impacts of dredging and marine traffic; impacts of sediment turnover; impacts of sediment physical changes; impacts of turbidity; and impacts of release of toxins from sediments.

APPENDIX D

IMPACTS OF OTHER WATER RESOURCES PROJECTS

Ahmad, Y.J., "Irrigation in Arid and Semi-Arid Areas", 1982, United Nations Environment Programme, Nairobi, Kenya.

The impact of irrigation projects is not exclusively beneficial, and when assessing the viability of such a project the adverse consequences must also be taken into consideration. The purpose of this document is to highlight the most serious of these consequences, which can be most damaging to the fragile and delicately balanced ecosystems that characterize the arid and semi-arid regions, so that full account may be taken of the necessary preventive or remedial actions that must be incorporated into any well-planned irrigation scheme, if the benefits gained are not to prove illusory. In this document the semi-arid region is defined as any region having a dry season of three to four months, regardless of the total annual precipitation. Such an area would need supplementary or seasonal irrigation. The arid region is taken as one in which the usual dry season will last longer than four months, and sometimes the whole year. Such regions must be designated for perennial irrigation.

Bradt, P.T. and Wieland, G.E., III, "The Impact of Stream Reconstruction and a Gabion Installation on the Biology and Chemistry of a Trout Stream", Completion Report, Jan. 1978, Department of Biology, Lehigh University, Bethlehem, Pennsylvania.

The purpose of this study was to evaluate the effect of a gabion installation and stream reconstruction in a 2 km section of a rechanneled stream. The Bushkill Creek, supporting a naturally reproducing brown trout population in Northampton County, Pennsylvania, was sampled bi-weekly biologically, chemically and physically for 16 months. Prior to the sampling, stream reconstruction efforts included both a gabion (rock current deflectors) installation to narrow and deepen the stream bed, and tree and shrub planting to cover bare banks and provide eventual shade. The stream bed was open to sunlight and primary productivity, as evidenced by larger algae populations, increased in the rechanneled area. The following benthic macroinvertebrate parameters significantly increased also through the rechanneled area: density index, biomass, total numbers, and number of taxa. The following chemical parameters increased significantly through the rechanneled area: conductivity, dissolved oxygen, percent oxygen saturation and alkalinity. Orthophosphate decreased significantly and flow velocity increased significantly. Limestone springs contributed to the increase in conductivity and alkalinity. Increased photosynthesis and turbulence also contributed to the increase in conductivity and alkalinity, and to the increase in dissolved oxygen and oxygen saturation. The gabions deepened and narrowed the stream channel resulting in a cooler stream in the summer.

Birtles, A.B. and Brown, S.R.A., "Computer Prediction of the Changes in River Quality Regimes Following Large Scale Inter Basin Transfers", Proceedings: Baden Symposium on Modeling the Water Quality of the Hydrological Cycle, IIASA (Laxenburg, Austria) and IAHS (United Kingdom), IAHS-AISH Publication No. 125, Sept. 1978, Reading, England, pp. 288-298.

Of the alternative proposed water resource development schemes to meet future demands in the southeastern part of England, one contender is a strategy to transfer water from the River Severn to the River Thames at appropriate times. Severn flows would be augmented when necessary from surface storages in Wales. Simulation experiments have been performed to estimate the water quality effects of interbasin transfers on the recipient system. A range of nearly conservative river quality determinants has been investigated using river quality models. Examples of the results obtained are presented in a form which facilitates estimating (1) the size and cost of any treatment plant which may be necessary to make transferred water compatible with receiving waters, and (2) the adverse effects of transfers on organisms with known concentration tolerance levels. The production of long simulated concentration sequences is not in itself particularly useful, because the great volume of data generated is not readily susceptible to manual interpretation; there is a clear requirement for some kind of condensation which will allow the essential characteristics of the data to be presented in a concise form that enables conformity with suitable defined standards or criteria to be judged quickly and accurately.

Darnell, R.M., "Overview of Major Development Impacts on Wetlands", Proceedings of the National Wetland Protection Symposium, June 6-8, 1977, Reston, Virginia, Department of Oceanography, Texas A&M University, College Station, Texas.

The primary categories of environmental impact of construction activities in wetlands, in order of importance, include habitat modification and loss, increase in suspended sediments, bottom sedimentation, and modification of the water flow regime. A need exists for a solid information base, synthesized into its critical elements and key processes, to provide the basis for informed action. The link between the environmental scientist and the environmental administrator must be strengthened to achieve the best, least-cost solutions to the complex wetland protection problems facing the nation.

deGroot, S.J., "The Potential Environmental Impacts of Marine Gravel Extraction in the North Sea", Ocean Management, Vol. 5, 1979, pp. 233-249.

The rapid increase in the mining of marine gravels in the North Sea offers a serious threat to the marine environment and especially to the herring populations of the southern North Sea and Channel. This paper discusses the extraction methods for gravel and the usefulness of applying programmed dredging to minimize the effects on the sea-bed. The impact of gravel extraction on the sea-bed and on the herring and sand-eel fisheries is discussed. It is estimated that the resources of marine gravel in the southern North Sea will be depleted within 50 years, whereas the fisheries will be practiced for many hundreds of years to come. Therefore, a careful evaluation, on a European level, is needed of the relative short-term benefits for the marine gravel industry and the long-term interests of the fisheries.

deGroot, S.J., "An Assessment of the Potential Environmental Impact of Large-Scale Sand-Dredging for the Building of Artificial Islands in the North Sea", Ocean Management, Vol. 5, No. 3, Oct. 1979, pp. 211-232.

The building of artificial islands is contemplated in the coastal waters of the Netherlands. The potential environmental impact of large-scale sand-dredging is evaluated. The most suitable mining methods, the effect on water quality turbidity, the disturbance of the balance between bottom sediment and water column in relation to heavy metals, nutrients, and PCB's are discussed, as well as the effect of dredging upon plankton, bottom fauna, fishes and on larvae and young stages of sea animals in general. The recovery time for the bottom fauna is estimated to be about three years. The consequences for the fisheries are mentioned, as the Netherlands coastal zone plays a very important role in the life of the North Sea species. This is related to the Dutch shrimp, plaice and sole fishery. An estimation of the yearly damage to the fisheries during the building phase is made. The construction time for the island would be about eight years.

Diamant, B.Z., "Environmental Repercussions of Irrigation Development in Hot Climates", Environmental Conservation, Vol. 7, No. 1, Spring 1980, pp. 53-58.

Irrigation projects intended to improve food supplies and the quality of life in third world countries have caused severe problems: spread of waterborne diseases and social upheavals among the peoples resettled by flooding impoundments. Malaria and/or schistosomiasis have appeared or increased in the areas near Lake Akosombo, Ghana; Lake Nasser, Egypt; the Indus River Basin, Pakistan; and the Helmand River project, Afghanistan. Environmental health aspects should be included in the planning, construction, and operation of large-scale irrigation projects. Vector control measures, which break the three-part cycle (sick carrier, transmitting vector, and recipient), can be accomplished by proper engineering to produce a habitat unfit for snails and mosquitos. Chemical, biological, and genetic control are not as effective or are still in the research stage. Some suggestions for destroying habitats for disease vectors are: controlling vegetation, straightening banks, producing frequent changes in water levels, reducing evaporation, eliminating wastewater influx, installing snail screens, using closed channels, sprinkling irrigation water, and choosing crops which do not need flooding. The complicated social and economic changes produced by these projects require a prior sociological survey. The population should be consulted before the fact on allocation of new farming plots, methods of cultivation, types of crops, and housing. Operation and maintenance of the water projects should be regularly surveyed by teams of ecologists, environmental engineers, entomologists, and health workers.

Elgershuizen, J.H., "Some Environmental Impacts of a Storm Surge Barrier", Marine Pollution Bulletin, Vol. 12, No. 8, Aug. 1981, pp. 265-271.

In 1985 the newly designed storm surge barrier (SSB) will be constructed across the mouth of the eastern Scheldt in the Netherlands to seal off most of the large inlets in the southwestern part of the land. The project was launched in answer to drastic flood losses in 1953. A complete closure of most of the inlets and also the Eastern Scheldt was envisaged, except for the New Waterway and the Western Scheldt, which are important to shipping. The original study had called for the construction of a dam, but this was changed in 1976 to the construction of a storm surge barrier to close the Eastern Scheldt estuary, and thus buy safety for the polders and preserve the ecological system of the Eastern Scheldt. The SSB will cause the formation of a new, unnatural ecosystem. To make the appropriate decisions to manage this area in as natural a way as possible, a sound knowledge of the Eastern Scheldt ecology is urgently needed. From the evaluation studies thus far conducted it appears that one closure per year for 12 hr would not have any significant effect, but if closure times are long, or if frequent complete closures occur, problems may arise. Thus it is suggested that, in place of complete closures, partly closing the barrier may avoid a concentrated attack of flood waters on sand and mud flats, dikes and salt marshes. This may result in the maintenance of a reduced tide in the Eastern Scheldt, and appears to offer a good compromise between safety and ecology.

Elkington, J.B., "The Impact of Development Projects on Estuarine and Other Wetland Ecosystems", Environmental Conservation, Vol. 4, No. 2, Summer 1977, pp. 135-144.

Despite various campaigns to publicize the value and plight of wetlands, pressures for development and economic growth are often such that ecological issues are obscured. The biology of the estuarine environment is discussed as a first step to evaluation of the impact of pollution and land use policies. The impacts of urban, agricultural and recreational use of wetland areas are reviewed with particular reference to wetland ecosystems of Third World countries. The importance of experiments in evaluating alternative land use options (such as intensive fish farming) is stressed, and a simple model for ecologically sound development is proposed which could apply to Lake Manzala in Egypt.

Environmental Resources Limited, "Environmental Health Impact Assessment of Irrigated Agricultural Development Projects", Dec. 1983, World Health Organization Regional Office for Europe, Copenhagen, Denmark.

This report addresses environmental health problems associated with irrigated agricultural development projects. This report provides outline guidance, based on past experience, on how environmental health impact assessment may be carried out. The report is organized into four main chapters, with additional background information provided in the form of annexes to the report. In Chapter 2 guidance is provided on identification of impacts on the environment, Chapter 3 focuses on prediction, and in particular on prediction of environmental health impacts, Chapter 4 focuses on mitigation of health impacts, while Chapter 5 discusses the organization and presentation of information for the decision maker, that is, the individual or agency who must take this information into account in deciding whether the development should proceed. The guidance developed in these chapters is based on an examination of numerous guidelines on EIA and on experience of application of the procedures and approaches recommended therein. A review of these guidelines is also presented.

Gysi, M., "Energy, Environmental, and Economic Implications of Some Recent Alberta Water Resources Projects", Water Resources Bulletin, Vol. 16, No. 4, Aug. 1980, pp. 676-680.

Four water projects in Alberta are discussed as examples of works which have been planned or built according to political decisions rather than economic analysis. This situation has led to energy, environmental, and economic losses in some cases. The Innisfail Regional Water Supply Plant pumps water uphill about 200 meters to the consuming communities.

Planners overlooked the fact that the Bow River, uphill from these communities, could adequately supply water by gravity flow. In addition, the pipe sizes and pump capacities at Innisfail appear to be inadequate for future expansion. The City of Calgary recently expanded its water supply facilities at a cost of $10 million. This would not have been required if the residential area had been metered to cut down the daily demand of 800 liters per capita. The Red Deer River Flow Regulation Dam benefit-cost ratio was not revealed to the public. The proponents were interested in boosts to industrialization and employment. In fact, a petrochemical plant was included among the benefits of the dam, even though the plant construction had been approved long before the dam was considered. Industrialization in this area is not totally dependent on the presence or absence of this dam. Benefits in constructing the dam are increased winter flow and raising of dissolved oxygen levels downstream. Flood control benefits are negligible although highly publicized. The Oldman River Flow Regulation Project, not yet approved, was the subject of two benefit-cost analyses. The Phase I ratio was 1.6:1. However, if questionable secondary benefits are subtracted, the ratio is 0.6:1. The Phase II study used inflated secondary benefits, a 5 percent discount rate, and some unlikely assumptions. Sensitivity analysis using a correct methodology would produce benefit/cost ratios less than 1.

Huber, W.C. and Brezonik, P.L., "Water Budget and Projected Water Quality and Proposed Man-Made Lakes Near Estuaries in the Marco Island Area, Florida", Proceedings of the National Symposium on Fresh Water Inflow to Estuaries, FWS/OBS-81-04, Vol. I, Oct. 1981, U.S. Fish and Wildlife Service, Washington, D.C., pp. 241-251.

The proposed Marco Island development plan calls for the excavation of a large group of interconnected lakes near Marco Island, Florida. The quality of the proposed lakes is of considerable importance, both to the riparian owners and to the nearby estuarine areas that will receive surface discharges. The deep lakes will be stratified due to the influx of hypersaline ground water below a depth of 2.0 m. Density differences are so great across the chemocline that the possibility of overturn is nil. The lakes receive water from surface runoff from the various land uses; from inflow, from regional ground water flow in the shallow fresh water layer, and from direct rainfall. Water is lost by surface runoff, ground water outflows, and evaporation. Annual precipitation in the area is about 50 inches. Residence times of 0.42 and 0.65 years have been calculated for two units of the development. On the basis of phosphorus loading rates, the lakes are expected to be mesotrophic with fair to good water quality. On the whole, predicted water quality is good with Secchi disk transparencies on the order of 1.2 to 1.5 m, and total nitrogen of about 1.3 mg/l. On the basis of nutrient loads, the urban development is expected to have little impact on the estuaries.

Livingstone, I. and Hazlewood, A., "The Analysis of Risk in Irrigation Projects in Developing Countries", Oxford Bulletin of Economics and Statistics, Vol. 41, No. 1, Feb. 1979, pp. 21-35.

Fluctuating water availability is an issue which arises in planning irrigation schemes in developing countries due to differences in rainfall. A recognized rule-of-thumb in engineering practice recommends that the volume of water assumed for planning purposes should be available in four years out of five. Only in one year out of five will water be insufficient for installed capacity. Such a rule provides an optimal solution only in particular circumstances, and its general application is inconsistent with standard net present value criterion for evaluating alternative investment projects. These rules should be determined, not on technical, but on economic and social grounds. A large mechanized farm may not accept frequent serious water deficiency, involving large financial losses. However, it is different for village irrigation with peasant production. If peasant farmers are poor with low incomes and food supply, they attach high values to extra output, even if available only in better years. Denying cultivators the full benefits of high river flow years so as to reduce risks of doing relatively worse in other years is not important. If crop storing is possible, a larger area could produce extra food in good years to be used as a reserve against subsequent crop failure. Therefore, the optimum solution shows that peasant farmers may follow a less cautious policy than a large mechanized farm.

Micklin, P.P., "International Environmental Implications of Soviet Development of the Volga River", Human Ecology, Vol. 5, No. 2, June 1977, pp. 113-135.

Hydrological alteration of major rivers may cause international environmental disturbances. The development of the Volga River by the Soviet Union is used to illustrate the problem. The most immediate victim of Volga development is Iran, which borders on the Caspian Sea. The Volga, a tributary of the sea, has had its natural flow changed by the development. This change has damaged the Caspian and thereby Iran. Proposed plans to alleviate problems in the Caspian Sea would clearly have global effects. The international problems associated with Soviet development of the Volga are not unique, and concern all governments. An American example is the use of the Colorado River by the United States to the detriment of Mexico. Suggested solutions include: (1) an "international environmental impact statement" modeled on the provisions of the United States National Environmental Policy Act; (2) an International Environmental Agency; and (3) involvement of United Nation's organizations.

Mulvihill, E.L. et al., "Biological Impacts of Minor Shoreline Structures on the Coastal Environment: State-of-the-Art Review, Volume I", FWS/OBS-77-51, Mar. 1980, U.S. Fish and Wildlife Service, Washington, D.C.

Information from 555 sources located in an information search was used to develop a computer data base for the analysis of the biological impacts of minor shoreline structures. Structures included were: breakwaters, jetties, groins, bulkheads, revetments, ramps, piers and other support structures, buoys and floating platforms, small craft harbors, bridges, and causeways. Data were compiled by type of structure and by coastal region including the following: structure functions; site characteristics; geographic prevalence; engineering, socio-economic, and biological placement constraints; construction materials; expected life span; environmental conditions; methodology of impact studies; physical and biological impacts; and alternatives. Results show that structure impacts on the environment are site-specific. Fourteen case studies are included. Small boat harbors, bridges and causeways, bulkheads, breakwaters, and jetties were found to have the most potential for coastal environment impact. Revetments, groins, and ramps have moderate impact potential, while buoys and floating platforms, piers, and other support structures have low impact potentials. Little information relative to the potential impacts of bridges, causeways, and small boat harbors was identified. Also, very little information on the quantitative impacts of specific structures was located.

National Oceanic and Atmospheric Administration, "Coastal Facility Guidelines: A Methodology for Development with Environmental Case Studies on Marinas and Power Plants", Working Paper, Aug. 1976, Rockville, Maryland.

This report provides state coastal zone management (CZM) agencies with information and recommendations for developing guidelines for facility development in the coastal zone. Section A of the report presents a methodology for identifying and initiating implementation procedures for management recommendations for specific facility types. Sections B and C apply the methodology to marinas and power plants in the states of Florida and Maryland, respectively. The two case studies from Florida and Maryland serve the dual purpose of (1) providing a useful CZM reference source on environmental mitigation techniques and relevant Federal authorities for the two facility types; and (2) further clarifying the format, intended information content, and applications envisioned in the methodology.

Pollard, N., "The Gezira Scheme--A Study in Failure", Ecologist, Vol. 11, No. 1, Jan.-Feb. 1981, pp. 21-31.

Water resources were developed for irrigation of cotton in the Gezira region, Sudan, starting with the construction of the Sennar dam and canals in 1914. The project has increased the incidence of waterborne disease and disrupted the cultural life of the population without producing much economic gain. Schistosomiasis, malaria, and yellow fever became major problems. Byssinosis affected many cotton gin workers. Simple irrigation has been in use here since 3000 B.C. Before British colonial intervention the population lived in balance with the environment. The family unit and the Muslim religion were the basis of cultural stability. Cotton had been grown, spun, and woven in family units. This practice declined as cotton was sent directly to gins. Land was divided into small plots of 12 to 16 hectares for working by tenant families. This was not successful because of burdensome government regulations on crop allocation and water management. Proceeds from the cotton crop rarely reach 100 pounds per tenant per year. Although the colonialists have left the area, the Gezira people have a legacy, a technology unsuitable for the cultural environment.

Ryner, P.C., "Chicago Lakefront Demonstration Project. Environmental Impact Handbook", 1978, Illinois Coastal Zone Management Program, Chicago, Illinois.

This handbook has been designed to serve as a guide in assessing the environmental effects of projects developed in accordance with the Lakefront Plan of Chicago. This handbook assumes that lakefront development has been identified by the City as being a potentially desirable activity, and that there is a fairly clear idea as to what public objectives such development would hope to meet. It presents a time and cost-effective approach to environmental analysis that can improve the environmental soundness of the planning and design process; facilitate the granting of city and state permits; and lead to the submission of clear, concise, and technically adequate environmental impact assessments to the U.S. Army Corps of Engineers as part of their required permit application process.

Shabman, L. and Bertelson, M.K., "The Use of Development Value Estimates for Coastal Wetland Permit Decisions", Land Economics, Vol. 55, No. 2, May 1979, pp. 213-222.

Readily available courthouse records are the data source for an inexpensive procedure for estimating development values of wetlands; the procedure provides economic information about preservation and development uses of marshland for granting of wetland alteration permits. A hedonic price equation, which includes a variable to measure the level of waterfront amenity created from filled coastal marsh, predicts the value derived from filling wetlands for residential development. The net value for development, which is calculated by subtracting development cost, can be compared to the value of natural wetland. The equation was tested using standard regression techniques and data for land parcel sales in Virginia Beach, Virginia. Neighborhoods selected for study were as socio-economically homogeneous as possible to reduce the number of variables. The transfer price of a parcel, which represents the present value of the flow of annual returns, was used instead of prices to account for changes over time. Transfer prices are treated as a function of: (1) location factors; (2) historical factors; (3) improvements; and (4) site amenities. A hypothetical situation applied to the Virginia Beach example results in a net development benefit of $8,232/acre. This value must then be compared to the more elusive value of wetland preservation, which is formatted in terms of rising marginal value, irreversibility, and uncertainty, to determine if a permit should be granted.

Smies, M. and Huiskes, A.H., "Holland's Eastern Scheldt Estuary Barrier Scheme: Some Ecological Considerations", Ambio, Vol. 10, No. 4, 1981, pp. 158-165.

The possible environmental consequences of construction and use of the Eastern Scheldt Barrier which is being built to protect Zeeland, the Netherlands, from flood waters, are discussed. The project includes the construction of a storm-surge barrier across the mouth of the estuary along with two secondary dams to separate the tidal basin of the Eastern Scheldt from the brackish-fresh water of the Rhine Scheldt shipping route. Turbulence and turbidity will decrease with decreasing mean tidal current velocities, and mean water residence time will increase. Chlorinity and nutrient concentrations depend on water residence time, so these values will be changing. There will probably be increases in net particulate carbon, which may result in increased sedimentation and resulting changes in depth. None of the effects of the barrier will be extreme. If the barrier should be completely closed for over a week, for example, to protect the estuaries from a tanker spill, problems may occur. The estuary may become anoxic, and severe ecological damage could result from lack of oxygen and flooding of the intertidal zone.

Smil, V., "China's Agro-Ecosystem", Agro-Ecosystems, Vol. 7, No. 1, 1981, pp. 27-46.

Regional lack of moisture is a critical environmental constraint to China's agro-ecosystem. The climate ranges from subtropical in the southeast to arid in the north and northwest. Mean annual precipitation varies from 577 mm to 1720 mm. In some areas evaporation exceeds precipitation by up to 4 times. Water resources development and control have been an integral part of China's agriculture for centuries. As of 1979 the irrigation system included 80,000 reservoirs, 5000 canal networks, and 2 million power-operated wells. Several large multipurpose water projects are under construction. Aquaculture has deteriorated in recent years because of pollution, dam building, and land reclamation. The capacity of agricultural pumps for drainage, and irrigation was 47 GW in 1978, a large increase from 0.13 GW in 1950. China's agricultural system faces many problems. The mechanical irrigation systems are not efficient. Leaks and seepage losses are 50-60 percent, power shortages idle the pumps for long periods, and sprinkler irrigation is in an early stage of development. Soil erosion is a wide-spread and serious environmental problem, caused by land mismanagement. A proposed water transfer from the Yangzi basin to the chronically dry north also poses environmental hazards.

Takahasi, Y., "Changes and Processes of Water Resources Development and Flood Control in Post-Second World War Japan", Water Supply and Management, Vol. 6, No. 5, 1982, pp. 375-386.

Water demand in Japan has increased dramatically since the end of World War II. Development of water resources to meet the increased household and industrial consumption, and to control floods, has resulted in significant environmental changes. Projects included construction of multipurpose reservoirs, saline water barriers in river mouths, river levee systems, flood control dams, and river works. At present 2060 dams exist, compared with 781 in 1930. Urbanization and industrialization have also had a great impact on flooding and water quality. Some of the negative impacts are: larger floods with increased peak discharges and flood wave velocities, more urban property vulnerable to flood damage, increased sedimentation rate in reservoirs, scouring of river beds near bridges and weirs, coastal erosion, river water quality deterioration (eutrophication and increased turbidity), damage to fisheries, changes in scenery, impairment of recreation, disappearance of brooks, and displacement of people by filling reservoirs.

Tucker, J.B., "Schistosomiasis and Water Projects: Breaking the Link", Environment, Vol. 25, No. 7, Sept. 1983, pp. 17-20.

Schistosomiasis, after malaria, is now the world's most widespread, serious infectious disease. Many water resource development projects in tropical areas have had the unintended consequence of markedly increasing the prevalence of schistosomiasis in the local population. Water contact is the most critical variable in the transmission of schistosomiasis. Irrigation of formerly arid regions creates additional habitats for the snail vectors beyond those already present in ponds and rivers. Defects in the design or engineering of water projects may also provide new habitats for the snail hosts. The water projects have provided new opportunities for villagers to come into contact with the infested water, particularly when the projects are located close to villages in which no alternative water sources are available. The advent of perennial irrigation, in which the irrigation canals are in use year-round, has facilitated the multiplication of the snail vectors of schistosomiasis. Perennial irrigation has removed natural checks on the snail population, enabling them to multiply out of control. Once snails have been introduced into an irrigation system it is impossible to eradicate them completely. If irrigation projects are supplied with efficient drainage, good water management and regular maintenance, it is possible to avoid an increase in the incidence of schistosomiasis.

Vendrov, S.L., "Interaction of Large Hydraulic Engineering Systems with the Environment", Hydrotechnical Construction, No. 2, Feb. 1980, pp. 175-181.

The history and present status of environmental protection during water resources development projects in the USSR are discussed. Present policies are based on party directives, the USSR constitution, and the constitution of union republics. Decisions affecting the environment are made throughout an entire project, from predesign through final construction. In several projects design plans were revised or projects dropped as changes in the natural and economic setting and goals for conservation changed. In general, the higher the level of development and installed power per worker, the more stringent the environmental protection requirements. As development increases, environmental protection becomes increasingly important. No land is so unimportant that it can be turned into a wasteland. There are territorial differences in policy, such as use of pipelines and underground structures rather than open channels for water transport in heavily populated and industrialized regions. The preliminary forecast of environmental impact uses file and literature sources and takes about a year in the predesign stage. The main forecast involves field and laboratory studies, mathematical modeling, and theoretical investigations over several years. These findings are successively refined as necessary. It is emphasized that forecasts must be individualized to the project. A forecast should include viewpoints of a variety of specialists and organizations, the effects of a project (such as interbasin transfer) on all the geographic zones, substantiation that adverse effects will not be irreversible, and unconditional provision of the medical-sanitary aspects of water.

Watling, L., "Artificial Islands: Information Needs and Impact Criteria", Marine Pollution Bulletin, Vol. 6, No. 9, Sept. 1975, pp. 139-141.

This is a review of the information needed before realistic forecasts can be made of the effects of new developments on the marine environment. The review resulted from the need to assess the impact of an artificial island off the east coast of the United States. Man-made structures in the sea, ranging from oil platforms to large anchored floating structures and reclaimed land in shallower water, are proliferating around the world with more foreseen in the future. It will be important to forecast the effect they have on the surrounding sea areas and marine resources as accurately as possible.

Watling, L., Pembroke, A. and Lind, H., "Environmental Assessment", Final Technical Report No. NSF-RE-E-054A, 1975, College of Marine Studies, University of Delaware, Lewes, Delaware, pp. 294-431.

The impact of the construction and operation of an artificial island on the marine ecosystem of the northeastern United States was examined. While this could not be done directly, this study allowed documentation of what is presently known and what needs to be known about the components of the system and the reaction of these components to pollutant additives so far as that had been determined; and finally, to develop criteria and recommendations to govern this kind of activity such that both the ecological system and the economic system are maintained in a healthy state. Some biologically important aspects of the biota from the region from Maine to Virginia are pointed out. It is not possible to offer a truly comprehensive view of this region due to the fact that levels of knowledge for the different biotic groups vary widely, and also, to the fact that the biota of the region is composed of several thousands of poorly documented species.

Witten, A.L. and Bulkley, R.V., "A Study of the Effects of Stream Channelization and Bank Stabilization on Warmwater Sport Fish in Iowa. Subproject No. 2. A Study of the Impact of Selected Bank Stabilization Structures on Game Fish and Associated Organisms", Report No. 76/12, May 1975, Iowa Cooperative Fishery Research Unit, Ames, Iowa.

Four types of bank stabilization structures (revetments, retards, permeable jetties, and impermeable jetties) were studied to determine their impact upon game fish habitat. Permeable jetties and retards deepened the channel near the structures 7 to 110 percent greater than the maximum depth in control sections. No other significant differences in either physical parameters or in mean body length or abundance of game fish were found between structured and nonstructured stream sections. Unusually high water during the sampling period may have allowed fish to remain dispersed throughout the streams rather than concentrating in the deep pools at or near stabilization structures. Rock revetments and impermeable jetties fostered the growth of some invertebrates, primarily mayflies and caddisflies. Revetments, which presented the most rock surface for invertebrate colonization, had the greatest impact on invertebrate abundance. A long rock jetty, extending far enough into the stream to produce a scour hole, would combine most of the advantages noted in the structures studied. For habitat improvement, rock was superior to steel as a construction material, and structures which cause the formation of scour holes superior to those that do not deepen the stream.

Yiqui, C., "Environmental Impact Assessment of China's Water Transfer Project", Water Supply and Management, Vol. 5, No. 3, 1981, pp. 253-260.

China's south-to-north water project is a broad concept employing numerous pipelines for diverting water of the Yangtze River in the south to the north of the Yellow River. In 1975 the North China Plain suffered an acute water shortage, and the eastern line was proposed by the relevant agencies in 1976 as the first phase of the plan to be established. Since the water supply is interrelated with the regional economy and agricultural production, these factors should be considered conjunctively. An environmental impact assessment must be performed as well; it has two major purposes. The first is to assess the impacts of the proposed projects on the environment, and the second is to select alternatives which might achieve the same goal as the proposed project, while minimizing any harmful effects. The material and technical data presently available for such assessments is quite limited. Thus the alternative to the eastern line, the middle eastern line, is being left in its preliminary form. However, the general trend is clear: by taking the middle eastern line as the first phase project for diverting water of the Yangtze River to the north of the Yellow River, both capital cost and annual operating cost can be economized, as compared with the eastern line alone. The cost/benefit ratio will be much lower than that of the middle line alone. Most of the environmental problems inherent in the eastern line can be avoided. The middle eastern line will be able to accommodate changes in the future.

APPENDIX E

IMPACTS OF NONPOINT SOURCES OF POLLUTANTS ON WATER ENVIRONMENT

Bailey, G.W. and Nicholson, H.P., "Predicting and Simulating Pesticide Transport from Agricultural Land: Mathematical Model Development and Testing", Symposium on Environmental Transport and Transformation of Pesticides, October, 1976, Tbilis, USSR, EPA-600/9-78-003, Feb. 1978, Environmental Research Laboratory, U.S. Environmental Protection Agency, Athens, Georgia, pp. 30-37.

The transport of pesticides from agricultural lands and other compartments of the environment has generated public apprehension concerning the fate and effects of these compounds. Legislative mandates require guidelines to be developed covering pesticide use in order to prevent or minimize water pollution resulting from pesticide transport from agricultural land. Computer simulation models of the dynamic multiple rainfall-event type are being developed and refined to describe and predict quantitatively transport of pesticides from soil as a function of agricultural management practices, watershed characteristics, climatic factors, and properties of soils and pesticides. This paper discusses the steps involved in development, testing and verification of such models.

Brookman, G.T. et al., "Technical Manual for the Measurement and Modelling of Non-Point Sources at an Industrial Site on a River", EPA 600/7-79-049, Feb. 1979, Industrial Environmental Research Laboratory, Research Triangle Park, North Carolina.

This manual provides a guide for the implementation of a measurement and modelling program for nonpoint sources at an industrial site on a river. Criteria for developing a field survey program and model selection are provided, along with program costs and manpower requirements. A sample list of equipment and computer costs is also provided. The development of a field survey includes sample site selection, selection of parameters to be measured, number and frequency of samples, collection methods, analytical methods, and data reduction and analysis. Included in the modeling section is a description of the SSWMM-RECEIV-II model which has been adapted to a coal-fired utility site. Application of the outlined procedures to the measurement of nonpoint sources from a coal-fired utility is also presented.

Burns, R.G., "An Improved Sediment Delivery Model for Piedmont Forests", Technical Completion Report No. ERC 03-79, June 1979, Environmental Resources Center, Georgia Institute of Technology, Atlanta, Georgia.

Two similar watersheds in the pine production area of central Georgia piedmont were studied to develop an improved sediment delivery model for use as a management tool for forest harvesting and site preparation. Eight watershed segments were instrumented in the treated basin and commercially clearcut and then prepared in two stages with drum roller choppers. Thirty-one sets of data were collected for rainfall, runoff, and fine sediment delivery. Other data collected include the occurrence, location, and condition of all bare soil in each area during each phase of harvesting and site preparation. The model, based in part on the "universal soil loss equation", has three components: energy, site and erosion hazard. Tested as indices of the energy component were the rainfall erosivity factor, R, gross precipitation, Pg, and the product of runoff volume and peak discharge rate, Qvp. Soil erodibility, K, watershed segment slope length, L, and gully density, G, were tested as site component indices. For the erosion hazard index, W, and its approximation, Wa, were used. Several model combinations were tested and all were superior to previous models. Use of the W hazard index indicates that minimal bare soil and maximum distance to stream channels are important management concerns for forest harvest practices.

Cluis, D.A., Couillard, D. and Potvin, L., "A Square Grid Transport Model Relating Land Use Exports to Nutrient Loads in Rivers", Water Resources Research, Vol. 15, No. 3, June 1979, pp. 630-636.

This transport model relates a nutrient-oriented land use data bank to mass discharges of total nitrogen and total phosphorus on an annual or seasonal basis. This bank is based upon already available Canadian statistics distributed upon drainage units of the Universal Transverse Mercator square grid system; it contains information on the intensities and spatial distributions of the produced loads for both types of sources (direct and diffuse). Stable sets of coefficients were obtained for each season, reflecting the hydrological regime and differences in mobility of phosphorus and nitrogen. Once it is calibrated, the model can be used to assess the relative contributions of each type of land use and to predict the effect of impoundments or changes in land use.

Davis, Jr., H.H. and Donigian, Jr., A.S., "Simulating Nutrient Movement and Transformations with the Arm Model", Transactions of the American Society of Agricultural Engineers, Vol. 22, No. 5, Sept.-Oct. 1979, pp. 1081-1086.

The plant nutrient section of the Agricultural Runoff Management (ARM) Model was developed for the EPA as a tool for predicting the amounts of plant nutrients (N and P) removed from agricultural lands by precipitation-induced runoff and erosion. First-order kinetics were used to model the transformations, plant uptake, and sorption of soil nitrogen and phosphorus. The influence of soil temperatures on the first-order rates was calculated with a modified Arrhenius equation. Initial testing was performed with data from two small (less than 1.5 ha) watersheds in Michigan and Georgia. Initial results showed that the model was able to represent soil nutrient data reasonably well. Monthly simulation results of N and P in the runoff were satisfactory; however, storm event simulation of soluble nutrients indicated further study of solution-based transport is needed.

Feller, M.C., "Effects of Clearcutting and Slash-Burning on Stream Temperature in Southwestern British Columbia", Water Resources Bulletin, Vol. 17, No. 5, Oct. 1981, pp. 863-867.

The effects of clearcutting and clearcutting plus slash-burning on stream water temperatures were studied in paired forest watersheds in British Columbia. The predominant trees were western hemlock, western red cedar, and Douglas fir. The streams were designated A (61% clearcut), B (19% clearcut and slash-burned), and C (undisturbed-control). The harvested watersheds were planted with Douglas fir seedlings. Summer water temperatures and daily temperature fluctuations were increased by both treatments. However, after 6 to 7 years the temperature of the clearcut stream returned to its pretreatment state, whereas the clearcut plus slash-burned stream showed no improvement. Although the control stream temperature never exceeded 17°C, the disturbed streams registered maximum temperatures of 21.8 and 20.3°C for A and B, respectively, and exceeded 19°C at other times. Winter temperatures recovered normal values after only 2 years. Clearcutting with slash-burning had greater effects than did clearcutting alone.

Fowler, J.M. and Heady, E.O., "Suspended Sediment Production Potential on Undisturbed Forest Land", Journal of Soil and Water Conservation, Vol. 36, No. 1, Jan.-Feb. 1981, pp. 47-50.

Estimations of suspended sediment production rates on undisturbed forest land are reviewed. The 614 million acres of undisturbed forest land in the United States annually contributes more than 100 million tons of suspended sediments to waterways, and these rates can be used for comparisons with rates on silvicultural lands. Some of the basic estimates involved in building a forest model which can be linked with the Iowa State University agricultural model include: parent soil material, drainage area, rainfall, runoff, slope, and elevation. Published research results for watersheds with over 95 percent of the area in undisturbed forest were used. Selected equations and statistics are presented by region. Nonpoint pollution from undisturbed forest land was found to vary widely, ranging from 0.001 to 0.009 tons per acre per year in scattered producing areas, to as much as 3.3 tons per acre per year on isolated forests in the southern and Pacific Coast areas. When results of sedimentation from forest areas are combined with results from agricultural lands, the Northwest and Southeast were found to have the heaviest suspended sedimentation rates in the contiguous United States.

Fusillo, T.V., "Impact of Suburban Residential Development on Water Resources in the Area of Winslow Township, Camden County, New Jersey", Water Resources Investigation 81-27, 1981, U.S. Geological Survey, Trenton, New Jersey.

Changes in land use as a result of large-scale residential development can have significant impacts on water resources. Data on the quality and quantity of surface water and ground water in the vicinity of the Winslow Crossing residential development, in Winslow Township, New Jersey, were collected from 1972 to 1979. Pumpage for water supply from the Cohansey Sand averaged 0.48 million gallons per day during 1978, and had little effect on water levels in the aquifer. Water quality was variable in the observation wells sampled. High levels of dissolved solids, nitrate-nitrogen, and phosphorus were found in the shallow ground water surrounding the effluent infiltration ponds of a wastewater treatment plant. A treatment process change in 1974 reduced nitrate-nitrogen levels. The development of 14 percent of a 1.64-square-mile drainage area resulted in an increase in the peak discharge of a 60-minute unit hydrograph from approximately 150 cubic feet per second to 270 cubic feet per second. Installation of a stormwater retention basin reduced this peak discharge to 220 cubic feet per second. Streams draining two highly developed drainage areas had significantly higher levels of calcium, magnesium, bicarbonate, and pH than streams draining less developed areas. Winslow Crossing's development had only a slight effect on Great Egg Harbor River in comparison with sources of contamination upstream from the study area.

Gaynor, J.D., "Phosphorus Loading Associated with Housing in a Rural Watershed", Journal of Great Lakes Research, Vol. 5, No. 2, 1979, pp. 124-130.

Phosphorus and biological indicator organisms were measured in a rural watershed to quantify housing and agricultural inputs to the drainage system. Total-P and orthophosphate-P (PO4-P) were significantly higher downstream from intensive housing areas than in areas with lower housing density. Housing areas discharged 40% particulate-P and 53% PO4-P, while rural areas discharged 62 and 29%, respectively. On the average, rural housing contributed 2 g P/house/day, which comprised 27% of the PO4-P discharged from the study area. For the rural subwatersheds, phosphorus from agriculture was not differentiated from biogeochemical and precipitation inputs. Assuming all phosphorus arose from fertilizer application, total-P losses represented 3.5% of that applied, while PO4-P losses represented less than 1%. It was concluded that housing in rural areas contributes PO4-P to nutrient enrichment of drainage systems, and that measures to control rural erosion may reduce total-P load but increase the proportion of PO4-P.

Gurtz, M.E., Webster, J.R. and Wallace, J.B., "Seston Dynamics in Southern Appalachian Streams: Effects of Clear-cutting", Canadian Journal of Fisheries and Aquatic Sciences", Vol. 37, No. 4, Apr. 1980, pp. 624-631.

Suspended particulate matter (seston) was studied from July 1977 to July 1978 in two second-order streams in the southern Appalachian Mountains. In the first stream, which drains an undisturbed hardwood forest watershed, seston concentrations fluctuated with season (lowest during winter high flows) and with storm flows. Most organic and inorganic particles were smaller than 105 micrometers in diameter. The second stream drains a watershed (formerly a hardwood forest) that was clearcut in early 1977. Increased levels of both organic and inorganic seston were found in the latter stream, especially beginning one year after clear-cutting (2 years after construction of logging roads). Particles larger than 234 micrometers in diameter accounted for most of the increases in inorganic seston. These increases were probably due to sediments deposited in the stream bed during road building and transported downstream during periods of peak flow. Increased levels of organic seston were probably related to breakdown of debris that entered the stream during logging, and reduced retention by leaf packs. It was hypothesized that eventual recovery of the stream will be limited by the rate of recovery of the surrounding terrestrial ecosystem.

Haith, D.A., "A Mathematical Model for Estimating Pesticide Losses in Runoff", Journal of Environmental Quality, Vol. 9, No. 3, July-Sept. 1980, pp. 428-433.

A simple mathematical model was proposed for the estimation of losses of dissolved and solid-phase pesticides in cropland runoff. The model was designed for use in water quality and pesticide screening studies for which field pesticide runoff data are unavailable. The model is based on commonly used methods for runoff and soil loss predictions and does not require calibration. A comparison of model predictions with measured atrazine (2-chloro-4(ethylamino)-6(isopropylamino)-s-trazine) losses from two small Georgia watersheds indicated that the model provides a reasonable means of estimating the approximate magnitudes of long-term (3 years for the two watersheds) total pesticide runoff losses. Although large errors were sometimes observed in estimates of losses from single runoff events, the observed variations in pesticide losses among events, and the distribution of these losses into dissolved and solid-phase components, were reflected in model predictions.

Hopkinson, Jr., C.S. and Day, Jr., J.W., "Modeling the Relationship Between Development and Storm Water and Nutrient Runoff", Environmental Management, Vol. 4, No. 4, July 1980, pp. 315-324.

The EPA Storm Water Management Model was applied to the des Allemands swamp forest system in the headwaters of the Barataria Basin, Louisiana. The effects of changing land use patterns on storm water runoff were modeled. By 1995 urban land on the uplands will increase by 321%, primarily at the expense of agricultural land. This urbanization will cause storm water runoff rates to increase by 4.2 fold. Nutrient runoff will increase 28% for N and 16% for P. Lake eutrophication will be a problem unless corrective actions are taken.

Interstate Commission on the Potomac River Basin, Proceedings of a Technical Symposium on Non-Point Pollution Control--Tools and Techniques for the Future, Technical Publication 81-1, Jan. 1981, Rockville, Maryland.

Thirty-two papers given at the symposium by experienced investigators in nonpoint water pollution are presented. Nonpoint water pollution, pollution from diffuse sources such as runoff from agriculture, forestry, and urban land development, accounts for up to 50 percent of the pollution in some rivers. Past water pollution control has emphasized point sources, and much progress has been made with these sources. Section 208 of the Federal Water Pollution Control Act Amendments has focused pollution control efforts on the identification and control of nonpoint sources. The papers are classified in five categories: (1) perspectives on nonpoint pollution control, (2) case studies on nonpoint sources of pollution, (3) modeling tools for evaluation of nonpoint pollutants, (4) control measures, and (5) planning and implementation strategies. A majority of the papers deal with the Potomac River basin and, in particular, the Occoquan watershed in Virginia.

Jewell, T.K., Adrian, D.D. and DiGiano, F.A., "Urban Storm Water Pollutant Loadings", Publication No. 113, 1980, Water Resources Research Center, University of Massachusetts, Amherst, Massachusetts.

The runoff quality portions of existing stormwater management models have not fostered confidence in their predictive capabilities. Lack of confidence has been engendered by the use of unverified predictive algorithms and by the variability of measured stormwater pollution data. This study was designed to produce improved stormwater pollutant washoff prediction techniques which would be verifiable and would take into account data variability. Available storm event pollution data is examined, and the catalogued data includes 261 storm events from 26 basins in 12 geographical areas. The data file is used to show that it is not possible to derive general pollutant washoff functions containing a given set of independent variables that would give reasonable results for most areas. This is true for either storm event total loadings or instantaneous fluxes. It is concluded that data should be gathered for each basin to be studied. A methodology was developed that permits the use of this data to predict stormwater pollution washoff for individual watersheds. It includes suggested trial formulations and guidelines for applying linear and intrinsically linear multiple regression analyses to pick the best model for predicting the washoff of each pollutant. This methodology represents a distinct improvement over existing stormwater quality predictive techniques. Researchers will not be able to predict stormwater pollutant loadings with a measurable degree of certainty.

Larson, F.C., "The Impact of Urban Stormwater on the Water Quality Standards of a Regulated Reservoir", Research Report No. 62, Mar. 1978, Water Resources Research Center, University of Tennessee, Knoxville, Tennessee.

Urban stormwater runoff may or may not have a significant effect on the water quality of a receiving stream. Changes in the water quality of Fort Loudoun Reservoir (near Knoxville, Tennessee) as a result of stormwater runoff were observed. A transient flow model was used to trace a slug of water for examination of water quality during runoff events. A sampling methodology was developed which, when used in conjunction with the flow model, allowed graphical predictions of water quality changes in the slug of water as it moved past the city. A grab sampling technique (used before, during and after runoff events) was also developed to increase the water quality data. Water quality parameters obtained were dissolved oxygen, pH, biochemical oxygen demand, conductivity, temperature, total solids and fecal coliforms. Over 30 test runs were included in the study. No significant water quality changes were observed between control values and those obtained after the addition of stormwater runoff in the reservoir. No general trends could be determined for any parameters which seemed to fluctuate according to conditions specific to each rainfall-runoff event. Total solids violated stream standards on two occasions, dissolved oxygen was in violation 19 times and standards for fecal coliform numbers were frequently violated.

Lusby, G.C., "Effects of Grazing on Runoff and Sediment Yield from Desert Rangeland at Badger Wash in Western Colorado, 1953-73", Water Supply Paper 1532-I, 1979, U.S. Geological Survey, Washington, D.C.

Four different systems of livestock management were compared hydrologically during a 20-year study (1953-73) in western Colorado. These systems were: (1) grazing by cattle and sheep from November 15 to May 15 each year, (2) complete elimination of grazing, (3) grazing by sheep from November 15 to February 15 each year, and (4) grazing by sheep from November 15 to February 15 every other year. Grazing by both cattle and sheep from November 15 to May 15 each year was the standard grazing practice in the area at the beginning of the study. Complete grazing exclusion resulted in a reduction in runoff of about 20% during the period 1953-65, and an additional 20% during 1966-73. During the same periods sediment yield was reduced by 35 to 28%, respectively, for a total of 63%. A change in grazing use from cattle and sheep, November 15-May 15 each year, to sheep only at approximately the same utilization rate, November 15-February 15 each year, was accompanied by a reduction in runoff and sediment yield of about 29%. The same change in use, except for grazing allowed every other year during the sheep grazing period, resulted in a reduction in runoff and sediment yield of about 20%.

Lynch, J.A., Corbett, E.S. and Sopper, W.E., "Evaluation of Management Practices of the Biological and Chemical Characteristics of Streamflow and Forested Water Sheds", 1980, Institute for Research on Land and Water Resources, Pennsylvania State University, University Park, Pennsylvania.

A 106-acre oak-hickory experimental watershed in central Pennsylvania was clearcut in three phases to evaluate the effects of the clearcuts on the physical, chemical, and biological properties of streamflow. Herbicides were used to control the regrowth of vegetation so that ground water conditions on the clearcuts would be similar for each phase analysis, and also to quantify maximum response and stress factors. The watershed response to these treatments was characterized by measuring changes in streamflow amounts and timing, stormflow parameters, streamwater temperature, nutrient concentrations, turbidity and sediment, and aquatic macroinvertebrate populations. Biologic implications for the aquatic ecosystem are presented on the basis that each aquatic organism has a particular set of environmental conditions and habitat preferences that are optimal for its maintenance.

Martin, C.W., Noel, D.S. and Federer, C.A., "The Effect of Forest Clear-Cutting in New England on Stream Water Chemistry and Biology", Research Report 34, July 1981, Water Resources Research Center, University of New Hampshire, Durham, New Hampshire.

Changes in stream chemistry following clear-cutting were sought in 56 streams at 15 locations throughout New England. Streams draining clearcut areas were compared with nearby streams in uncut watersheds over periods of up to 2 years. In general, concentrations of all elements studied (inorganic N, SO4-S, Cl, Ca, Mg, K, Na), as well as pH and specific conductivity, varied as much among uncut streams at a location as between uncut and cutover streams. However, at most locations, at least one of these variables differed between uncut and cutover streams. The greatest differences occurred with nitrogen in northern hardwood forests in the White Mountains of New Hampshire. At four of the locations the effect of cutting on algae and invertebrates in the streams were also examined. Both algal and invertebrate densities were greater in cutover streams by factors of 2 to 4, probably because of increased light and temperature. The taxonomic composition of both algal and invertebrate populations was also changed by cutting. Partial cuts and sufficiently wide buffer strips can minimize both chemical and biological changes.

Mather, J.R., "The Influence of Land Use Change on Water Resources", June 1979, Water Resources Center, University of Delaware, Newark, Delaware.

This report provides detailed examples of the hydrologic consequences of two specific land-use changes. The first involves the effect on streamflow of a gradual change from cultivation to suburbanization in a basin near Wilmington, Delaware, while the second involves the effect on recharge to the water table of introducing a suburban housing development in a formerly agricultural and wooded tract. Both examples make use of information developed from the climatic water budget. In the first example, it was found that the gradual change from forest and cultivation to suburbanization has resulted in a progressive increase in stream runoff averaging about 0.04 inch per year over the past 30 years. In the second example, increasing the amount of impervious surfaces decreases ground water recharge. In order to have no detrimental effect on the water table recharge, detention (by means of some sort of infiltration basins or ponds) of about one-half of the water falling on impervious surfaces in the housing development is necessary in order for the water table recharge after development to equal the recharge before development.

McCuen, R.H. et al., "Estimates of Nonpoint Source Pollution by Mathematical Modeling", Technical Report No. 43, Mar. 1978, Maryland Water Resources Research Center, University of Maryland, College Park, Maryland.

The environmental consequences of the rapid expansion of urban communities into areas that had previously been dominated by rural land use has been the cause of increasing concern over the past decade. Techniques for the rational analysis of the environmental quality of the affected region are needed to examine the environmental consequences of existing and future projects in the planning community. A computer simulation model was developed to estimate the accumulation of pollutants in urban subbasins. A model component is also used to estimate the removal of the pollutants for any rainfall event. The simulation model could be used by planners to estimate the pollution loading from an urban subbasin that enters streams for all or part of that region.

Olivieri, V.P., Kruse, C.W. and Kawata, K., "Micro-organisms in Urban Stormwater", EPA-600/2-77-087, July 1977, U.S. Environmental Protection Agency, Cincinnati, Ohio.

Microbiological quantitative assays of Baltimore City urban runoff were conducted throughout a 12-month period to show the relationship to several factors such as separate or combined sewer flow, urban characteristics of drainage area, rainfall, and quantity of flow during and between rain storms. In general, there were consistent indicator organisms throughout the study except for Shigella sp., which is believed to have been present but could not be isolated due to interferences during the culture procedure. There appeared to be little relationship between pathogen recovery and season of the year, amount of rainfall, period of the antecedent rainfall, and stream flow. The most concentrated pathogens were Pseudomonas aeruginosa and Staphylococcus aureus. The background samples (sewage, urban streams and reservoirs) between storms gave good positive correlation between indicators and pathogens at a 95 to 99 percent level of confidence, whereas, the stormwater had no or poor correlation. The logical solution would point to the removal of sanitary sewage over-flows rather than the disinfection of all urban runoff for removing the health hazard and improving the quality of urban runoff.

Ongley, E.D. and Broekhoven, L.H., "Data Filtering Techniques and Regional Assessment of Agricultural Impacts Upon Water Quality in Southern Ontario", <u>Progress in Water Technology</u>, Vol. 11, No. 6, 1979, pp. 551-577.

This study examined the effect of agriculture relative to other land uses upon river water quality at a regional level using 49 nonpoint (diffuse) source and 52 point source drainage basins in southern Ontario, and which are tributary to Lakes Ontario, Erie, Huron, and connecting waterways. Methods of data filtering were examined, and substantive relationships among river water quality and categories of land use were identified. Water quality data were mean annual concentrations for the period 1968 to 1972 for nutrients and other physical and chemical variables. Land use variables employed a reduced set of Canada Land Inventory categories, which included agricultural and urban variables. Filtering techniques included correlation, factor, and discriminant analyses. Factor analysis did not effectively group these variables. Alternatively, discriminant analysis offered a rapid and convenient method for identifying those independent variables responsible for variations in water quality. Correlation was used to examine those relationships among variables which were identified by discriminant analysis. Although urban variables appear to be responsible for extremely high levels of total nitrogen, and total soluble phosphorus found in a few basins, croplands appear to influence levels of those attributes in basins representing the top one-third of ranked mean annual values. Cropland is the single discriminating variable for basins having large suspended sediment concentrations, and is strongly related to nitrate levels.

Robbins, J.W.D., "Environmental Impact Resulting from Unconfined Animal Production", EPA-600/2-78-046, Feb. 1978, U.S. Environmental Protection Agency, Washington, D.C.

Knowledge related to environmental effects of unconfined animal production is outlined and evaluated. Animal species include cattle, sheep, and hogs. All available data indicate that pollutant yields from pasture and rangeland operations are not directly related to the number of animals or amount of wastes involved. Rather, these nonpoint source problems are intimately related to hydrogeological and management factors, and are best described as the results of the erosion/sediment phenomenon. Unconfined livestock production can cause changes in vegetative cover and soil physical properties that may result in increased rainfall runoff and pollutant transport to surface waters. The most common stream water quality result is elevated counts of inorganic and organic sediments with associated plant nutrients and oxygen demands. Generally, the pollutant levels from the remainder of the production site are not discernible from background levels. If other changes, such as those affecting ground water quality, occur, they are of no environmental consequence. A major challenge remaining is to demonstrate cost-effective routes toward achievement of various levels of pollution control from unconfined animal production.

Ross, B.B., Shanholtz, V.O. and Contractor, D.N., "A Spatially Responsive Hydrologic Model to Predict Erosion and Sediment Transport", Water Resources Bulletin, Vol. 16, No. 3, June 1980, pp. 538-545.

A finite element numerical model has been developed which routes overland and channeled flows in a watershed, given soils, land use, topographic descriptors, and rainfall as input. Such processes as infiltration, canopy interception, seasonal growth of vegetation, and depression storage are described in the hydrologic context of the model. These capabilities, along with the spatial detail and responsiveness of the model, allow a ready adaptation of the model to provide for the prediction of sediment transport and yield. It is assumed that the best results can be obtained by a technique which utilizes the following procedures. Sediment yield to the channel is described by functions describing soil detachment by rainfall and overland flow and transport by overland flow.

Since the model description of the channel flow processes involves a more realistic representation of the physical drainage system, an attempt was made to define sediment transport in the channel by erosion and sedimentation mechanics. A conceptual framework is provided whereby the integrated effects of various land use activities on sediment transport and yield can be evaluated. Inherent in this provision of the model is the capability of determining the effects of any control measures to be implemented on a watershed.

Schillinger, J.E. and Stuart, D.G., "Quantification of Non-Point Water Pollutants from Logging, Cattle Grazing, Mining, and Subdivision Activities", Report No. 93, 1978, Water Resources Research Center, Montana State University, Bozeman, Montana.

Water quality data were collected over a two-year period at 57 points in the Hyalite Creek and Bozeman Creek watersheds in southwestern Montana to determine the contribution of different activities to changes in the water chemistry, suspended sediment levels, and bacterial indicator counts in streamflow. Cattle grazing increased fecal coliform; a cattle grazing index is formulated and correlated with the fecal coliform concentration and is shown as a function of a grazing management rating to relate a cattle management system to water quality. Clearcut logging did not significantly effect water quality. Although soil erosion occurred in clearcut areas, little sediment reached the streams from disturbances more than 30 m away. Stormwater runoff, septic tank drainfields and construction in subdivisions in urban and suburban areas significantly altered the bacteriological and chemical quality and amount of suspended sediment in streams. Effects of ground water seeping through a lead mine operated intermittently adjacent (300 m) to a stream were insignificant on water quality. A factor analysis indicates natural processes (1) correlate significantly with changes in water quality in high mountain streams rather than logging and grazing, and (2) do not correlate significantly in urban and suburban developments. Bacteriological analysis indicates Klebsiella and Enterobacter may indicate pollution from storm sewers and septic tanks better than total coliform. Methods for in situ and in vitro examination of leptrospires are presented.

Schreiber, J.D., Duffy, P.D. and McClurkin, D.C., "Aqueous and Sediment-Phase Nitrogen Yields from Five Southern Pine Watersheds", Soil Science Society of America, Vol. 44, No. 2, Mar.-Apr. 1980, pp. 401-407.

Nitrogen (N) in solution and associated with suspended sediments in stormflow from five reforested watersheds (1.5-2.8 ha) in northern Mississippi was determined during the 1975 water year. Samples were collected using Cohocton wheel samples set below 0.91-m H-flumes. Mean yearly solution NH4-N and NO3-N concentrations, 0.20 and 0.01 mg/l, respectively, did not differ among watersheds. Sediment-N concentrations ranged from 2410 to 6080 micrograms/gram, and were 5.4 to 10.0 times those in the watershed soils. The N enrichment of sediment relative to soil was attributed to selective erosion of fine sediments (clay) and/or deposition of coarse sediments in transport. Significant differences in sediment N yields among the five watersheds were related to stormflow volume, sediment concentration, and sediment N concentration. Nitrogen was transported from the watersheds about equally in the sediment and solution phases. Mean solution NH4-N and NO3 yields for the water year were 422 and 28 g/ha, respectively, for the five watersheds mean sediment N yield was 3915 g/ha. For the year, 8 to 17 percent of the precipitation was measured as stormflow; the remainder was deep seepage and evapotranspiration since there was essentially no baseflow.

Smith, R. and Eilers, R.G., "Stream Models for Calculating Pollutional Effects of Stormwater Runoff", EPA-600/2-78-148, Aug. 1978, U.S. Environmental Protection Agency, Cincinnati, Ohio.

Three related studies for developing better methods for quantifying the pollutional effects in streams caused by stormwater overflows are described. Mathematical models developed simulate the biological, physical, chemical, and hydraulic reactions that occur in a flowing stream. The relationships are presented as differential or difference equations with the two independent variables: time and distance along the stream. Analog or digital computers, or calculus, can be used to find solutions to the differential equations. This solution is the concentration of all species of pollutional interest within the stream as a function of time and distance. There are two regimes of solutions--steady-state and transient; the pollution loads on a stream can also be specified as steady-state or transient. Two digital computer programs were developed: one computes the dissolved oxygen (DO) deficit in a stream as a function of time and distance along the stream caused by any specified overflow; the second computes the hydraulic effect on the stream of large volume overflows. A method for summarizing the results of computations in an easily used manner was also developed. Because of advanced technology, it is now possible to estimate the impact of any stormwater overflow on the dissolved oxygen resources of a stream by hand computation.

Tubbs, L.J. and Haith, D.A., "Simulation Model for Agricultural Non-Point Source Pollution", Journal of the Water Pollution Control Federation, Vol. 53, No. 9, Sept. 1981, pp. 1425-1433.

A mathematical simulation model, referred to as the Cornell Nutrient Simulation (CNS) model, is presented which is suitable for estimating dissolved and solid-phase nitrogen and phosphorus in runoff, and dissolved N in percolation from cropped fields with well-drained soils. The CNS model consists of two basic components: a daily soil moisture/erosion submodel that predicts runoff, percolation and soil loss; and a monthly submodel of soil nutrient levels. Validation studies for several small fields in Georgia and New York are included. The nutrient component has a monthly time step. The nutrient balance model estimates monthly losses of dissolved and solid-phase N in runoff, dissolved N in percolation, dissolved P in runoff, and solid-phase P in runoff. Validation studies in Georgia were performed in two small fields that were monitored for runoff, sediment and nutrient loss in runoff from May, 1974 through September, 1975. The New York testing sites are six 0.3 ha plots in Aurora, New York, from which runoff, percolation, and nutrient loss data were collected from January, 1972 through December, 1973. Measured nutrient, water, and sediment losses were compared with those predicted by the CNS model. Runoff predictions exceeded observations by substantial amounts from both fields in Georgia, although errors were smaller on one. Dissolved N and P were over-predicted by about the same degree as runoff on one field. The most critical problem is in the simulated losses of dissolved P in runoff. In New York there were additional problems which made comparisons of the predictions and the observations meaningless. The principal weakness of the model is in its handling of organic matter such as manures and fresh plant residues.

Turner, F.T., Brown, K.W. and Deuel, L.E., "Nutrients and Associated Ion Concentrations in Irrigation Return Flow from Flooded Rice Fields", Journal of Environmental Quality, Vol. 9, No. 2, Apr.-June 1980, pp. 256-260.

A 3-year field study of nutrient and common ion concentrations in irrigation return flow (IRF) from flooded rice fields utilized replicated plots that received either recommended or excessive fertilizer rates with continuous flow or intermittent flood irrigation. All P and K and 40% of the ammonium sulfate nitrogen were applied preplant and incorporated. The remaining N was applied just prior to permanently flooding (40%) and at 7-mm panicle stage (20%). The irrigation and plot waters were analyzed for NH4(+)-N, NO3(-)-N, NO2(-)-N, PO4(3-)-P, K(+), Ca(2+), Mg(2+), Na(+), Cl(-), and SO4(2-). Highest nitrate-N concentrations occurred early in the season before the permanent flood period, but did not exceed drinking water standards. Nitrite-N concentrations in the IRF were low at all times. Maximum NH4(+)-N concentrations occurred following B fertilizer applications which were not incorporated into the soil, and persisted in the floodwater for 5 to 7 days. These peak NH4(+)-N concentrations exceeded acceptable drinking water standards by a factor of 10 or greater. Concentrations of PO4(3-)-P and K(+) in the floodwater were similar to those in the surface water used for irrigation. The concentration of the other common ions in the floodwater did not greatly exceed those in the irrigation water, and all were within concentrations acceptable for drinking water. Method of fertilizer application has more influence on IRF nutrient concentrations than did fertilizer rates or irrigation management practices. Under the conditions of this study, it appears that only the ammonium concentrations may have a detrimental impact on the quality of IRF from flooded rice fields. This potential problem could be minimized by preventing IRF for a period of one week following surface applications of ammonium sulfate fertilizer.

Unger, S.G., "Environmental Implications of Trends in Agriculture and Silviculture. Volume II: Environmental Effects of Trends", EPA-600/3-78-102, Dec. 1978, Development Planning and Research Associates, Inc., Manhattan, Kansas.

This study assesses those trends in U.S. agriculture and silviculture that will have the most significant environmental implications, either beneficial or adverse, in the short term (1985) and in the long term (2010). Volume II identifies the major ecological impacts of the major trends on aquatic life, terrestrial life, and human health. The second volume also contains an assessment of continuing research needs and prospective policy issues involving environmental quality management.

U.S. Bureau of Reclamation, "Prediction of Mineral Quality of Irrigation Return Flow: Volume I. Summary Report and Verification", Report No. EPA-600/2-77-179a, Aug. 1977, Denver, Colorado.

This volume outlines the purpose and scope of return flow research and explains the capabilities of the conjunctive use model for predicting the mineral quality of irrigation return flow. The purpose of the research was to develop a conjunctive use model which would (1) predict the salinity contribution from new irrigation projects, and (2) predict the change in return flow salinity that would result from operational changes on existing projects. The model describes the chemical quality in terms of eight ionic constituents and total dissolved solids. A nodal concept has been used to facilitate subdividing the project area along physical or hydrologic boundaries as desired. The study may be limited to one or as many as 20 nodes.

U.S. Environmental Protection Agency, "Areawide Assessment Procedures Manual, Volume I", EPA-600/9-76-14-1, July 1976, Municipal Environmental Research Laboratory, Cincinnati, Ohio.

This manual provides an environmental management statement of procedures available for water quality management with particular emphasis on urban stormwater. The manual summarizes and presents in condensed form a range of available procedures and methodologies for identifying and estimating pollutant load generation and transport from major sources within water quality management planning areas. The major emphasis of the manual is directed toward the assessment of problems and selection of alternatives in urban areas, with particular concern for stormwater related problems. Also included in the manual are methodologies for assessing the present and future water quality impacts from major sources as well as summaries of available information and techniques for analysis and selection of structural and nonstructural control alternatives.

U.S. Environmental Protection Agency, "Areawide Assessment Procedures Manual, Volume II", EPA-600/9-76-014-2, July 1976, Municipal Environmental Research Laboratory, Cincinnati, Ohio.

This manual provides an environmental management statement of procedures available for water quality management with particular emphasis on urban stormwater. The manual summarizes and presents in condensed form a range of available procedures and methodologies for identifying and estimating pollutant load generation and transport from major sources within water quality management planning areas. This volume presents rainfall runoff and water quality model applicability, and land use and rainfall data analysis.

U.S. Environmental Protection Agency, "Areawide Assessment Procedures Manual, Volume III", EPA-600/9-76-014-3, July 1976, Municipal Environmental Research Laboratory, Cincinnati, Ohio.

This manual provides an environmental management statement of procedures available for water quality management with particular emphasis on urban stormwater. The manual summarizes and presents in condensed form a range of available procedures and methodologies for identifying and estimating pollutant load generation and transport from major sources within water quality management planning areas. Although an annotated chapter is provided for the assessment of nonurban pollutant loads, the major emphasis of the manual is directed toward the assessment of problems and selection of alternatives in urban areas, with particular concern for stormwater-related problems. Also included in the manual are methodologies for assessing the present and future water quality impacts from major sources as well as summaries of available information and techniques for analysis and selection of structural and nonstructural control alternatives.

Walker, W.R., "Assessment of Irrigation Return Flow Models", EPA-600/2-76-219, Oct. 1976, Department of Agricultural and Chemical Engineering, Colorado State University, Fort Collins, Colorado.

Throughout the western United States irrigation return flows contribute to the problem of water quality degradation. Evaluating the effectiveness of alternative management strategies involves models which simulate the processes encompassed by irrigated agriculture. The development and application of these models require multidisciplinary expertise. A workshop involving 15 specialists in the varied aspects of irrigation return flow modeling was held to review the status of these models. Irrigation return flow and conjunctive use models recently developed by the U.S. Bureau of Reclamation served as focal points for the workshop. As the field verification and potential applications of these models were discussed, several general problems were identified where further investigation is needed. Particular emphasis was given to the description of the spatially varied aspects of soil, crop, and aquifer systems, and the proper alignment of model objectives with available data. The large number and diversity of existing models illustrate the individualistic nature of irrigation return flow modeling. In order to effect more widespread utilization of existing models, a systematic procedure should be developed to update and disseminate this modeling technology.

Watson, V.J. et al., "Impact of Development on Watershed Hydrologic and Nutrient Budgets", Journal of the Water Pollution Control Federation, Vol. 51, No. 12, Dec. 1979, pp. 2875-2885.

This study assessed the impact of development on the hydrologic and nutrient budgets of contrasting watershed systems. Nutrient yields from the watersheds of oligotrophic Lake George (New York) and eutrophic Lake Wingra (Wisconsin) were estimated from measured nutrient concentrations and hydrologic flows, and from hydrologic budget computations, and were compared to reconstructed presettlement yields. Development-induced changes within the watersheds have altered phosphorus budgets more than nitrogen budgets, but the greatest impact on the Lake George budget comes from nitrogen-enriched precipitation. The Lake Wingra watershed appears to have been most affected by the alteration of its hydrology and morphology.

APPENDIX F

TRANSPORT AND FATE OF POLLUTANTS IN THE WATER ENVIRONMENT

Anderson, J.W., "An Assessment of Knowledge Concerning the Fate and Effects of Petroleum Hydrocarbons in the Marine Environment", Marine Pollution, Functional Responses, Vernberg, W.D. et al., Editors, 1979, Academic Press, New York, New York, pp. 3-21.

The amount of data accumulated from laboratory and field investigations regarding the environmental effects of petroleum hydrocarbon spills allows a good understanding of this problem. Considerable knowledge is available regarding the acute toxicity and the effects of sublethal concentrations of petroleum on a variety of vertebrate and invertebrate organisms. The uptake and release of petroleum hydrocarbons by marine and estuarine molluscs suggests their use for field monitoring of sediment-bound petroleum hydrocarbons by aquatic organisms.

Buikema, Jr., A.L., McGinniss, M.J. and Cairns, Jr., J., "Phenolics in Aquatic Ecosystems: A Selected Review of Recent Literature", Marine Environmental Research, Vol. 2, No. 2, Apr. 1979, pp. 87-181.

This review surveys the pertinent literature on phenolics in the aquatic ecosystem. Approximately 2 percent of the total organics manufactured in the United States are phenols. Of the total phenolics produced in the U.S., 96 percent were synthetic and 4 percent were naturally occurring. Synthetic phenols arise from coking of coal, gas works and oil refineries, chemical plants, pesticide plants, wood preserving plants, and dye manufacturing plants. Natural phenolics occur from aquatic and terrestrial vegetation, and much of this is released by the pulp and paper industry. Toxicity of phenolics has been studied on selected microbes (e.g., protozoa, yeast and bacteria), algae, duckweed, and numerous invertebrates and vertebrates. Studies on the biological effects of phenolics are limited and varied. Fish development and embryo survival were not affected by phenol levels less than 25 mg/l. Amphibian embryos were sensitive to 0.5 mg/l phenol. Pentachlorophenols inhibited fish growth at certain levels. Many aspects of behavior are affected by phenolics, and the phases of "intoxication" leading to death have been described for fish and invertebrates. Little research has been done on the cycling of phenol and phenolics (other than pesticides) in aquatic ecosystems.

Damman, W.H., "Mobilization and Accumulation of Heavy Metals in Freshwater Wetlands", Research Project Technical Report, 1979, Institute of Water Resources, Connecticut University, Storrs, Connecticut.

The vertical distribution of Zn, Pb, Mn, Cu, Ni, K, Na, N, Fe, and Al was studied in a series of cores taken from two Connecticut wetland types; a mesotrophic red maple swamp, and an oligotrophic Sphagnum bog. Parts of the red maple swamp affected by spring and seepage water contain much higher concentrations of Cu, Ni, Mn, Fe, Na and K than the rest of the swamp or bog. Lead, Mn and Fe concentrations are highest above the anaerobic level. Of the heavy metals, only Mn and Zn are enriched by nutrient cycling in the surface horizons of the swamp. This is best expressed in the higher parts of the swamp. The annual growth of a low Sphagnum flavicomans hummock contains amounts of Zn and Cu roughly equal to the annual atmospheric input, but only 1/2 of the Pb and 1/3 of the Ni. Apparently, Ni, Cu, and Mn are leached from the surrounding uplands into the swamp, but this affects only the border areas. The red maple swamp does not act as a sink for Pb and Zn. Atmospheric input appears to be the only source of heavy metals in the surface of the bog.

Drill, S. et al., "The Environmental Lead Problem: An Assessment of Lead in Drinking Water from a Multi-Media Perspective", EPA-570/9-79-003, May 1979, Mitre Corporation, McLean, Virginia.

Human exposure to lead has been shown to be cumulative in nature. In order to assess the toxicological significance of environmental lead exposures, it is necessary to define the contributions to an individual's daily lead uptake from all possible exposure pathways. This paper defines and quantifies the major environmental sources of lead exposure, describes the absorption characteristics of lead compounds in man via each exposure route, determines the source contribution factors for daily lead uptake by each exposure pathway, and relates those contributions to an individual's blood-lead level.

Fenet-Robin, M. and Ottmann, F., "Comparative Study of the Fixation of Inorganic Mercury on the Principal Clay Minerals and the Sediments of the Loire Estuary", *Estuarine and Coastal Marine Science*, Vol. 7, No. 5, Nov. 1978, pp. 425-436.

Samples of kaolinite, illite, and montmorillonite were agitated with solutions of mercuric chloride at different salinities, clay turbidities, and concentrations of mercuric ions. The rates of adsorption and maximum quantities adsorbed were obtained in terms of these factors. Clays with known quantities of adsorbed mercury were agitated with fresh and salt water to measure rates of desorption. The values obtained were compared with an analysis of the water and sediments of the Loire Estuary, based on a large number of samples taken over the period 1972 to 1975. Mercury pollution has decreased considerably over this period.

Hansen, D.J., "Impact of Pesticides on the Marine Environment", *First American-Soviet Symposium on the Biological Effects of Pollution on Marine Organisms, 20-24 September 1976, Gulf Breeze, Florida*, EPA-600/9-78-007, May 1978, Environmental Research Laboratory, U.S. Environmental Protection Agency, Gulf Breeze, Florida, pp. 126-137.

Laboratory bioassay experiments were conducted to test effects of toxicants on estuarine animals: (1) acute bioassays with flowing water at constant salinity and temperature, and with measurement of toxicant concentration in water and test organism and statistical analysis of mortality data; (2) bioassays on sensitive larval stages of crabs and shrimp; (3) bioassays over the reproductive period or entire life cycle of grass shrimp (Palacemonetes pugio) and sheepshead minnow (Cyprinodon variegatus); and (4) bioassays on communities of benthic macroinvertebrates. Bioassaying, perhaps the most useful technique available for predicting a chemical's potential hazard, includes acute toxicity experiments (either static or flow-through), usually used to determine the concentration of a chemical lethal to a certain percentage of test organisms over a short time; long-term bioassays on sensitive life stages, used to determine the concentration which can be tolerated by a species through all or part of its life cycle; and community bioassays, for predicting community response to a toxicant. The findings were: (1) C. variegatus typically develops from embryo to maturity in 10-14 weeks with about 70 percent survival in the laboratory, females produce an average of eight eggs/day, and fertilization exceeds 90 percent; (2) the pesticides endrin and heptachlor reduced spawnings and eggs of C. variegatus, decreased egg fertility, and increased mortality of fry and spawning females. Effects of the pesticide toxaphene and the PCB Aroclor 1254 were also tested.

Hoover, T.B., "Inorganic Species in Water: Ecological Significance and Analytical Needs, A Literature Review", EPA-600/3-78-064, July 1978, U.S. Environmental Protection Agency, Environmental Research Laboratory, Athens, Georgia.

Representative studies of the environmental significance of inorganic species (as opposed to total-element content) in water are reviewed. The effects of chemical forms on human health and on plant and animal life, and the roles of valence state, ionization, complexation, and adsorption in the transport and cycling of elements are considered along with factors affecting the distribution of elements and species in freshwater streams and impoundments in estuaries, and in the sea. Information on the chronic effects on human health of trace inorganic pollutants in water is almost entirely limited to total elements because of an inability to distinguish among forms of an element. The elements of greatest concern with respect to the toxicity of different species are arsenic, chromium, lead, mercury, and selenium. In the toxicology of aquatic biota, there is a rapidly growing appreciation that both acute and chronic effects are strongly related to chemical species. The movement of inorganic pollutants in the aquatic environment is strongly influenced by adsorption of particular species on both mineral and organic particulates. No broadly applicable analytical techniques of adequate sensitivity are available for elemental speciation. This deficiency in analytical ability prevents the evaluation of research on toxicology and on transport of these chemical forms.

Iwamoto, R. et al., "Sediment and Water Quality: A Review of the Literature Including a Suggested Approach for Water Quality Criteria with Summary of Workshop and Conclusions and Recommendations", EPA-910/9-78-048, Feb. 1978, U.S. Environmental Protection Agency, Washington, D.C.

A review of the literature dealing with physical and biological aspects of sedimentation in the lotic environment was compiled. Over 300 annotated and 100 unannotated references are included. These were used as a resource for a workshop on needs for an alternative to turbidity measurements as a water quality criterion. Also included are a summary of the current state of investigations on stream sediments and turbidity measurements as a criterion. These include: composition of bed material; behavioral aspects of aquatic fauna; and clinical measurements of physiological functions as a measure of stress. The more significant conclusions are that sedimentation of the stream's substrate, particularly the gravel used for spawning, produces significant detrimental effects on the salmonid resources. Turbidity measurements are useful indicators of general quantities of suspended sediment; but they are difficult to relate to biological significance. In relation to logging, the Best Management

Practices (BMP) concept appears to be the most practical, but BMP should be monitored by a set of criteria that are suggested.

Leland, H.V., Luoma, S.N. and Fielden, J.M., "Bioaccumulation and Toxicity of Heavy Metals and Related Trace Elements", Journal of the Water Pollution Control Federation, Vol. 51, No. 6, June 1979, pp. 1592-1616.

This review is confined to a discussion of bioaccumulation and toxicity of heavy metals and related trace elements. The decision to narrow the scope of the review reflects a large body of literature now available on trace element distributions and their environmental effects. Included in this review are reports dealing directly with concentrations or activities of trace elements in aquatic ecosystems and the impact of these trace constituents on aquatic life. Included is a bibliography containing 199 literature references.

Malins, D.C., Effects of Petroleum on Arctic and Subarctic Environments and Organisms, 1977, Academic Press, New York, New York.

This book, in 2 volumes, includes 12 papers that deal with properties and analyses of petroleum in biotic and abiotic systems; inputs, transport mechanisms, and observed concentrations of petroleum in the marine environment; alterations in petroleum resulting from physico-chemical and microbiological factors; acute toxic effects of petroleum on arctic and subarctic marine organisms; marine fish and invertebrate diseases, host disease resistance, and pathological effects of petroleum; accumulation and biotransformation of petroleum hydrocarbons in marine organisms; sublethal biological effects of petroleum hydrocarbon exposures; effects of petroleum on ecosystems; biological effects of petroleum on marine birds; and effects of spills in arctic and subarctic environments.

APPENDIX G

BASELINE STUDIES OF THE WATER ENVIRONMENT

Adrian, D.D. et al., "Cost Effective Stream and Effluent Monitoring", Publication No. 118, Sept. 1980, Water Resources Research Center, University of Massachusetts, Amherst, Massachusetts.

The location of sampling stations is a critical factor in the design of a water quality monitoring network in a river basin. There are three levels of design criteria associated with sampling station location. They are: the macrolocation--river reaches which will be sampled within the river basin; the microlocation--station location relative to outfalls or other unique features within a river reach; and representative locations--points in the river's cross section from which grab samples will provide a lateral profile of the stream. This report addresses the determination of macrolocations and representative locations. In addition, economic aspects of monitoring are considered so that the costs and benefits associated with a monitoring design can be used to determine the optimal level of monitoring. The procedure is applied to the Massachusetts section of the Connecticut River Basin. The benefits resulting from water quality monitoring depend upon the reliability of the monitoring. The reliability in turn depends on factors such as sampling frequency and station location, among others. Multiple regression analyses were performed, using streamflow data, to observe how the reliability of monitoring data varies with location and frequency. A procedure using Bayesian analysis was developed to determine the value of water quality monitoring.

Bingham, C.R. et al., "Grab Samplers for Benthic Macroinvertebrates in the Lower Mississippi River", Misc. Paper E-82-3, July 1982, U.S. Army Engineer Waterways Experiment Station, Vicksburg, Mississippi.

The use of any one single type and size of existing grab sampler for gathering representative benthic macroinvertebrate samples from the various habitats within rivers is impractical, if not impossible. Adjusting one sampling gear (grab sampler) to the existing habitat and program objectives appears to be the best approach. Although macroinvertebrate relative catch efficiency (catch by area inscribed by cocked sampler) varies among sampler types, the catch efficiency of a single type varies with substrate, current regime, and bottom contours to a greater extent. Data from 24 grab samples, eight each taken with the Standard Ponar, Petite Ponar, and Shipek grabs from a backwater habitat were statistically compared using one-way analysis of variance. Total grabs produced 5696 organisms and 24 distinct taxa. Results showed no difference in catch efficiency among grab samplers for distinct taxa, total densities, and densities of the dominant taxa: Lirceus sp., Ilyodrilus templetoni, Hexagenia sp., Limnodrilus cervix, Limnodrilus hoffmeisteri, Sphaerium sp., and immature Iliodrilus. The Petite Ponar grab captured significantly fewer immature Limnodrilus. This was attributed to clumped distribution of these worms rather than difference in gear type. Analysis of the cumulative percent composition of newly-acquired species showed that second and additional replicates of each grab type accounted for 10 percent or less of the total standing crop, as also found in marine substrate types.

A single grab per station is sufficient to characterize the dominant benthic macroinvertebrate standing crop community and is, therefore, recommended for survey level studies. The Shipek grab is the preferred grab in habitats with strong currents; rough bottom morphology; and sand, gravel, or firm clay substrates. Ponar-type grabs are preferred for softer substrates under lower current velocities. This includes most backwater-type habitats, e.g., river borders, abandoned channels, oxbow lakes, and dike fields under low-flow conditions. The choice between Petite Ponar and the Standard Ponar grabs depends upon the project design, but should be made with the following facts in mind. A single Standard Ponar grab samples a larger surface area than a single Petite Ponar grab; therefore, it provides a better representation of the immediate (station) benthic macroinvertebrate community than does the Petite Ponar grab. However, almost twice as many Petite Ponar samples can be taken and processed as Standard Ponar samples with similar effort. Greater numbers of stations dispersed over an area gives a better areal representation. Greater number of replicates provide smaller experimental error and, therefore, better support data for statistical inferences.

Burke, H.D., "Bibliography of Manual and Handbooks from Natural Resource Agencies", FWS/OBS-78/22, Mar. 1978, Thorne Ecological Institute, Boulder, Colorado.

This bibliography locates handbooks and manuals relating to wildlife (produced by states and Federal agencies) which are not cited in libraries or information retrieval systems. These are normally in the form of instructions to field workers or planning personnel and not considered as publications. The handbooks and manuals were collected from 17 western states (North Dakota to Texas, and states to the west of these) and from Federal agencies such as the U.S. Forest Service, Soil Conservation Service, and Bureau of Land Management.

Burns, E.A., "Symposium Proceedings of Process Measurements for Environmental Assessment Held at Atlanta, on February 13-15, 1978. Final Task Report, April 1977-February 1978", EPA-600/7-78-168, Aug. 1978, TRW Systems Group, Redondo Beach, California.

This report documents the 26 presentations made at the subject Symposium. The objective of the Symposium was to bring together people who were responsible for planning and implementing sampling and analysis programs for multimedia environmental assessment. The program consisted of sessions defining the uses of environmental assessment data, the techniques for acquiring information, and recent user field experience with environmental assessment measurement programs.

Bogucki, D.J. and Gruendling, G.K., "Remote Sensing to Identify, Assess, and Predict Ecological Impact on Lake Champlain Wetlands", Final Report, 1978, State University of New York at Plattsburgh, Plattsburgh, New York.

The objectives of this study were to (1) utilize remote sensing techniques (color and color infrared aerial photography) to map aquatic vegetation for 12 priority Lake Champlain wetlands (9135 hectares) at a 1:2500 scale, (2) analyze the ecological effects of naturally fluctuating lake levels on the composition and distribution of major aquatic plant populations, and (3) assess the possible effects of water level regulation from the proposed Richelieu River dam construction. Evaluation of the effects of naturally fluctuating water levels on selected aquatic vegetation over a three-year period indicated that certain floating, emergent, and shrub species (particularly Typha angustifolia, Sparganium curycarpum, Scirpus fluviatalis, Cephalanthus occidentalis, Nuphar variegatum, and Nymphaea tuberosa) benefit from high water levels. All showed decreases in distribution, density, and/or vigor during the 1975 and 1977 low water years. Low water conditions appear most favorable for the emergent Zizania aquatica and for the maintenance of the green timber areas. If Lake Champlain should be regulated, each of the Lake Champlain wetlands would react in a different manner. The varied present day wetland characteristics with respect to vegetation composition, slope, exposure, soils, and basin configuration all combine to result in a wide range of possible ecological impacts.

Brown, R.L., "Monitoring Water Quality by Remote Sensing", Final Report No. NASA CR 154 259, July 1977, California State Department of Water Resources, Sacramento, California.

The results of a study to determine the applicability of remote sensing for evaluating water quality conditions in the San Francisco Bay-Delta area and Lake Tahoe, California, are presented. Coincident ground truth was obtained during LANDSAT and U-2 flights and correlated with the remote sensing images to establish a data comparison base line. Images were analyzed for apparent surface anomalies which might indicate water quality problems. It is concluded that: (1) for most water quality monitoring applications, LANDSAT imagery is too infrequent and of too small a scale to be useful in routine monitoring programs; (2) imagery from U-2 and conventional aircraft can be effectively used to monitor gross water quality changes; (3) with the present state-of-the-art in image analysis and the large amount of ground truth needed, remote sensing has only limited application in monitoring water quality; (4) California water quality conditions are improving as a result of the Porter-Cologne Water Quality Act and provisions of P.L. 92-500; and (5) in complex and dynamic systems such as the San Francisco Bay and Delta, large amounts of ground truth data must be collected to support remote imagery; spatial and temporal variations of the parameters are so great that approaches other than synoptic (synchronized multi-point sampling) do not provide enough information to evaluate patterns observed in specific images.

Cairns, Jr., J. and Gruber, D., "A Comparison of Methods and Instrumentation of Biological Early Warning Systems", Water Resources Bulletin, Vol. 16, No. 2, Apr. 1980, pp. 261-266.

Rapid biological information systems using aquatic organisms to monitor water and wastewater quality have only recently begun to develop technologically for practical on-site applications. One approach which has been demonstrating its feasibility recently monitors the ventilatory behavior of fish to assess, for example, the quality of drinking water supplies and industrial wastewater discharges. A comparison of the basic strategies of the various biological monitoring systems making use of this concept was presented. In addition, the applications of these systems were discussed.

Cermak, R.J., Feldman, A.D. and Webb, R.P., "Hydrologic Land Use Classification Using Landsat", Technical Paper No. 67, Oct. 1979, U.S. Army Engineers Hydrologic Engineering Center, Davis, California.

This report describes the Hydrologic Engineering Center's experience with land use classification from LANDSAT multispectral imagery. Land use is required for the estimation of hydrologic model parameters. The land use classification procedure used, developed at the University of California, Davis, for the Corps of Engineers, is an unsupervised, noninteractive approach requiring no special image processing equipment. Watershed land use was determined from LANDSAT digital data, entered into a geographic data bank, and compared with a conventional land use classification. Hydrologic simulation model parameters were estimated from land use and other basin characteristics. The generated discharge frequency curves, corresponding to the alternative land use classifications, permitted the hydrologic significance of accuracy in land use identification to be assessed.

Collotzi, A.W. and Dunham, D.K., "Inventory and Display of Aquatic Habitat", Classification, Inventory, and Analysis of Fish and Wildlife Habitat--The Proceedings of a Natural Symposium, January 24-27, 1977, Phoenix, Arizona, FWS/OBS-78/76, 1978, U.S. Forest Service, Washington, D.C., pp. 533-542.

A systematic and uniform approach to the inventory and display of aquatic habitat has been developed in the Intermountain Region of the U.S. Forest Service. A line transect method constitutes the inventory of aquatic habitat by measuring habitat variables that are grouped into three habitat categories: water surface, streambottom, and streambanks. A rating score for stream habitat variables is displayed for interpretation and analysis based on pool measure, pool structure, stream bottom, and stream environment. A land classification system for the valley bottom lands provides for the display of the aquatic resource at various levels of planning, and for the interaction of all resources within the valley bottom. This classification consists of four hierarchical levels: subsection, valley bottom association, valley bottom type, and valley bottom phase--the association level being the most important.

Colwell, J.E. et al., "Use of Landsat Data to Assess Waterfowl Habitat Quality", Jan. 1978, Environmental Research Institute, University of Michigan, Ann Arbor, Michigan.

The feasibility of using Landsat data is examined for (1) the analysis of annual waterfowl production by monitoring the number of breeding and brood ponds that are present, and (2) the ability to assess waterfowl habitat based on various relationships between ponds and surrounding upland terrain types. A large coarse Landsat census can improve the estimate of the number of ponds in a population with respect to an accurate small sample done by lowflying aircraft. One reason is the large semi-random variability in pond area per small sample unit. In a multitemporal Landsat classification map, marsh categories were reasonably well recognized, although there was some misclassification between shallow and deep marshes. Bare soil was generally well recognized as were the upland terrain classes. Landsat data may also determine waterfowl habitat quality, as related to polluted terrain conditions.

Fry, J.P. and Pflieger, W.L., "Habitat Scarcity, a Basis for Assigning Unit Values for Assessment of Aquatic Wildlife Habitat", Classification, Inventory, and Analysis of Fish and Wildlife Habitat--The Proceedings of a National Symposium, January 24, 1977, Phoenix, Arizona, FWS/OBS-78/76, 1978, U.S. Forest Service, Washington, D.C., pp. 491-494.

Current methods, for numerical evaluation of wildlife habitat, measure habitat suitability for selected species, weighted toward those important for uses such as hunting and fishing. Evaluation should address future optional needs as well as identified uses. The security of a species is inversely related to the scarcity of its habitat, thus relative scarcity of a habitat type is a measure of the types important for future management options. A system for describing and assessing aquatic habitat in terms of relative scarcity and condition is being developed in Missouri. The fish fauna and a few physical factors will be used to delimit stream habitat types. Index values will be computed for each type based on its abundance and condition. Unit values of degraded habitat could be increased by habitat restoration as an exchange, for unit value losses caused by water resources projects, using habitat scarcity and condition as the primary basis for determining the location and amount of exchange.

Gonor, J.J. and Kemp, P.F., "Procedures for Quantitative Ecological Assessments in Intertidal Environments", EPA-600/3-78-087, Sept. 1978, School of Oceanography, Oregon State University, Corvallis, Oregon.

This report is an effort to assemble the best available published procedures for quantitative ecological studies in marine intertidal benthic environments, applicable to evaluating existing or potential pollution effects. Considerable material on how to rigorously devise sampling programs is included. Methods and procedures recommended are synthetic and compiled from a variety of sources. Guidelines are intended for application in marine intertidal ecological impact assessments, pre- and post-pollution baseline surveys, and long-term monitoring programs intended for the detection and forecasting of potential and human impacts in the intertidal environment.

Groves, D.H. and Coltharp, G.B., "Remote Sensing of Effects of Land Use Practices on Water Quality", Final Report, May 1977, Department of Forestry, University of Kentucky, Lexington, Kentucky.

An intensive 2-year study of 6 watersheds in the Cumberland Plateau region of eastern Kentucky determined the utility of manual densitometry and color additive viewing of aircraft-land LANDSAT transparencies for monitoring land use and land use change. The study area was comprised of reclaimed surface-mined land and forestland. Manual photo interpretation techniques stratified the study area into vegetative types. An intensive ground survey was undertaken to ascertain kind, size and extent of vegetation present in each. Values obtained from subsequent densitometric sampling of NASA research aircraft and LANDSAT imagery were examined for correlation and predictability of corresponding vegetation types. Densitometer filter-aperture combinations were examined for determination of vegetation classification success. Densitometric values were compared with sample water quality values obtained from the watersheds. Linear regression equations were derived for water quality-densitometry and water quality-percent disturbed land relationships. Manual densitometry of LANDSAT imagery merely provided discrimination between wholly forested and partially surface-mined watersheds, while that of seasonal color infrared imagery yielded good classification of 8 broad vegetation categories of forested and reclaimed surface mines. Some correlation between densitometric data and water quality parameters appears to exist.

Haugen, R.K., McKim, H.L. and Marlar, T.L., "Remote Sensing of Land Use and Water Quality Relationships--Wisconsin Shore, Lake Michigan", Report No. 76-30, Aug. 1976, Cold Regions Research and Engineering Laboratory, U.S. Department of the Army, Hanover, New Hampshire.

Various types of aerial imagery were evaluated for their usefulness in assessing effects of land use on sediment loading of streams on the Wisconsin shore of Lake Michigan. High altitude color infrared imagery acquired at 60,000 ft with a 9-in format RC-8 and Zeiss cameras was the most acceptable remote sensing technique for the mapping and measurement of land use types. Other sensors evaluated included: (1) high altitude RS-7 thermal scanner--useful for mapping major circulation patterns in large water bodies; (2) RS-14 thermal scanner--recommended for mapping effluent plumes and differentiating within wetlands or impervious urban settings, although lack of geometric control, narrow field of view, and cost largely preclude its use; (3) high altitude multispectral 70 mm photography (Hasselblad)--the small format was difficult to work with, and data was more easily obtained with the 9-in format cameras; (4) high-altitude color (SO 397)--although applicable to small area, single flight line surveys, this method does not appear feasible for regional area surveys or with high altitude imagery due to wide variation in color balance and other factors; and (5) PMIS (Passive Microwave Imaging System)--severe limitations were encountered, and it was not possible to consistently identify major patterns on a 64 color photographic rendition of PMIS scanning data. Sampling mapping was done for the Manitowoc, East Twin, and Oconto watersheds.

Hellawell, J.M., <u>Biological Surveillance of Rivers: A Biological Monitoring Handbook</u>, 1978, Water Research Center, Stevenage, England.

This 10-chapter monograph deals with water quality and pollution surveys, including monitoring objectives, water standards and criteria, field surveys, sampling strategies (for macroinvertebrates, fish, benthic organisms, macrophytes, etc.), data analysis and biotic indices, comparisons of data-handling methods, presentation and interpretation of survey results, and sources of additional help (keys for taxa identification, mathematical tables, bibliographic references, etc.).

Hundemann, A.S., "Remote Sensing Applied to Environmental Pollution Detection and Management (A Bibliography with Abstracts)", NTIS/PS-78/9789/4WP, Aug. 1978, National Technical Information Service, U.S. Department of Commerce, Springfield, Virginia.

Application of remote sensing methods to air, water, and noise pollution problems is discussed. Topical areas cover characteristics of dispersion and diffusion by which pollutants are transported, eutrophication of lakes, thermal discharge from electrical power plants, outfalls from industrial plants, atmospheric aerosols under various meteorological conditions, monitoring of oil spills, and application of remote sensing to estuarine problems.

Hyman, M.A.M., Lorda, E. and Saila, S.B., "A Standard Program for Environmental Impact Assessment: Phase I--Ichthyoplankton Sampling", Proceedings of Program Review of Environmental Effects of Energy Related Activities on Marine/Estuarine Ecosystems, EPA-600/7-77-111, Oct. 1977, U.S. Environmental Protection Agency, Washington, D.C., pp. 153-159.

The complexity of estuarine ecosystems and the growing need for proper management decisions and regulations have placed a high priority on large-scale integrated models. However, the practical effectiveness of predictions from such a model is limited. It is suggested that a successful environment impact assessment program is predicated on a sequence of carefully planned experiments designed to answer specific questions, rather than on a general measurement system. This carefully planned standard program can then be used for establishing suitable guidelines for more routine types of assessment. To assign sources of variability in ichthyoplankton surveys more accurately, an intensive sampling scheme was proposed for a single site. The purpose of this study was to work toward producing a standard program for ichthyoplankton sampling. If variability is assigned to various effects, as opposed to being confined into an overall error term, data will be of an overall higher consistency and thus, generally more useful. Specifically, once this variability is better understood, any modeling effort should prove more productive.

Jacobs, F. and Grant, G.C., "Guidelines for Zooplankton Sampling in Quantitative Baseline and Monitoring Programs", EPA-600/3-78-026, Feb. 1978, Virginia Institute of Marine Science, Gloucester Point, Virginia.

Methods of zooplankton sampling and analysis for quantitative baseline and monitoring surveys are described and evaluated. Zooplankton exhibit wide spatial, diurnal, and seasonal variations which, along with gear bias and capture avoidance, complicate data collecting and subsequent assessment of relationships. This study concluded: (1) baseline studies require more frequent sampling and closely spaced stations than do monitoring studies; (2) sampling locations can be further apart in homogeneous waters, while in heterogenous coastal or estuarine water sites should be more closely spaced and sampling conducted more frequently; (3) sampling sites can be selected by means of a grid overlaid on the study area, though transects may be used with study areas which cover great distances and when ship time is limited; (4) in pollution studies a series of transect lines radiating from a single source may be advisable; (5) pumping systems are an expensive but efficient means of capturing microzooplankton; a rate exceeding 150 liters/min is necessary to minimize avoidance; (6) nets with mouth openings of 50-100 cm diameter are recommended for most groups of mesozooplankton; (7) in areas of high plankton density, a 333-micrometer mesh is preferable; (8) specimens are generally best preserved in 4 percent buffered formaldehyde; and (9) the Folsom and Burrell splitters are suggested. Statistical methods of analysis are discussed, and a bibliography is provided.

James, W.P., Woods, C.E. and Blanz, R.E., "Environmental Evaluation of Water Resources Development", Completion Report TR-76, July 1976, Texas Water Resources Institute, Texas A and M University, College Station, Texas.

A methodology for the utilization of LANDSAT-1 imagery and aerial photography on the environmental evaluation of water resources development is presented. Environmental impact statements for water resource projects were collected and reviewed for the various regions of Texas. The environmental effects of channelization and surface impoundments are discussed for 12 physiographic regions of the state as delineated on black and white satellite (LANDSAT-1) mosaic of band 7. With the aid of LANDSAT-1 imagery, representative or typical transects were chosen within each region. Profiles of each site were constructed from topographic maps, and environmental data were accumulated for each site and related to low altitude aerial photography and enlarged LANDSAT-1 false color composites. Each diagrammatic transect, with accompanying data and photographs, provides significant information for input of environmental amenities on a local and regional scale into preliminary water resources development studies. The utilization of the transects provides a visual display of available information, aids in the identification and inventory of resources, assists in the identification of data gaps, and provides a planning tool for additional data acquisition. Remote sensing techniques are readily adapted to water resources planning. LANDSAT-1 imagery as well as conventional low altitude aerial photography provides the planner with a synoptic overview of the resource area. The delineation of physiographic regions by LANDSAT-1 imagery will be helpful in defining delicate border areas and delineating broad environmental areas. Satellite imagery is applicable for transect siting in aerial river basin studies or regional analysis.

Liebetrau, A.M., "Water Quality Sampling: Some Statistical Considerations", Water Resources Research, Vol. 15, No. 6, Dec. 1979, pp. 1717-1725.

Typically, a state, regional, or national water quality (WQ) monitoring system has many objectives. Examples include: (1) providing a systemwide synopsis of WQ; (2) determining whether selected WQ parameters show gradual changes over time; (3a) detecting actual or potential WQ problems, (3b) determining the specific causes of actual problems, and (3c) assessing the effect of any corrective action; and (4) enforcing the law. While these purposes have different data requirements, they are interrelated, and data collected for any one can have value for others; for example, data collected for the listed reasons could be used for future long-range planning. Each objective makes certain demands of a WQ sampling network, and these in turn have implications concerning sampling design and statistical analysis of resulting data. In a sampling scheme flexible enough to encompass multiple objectives, data sets collected for answering questions arising under points 1-4 need not all have similar characteristics; consequently, a variety of statistical methods are needed. A good sampling design is important for achieving the first two objectives. A working definition of "water quality" was given initially, and this was followed by the definition of a sampling population. The set of all segments of the drainage network serves as the population for several sampling plans. In particular, the use of probability sampling to design a network of synoptic sampling stations was discussed in detail. Finally, sequential statistical procedures were applied to a specific problem of type 3c. In terms of the amount of data required, sequential methods are quite efficient for detecting changes of a specific magnitude in some WQ parameters.

Loftis, J.C. and Ward, R.C., "Sampling Frequency Selection for Regulatory Water Quality Monitoring", Water Resources Bulletin, Vol. 16, No. 3, June 1980, pp. 501-507.

The selection of sampling frequencies in order to achieve reasonably small and uniform confidence interval widths about annual sample means or sample geometric means of water quality constituents was suggested as a rational approach to regulatory monitoring network design. Methods were presented for predicting confidence interval widths at specified sampling frequencies while considering both seasonal variation and serial correlation of the quality time series. Deterministic annual cycles were isolated, and serial dependence structures of the autoregressive, moving-average type were identified through time series analysis of historic water quality records. The methods were applied to records for five quality constituents from a nine-station network in Illinois. Confidence interval widths about annual geometric means were computed over a range of sampling frequencies appropriate in regulatory monitoring. Results were compared with those obtained when a less rigorous approach, ignoring seasonal variation and serial correlation was used. For a monthly sampling frequency the error created by ignoring both seasonal variation and serial correlation was approximately 8 percent.

Loftis, J.C. and Ward, R.C., "Water Quality Monitoring--Some Practical Sampling Frequency Considerations", Environmental Management, Vol. 4, No. 6, Nov. 1980, pp. 521-526.

Relationships between water quality sampling frequency in a monitoring network and the effect of statistical methods and assumptions on sampling frequency were examined. Detailed data from the Illinois State Water Survey network was studied using three methods: (1) considering neither seasonal variation nor serial correlation (the most widely used method), (2) considering only seasonal variation, and (3) considering both seasonal variation and serial correlation (assumed to be the most accurate method). Confidence level widths for each were plotted as a function of sampling frequency, which were divided into 3 regions with zones of uncertainty between each region. In Region 1 (sampling interval, 1-12 days), Method 3 is most accurate; in Region 2 (sampling interval, 12-34 days), Method 1 or 3 apply; and in Region 3 (sampling interval greater than 34 days), Method 1 is pertinent.

McNeeley, R.N., Neimanis, V.P. and Dwyer, L., "Water Quality Sourcebook: A Guide to Water Quality Parameters", 1979, Water Quality Branch, Department of the Environment, Ottawa, Ontario, Canada.

A broad spectrum of water quality parameters that are frequently encountered in the aquatic environment are discussed in general terms. Seventy parameters ranging from alkalinity to zinc are outlined under the following headings: (a) general information, (b) environmental range, (c) natural and man-made sources, (d) water quality guidelines, and (e) effects on use. A glossary is included to further explain the scientific terminology which is essential to the text. An appendix of "Specific Use Guidelines" draws together and summarizes in a tabular form all the water quality guidelines presented in the text.

Persoone, G. and DePauw, N., "Systems of Biological Indicators for Water Quality Assessment", Biological Aspects of Fresh Water Pollution, O. Ravera, Editor, Pergamon Press, New York, New York, 1978, pp. 39-75.

The paper reviews and comments on the major methods which have been worked out at the international level for the assessment of the quality of surface waters with the aid of biological-ecological criteria. The endeavours of the Commission of the European Economic Communities aiming at the intercalibration of methods used in different countries, and the trends which emerged from recent exercises carried out in different hydrographic basins, are outlined briefly.

States, J.B. et al., "A Systems Approach to Ecological Baseline Studies", FWS/OBS-78/21, Mar. 1978, U.S. Fish and Wildlife Service, Fort Collins, Colorado.

This handbook describes a systematic approach to planning and conducting a "holistic" study of selected ecosystem components and functions, which are significant with regard to energy development projects in the western United States. Techniques of ecological systems analysis are described, and the manual explains how to build a conceptual ecosystem model and use it to plan a baseline study. A glossary of key terms and an annotated bibliography are included.

Stout, G.E. et al., "Baseline Data Requirements for Assessing Environmental Impact", IIEQ-78-05, May 1978, Institute for Environmental Studies, University of Illinois, Urbana-Champaign, Illinois.

This study has developed a guide that may be used by technical personnel to perform an integrated baseline evaluation of changes in the total environment--in plants, soils, and animals (including man)--that is needed for a factual pinpoint assessment. The methodology outlined in this guide requires substantial resources both in manpower and funds. The management and evaluation of the survey instrument should be performed by a qualified organization. Comprehensive ecosystem evaluation requires an interdisciplinary team of scientists. As a result, the execution of a baseline impact assessment requires considerable planning, funding, and evaluation.

Stofan, P.E. and Grant, G.C., "Phytoplankton Sampling in Quantitative Baseline and Monitoring Programs", EPA-600/3-78-025, Feb. 1978, Virginia Institute of Marine Science, Gloucester Point, Virginia.

An overview of methods for phytoplankton sampling, sample treatment and sample analysis is presented, along with an extensive bibliography. The complexity of a dynamic oceanic and estuarine ecosystem precludes prescription of a detailed and concise guideline for baseline or monitoring surveys. However, various general recommendations include: (1) divide the investigative area into hydrographic subareas, and overlay them with a grid to provide a statistically valid sampling basis; (2) sample at weekly to monthly intervals in estuarine systems, and biweekly to bimonthly in ocean environments; (3) intensify sampling following ecosystem disturbances; (4) subsurface bottle sampling is most efficient in subtrophic waters with high phytoplankton densities, a net or pump sampler may be preferable for oligotrophic ocean areas, while screens are well-suited for surface sampling; (5) the Utermohl enumeration method is the most widely used because it provides a wide range of adaptability at medium cost; conventional live analysis should also be performed on sample aliquots; (6) phytoplankton analysis in a general baseline survey should be accompanied by a wide array of ancillary data, including delineation of zooplankton, ichthyoplankton, and benthic communities; (7) Carbon-14 is the usual productivity estimate; and (8) Chlorophyll-a is a good measure of standing stock. Several data storage banks were searched for this study to determine the current status of phytoplankton surveys.

Villeneuve, J.P. et al., "Kriging in the Design of Streamflow Sampling Networks", Water Resources Research, Vol. 15, No. 6, Dec. 1979, pp. 1833-1840.

Kriging is of particular interest in network design because of its ability to estimate streamflow values using existing stations. Another possibility offered by kriging is the estimation of variance reduction gained by addition of fictitious stations in regions of high variance. A brief description was given of kriging theory as developed at the Ecole des Mines de Paris. In order to improve and optimize the Quebec streamflow recording network design the authors kriged specific streamflows with a given return period over the Quebec province, using the data observed at existing stations. For the evaluation of the given return period flow and its sampling variance at each gaged site the authors used the log Pearson type 3 distribution model. Kriging is an optimal estimation technique, in terms of minimum variance, and contrary to other methods, it gives an estimation variance for any point in the kriged domain, which is esential in network design. Network analysis and optimization are presently expensive and time-consuming undertakings. In this study of flows with given return periods, the authors demonstrated the aptitude of kriging to transpose information from gaged to ungaged sites, and demonstrated how the different elements of kriging can be used for network design.

Ward, D.V., Biological Environmental Impact Studies: Theory and Methods, 1978, Academic Press, New York, New York.

This book shows how the time resources presently devoted to descriptive empirical efforts and guessing of impacts can be redirected to focused manipulative experimentation so that the predictability of biological changes is greatly improved. The author specifies how this type of biological environmental impact can be approached and accomplished. The book makes three major contributions: (1) to present the idea and some examples of manipulative rather than descriptive ecological studies to managers and government agents concerned with requesting and reviewing environmental impact assessment studies; (2) to aid the biologist performing the impact study to review rapidly a battery of approaches and methods that are applicable to manipulative impact assessment studies; and (3) as an aid to training biology students, many of whom will be employed to perform environmental impact studies. Chapter headings include Environmental Impact Analysis, The Field Survey: Preliminary System Analysis, Modelling the System, The Field Experiment, Laboratory Studies, Some Examples, and Conclusions.

Wiederholm, T., "Use of Benthos in Lake Monitoring", Journal of the Water Pollution Control Federation, Vol. 52, No. 3, Mar. 1980, pp. 537-547.

This paper discusses the rationale behind the use of biological variables in environmental monitoring of lakes, as well as the principles of variable selection and the limitations of data usability. Profundal benthic communities were suggested to be an integral measure of autotrophic and heterotrophic lake processes. Measures of community structure and their relationship to morphometric and epephic factors were presented and discussed, including indicator species/communities, diversity/species richness, oligochaete/chironomid ratio, and oligochaete abundance. It was demonstrated that each of these measures could be related to lake characteristics such as concentration of phosphorus or chlorophyll. To reach a logical ordering of data from different types of lakes or from different depths in one lake, it is necessary, however, to include the depth factor in some way. This was done simply by dividing or multiplying primary data by the factual sampling depth depending on whether the size of a specific community measure is positively or negatively correlated with depth. Although rather crude, this operation was felt to be intuitively sound in that it accounted for the combined effect of important factors such as food supply (greater degree of mineralization but more accumulation of sediment at greater depth), predation pressure (generally declining with increasing depth), and unfavorable abiotic factors (for example, oxygen deficit).

APPENDIX H

ENVIRONMENTAL INDICES AND INDICATORS

Ball, R.O. and Church, R.L., "Water Quality Indexing and Scoring", Journal of the Environmental Engineering Division, American Society of Civil Engineers, Vol. 106, No. EE4, Aug. 1980, pp. 757-771.

The desirability of developing water quality indices, which has been frequently cited, was examined. The development of currently available indices was reviewed, and a distinction was made between the indexing and scoring of water quality samples. It was pointed out that many indexes are, in fact, scoring methods, and that the mathematical basis for the scoring methods are not always consistent with the objectives of scoring. A general mathematical approach to scoring was described, and the necessity of considering the uniformity as well as the average quality of the water was described. The development and use of "true" indices was briefly reviewed, and recommendations for future indexing approaches were presented.

Booth, W.E., Carubia, P.C. and Lutz, F.C., "A Methodology for Comparative Evaluation of Water Quality Indices", 1976, Worcester Polytechnic Institute, Worcester, Massachusetts.

The relative usefulness of water quality assessment indices is examined by use of a methodology that compares and evaluates their performance characteristics. The method so devised demonstrates the evaluation of two indices by identifying their assumptions and limitations. Evaluation of the National Sanitation Foundation (NSaF) index shows that it is only suitable to assess the overall water quality as characterized by nine physicochemical parameters. It is somewhat insensitive to pollution problems exhibited in only one of the parameter's values. Since the parameters quality ratings, and weights that are used are fixed, they cannot vary to account for the ultimate use of the index, data availability, new discoveries, or conditions associated with different geographic locations. Applications of the NSaF index are extremely limited when attempting to identify or analyze specific pollution problems as they may affect a water use. Conversely, the Harkins index is best used when assessing trends in a specific pollution problem area. The index has no set parameter requirements and is therefore very flexible in application since only those parameters of interest to the user need be included in the computations.

Chiaudani, G. and Pagnotta, R., "Ratio of ATP/Chlorophyll as an Index of Rivers' Water Quality", Proceedings: Congress in Denmark 1977 Part 3; Internationale Vereinigung fur Theoretische und Angewandte Limnologie, Instituto di Ricerca sulle Acque, Rome, Italy, Vol. 20, 1978, pp. 1897-1901.

The September 1975 to April 1976 study of a segment of the lower Tiber River, Italy, confirmed that ATP/chlorophyll (Ch) ratios are useful for rapidly detecting water quality shifts in rivers where eutrophication or organic wastes are involved. The ATP/Ch method more accurately assesses water quality than biomass-chlorophyll ratios. Water quality is classified by the burden of organic, biologically decomposable substances; low to moderate, moderate to critical, and severely polluted. Main chemical parameters used as references are COD, ammonium content, and oxygen content. The ATP/Ch ratio increases with ammonium concentration and COD, while DO decreases with relation to increased organic load. Correlation coefficients are all highly significant for chemical parameters and mean ATP/Ch. In free-flowing rivers, ATP/Ch ratios (x 100) less than 10 indicate low to moderate pollution. Moderate to critical pollution is characterized by a ratio of 10-20. Severely polluted rivers have ATP/Ch ratios greater than 20. In cultured algae, values less than 10 characterize autotrophic conditions; with values more than 20, heterotrophic organisms prevail. Values ranging from 10 to 20 indicate a succession of autotrophic conditions; with values more than 20, heterotrophic organisms prevail.

Dunnette, D.A., "A Geographically Variable Water Quality Index Used in Oregon", Journal of the Water Pollution Control Federation, Vol. 51, No. 1, Jan. 1979, pp. 53-61.

An Oregon Water Quality Index has been developed which takes into consideration differences in water quality resulting from geographical characteristics of separate basins. The index was developed for the purpose of providing a simple, concise, and valid method for expressing the significance of regularly generated laboratory data. The trend-monitoring value of the index was demonstrated for two quite different Willamette River stations. Correlations among this and several other proposed indexes averaged 0.87. Yearly and seasonal variations in water quality were quantitized and found to average 88.9 and 78.9 Oregon Water Quality Index units for the higher and lower water quality stations, respectively, over the period 1971 to 1976. Calculated rates of change in water quality were +0.68 and +0.91 Oregon Water Quality Index units/year for the two stations for 1971 to 1976. The Oregon Water Quality Index is now used routinely in that state's primary station sampling program to recognize water quality trends.

House, M. and Ellis, J.B., "Water Quality Indices: An Additional Management Tool", Water Science and Technology, Vol. 13, No. 7, 1981, pp. 413-423.

A comparison of the geometric weighted formulation of the Scottish Development Department (SDD) water quality index with the National Water Council (NWC) and Thames Water Authority (TWA) classifications showed 76 percent and 81 percent agreement, respectively. This suggests that, with modifications, the SDD index could be successfully used to monitor changes in water quality. In its present form it is most accurate in good quality waters and less accurate in low quality waters. There are several advantages to adopting a water quality index: (1) it reduces a large amount of data to a single index; (2) it reduces subjective judgments; (3) it provides more information on water quality; (4) it allows greater detail in describing specific river reaches and in showing small changes in quality; (5) it reports specific quality rather than qualitative approximations; and (6) it is more easily understood by laymen.

Ibbotson, B. and Adams, B.J., "Formulation and Testing of a New Water Quality Index", Water Pollution Research in Canada 1977, Proceedings of Twelfth Canadian Symposium on Water Pollution Research, 1977, Department of Civil Engineering, Toronto University, Toronto, Ontario, Canada, pp. 101-119.

The need to protect the environment has promoted the development of new ways to communicate environmental information to policy makers and the general public. Presented is a new formulation which translates water quality parameter values into simple numerical results which can then be summed to give water quality index scores. The mechanism uses the matrix format to organize the display results with water quality parameters on one axis and common water activities on the other. The mechanism's task is then to assess the suitability of each parameter to each activity, and subsequently to render the results into simple scores. The index can be applied to any situation in which water quality is a consideration. Its applicability lies in its offering to the perspective user the opportunity to calibrate the index to the needs of a specific situation. The procedure offers a systematic approach that is highly flexible and that generates results which are easily understood. The principal reasons for developing this index mechanism are to enable presentation of a simplified measure of water quality, and to augment expert assessment of raw water quality data, not to supplant this important aspect of water quality management.

Inhaber, H., Environmental Indices, 1976, John Wiley and Sons, Inc., New York, New York.

This book provides a general summary of environmental indices, beginning with a discussion of concepts and economic indices. Chapters are included on air quality indices, water indices, land indices, biological indices, aesthetic indices, and other environmental indices.

Keilani, W.M., Peters, R.H. and Reynolds, P.J., "A Water Quality Economic Index", Proceedings of the 9th Canadian Symposium on Water Pollution Research, 1974, Department of the Environment, Ottawa, Ontario, Canada, pp. 1-24.

A methodology is presented for producing a water quality economic index which will provide information on the direction and amount of change in water quality on a regional and national basis. It will be useful for evaluation of water quality trends and for economic decision-making regarding preventive measures and treatment programs. The methodology consists of statistical and mathematical models, supported by tables, for the different steps in calculating the indices. Two sample indices derived from actual and estimated data are also given. The method utilizes an adaption of Laspeyre's weighted aggregative formula which employs base year weights for construction of indices. Five water quality parameters are assigned to each of 12 water uses. Each parameter value, measured at the actual or potential point of use, is rated between a "desired objective" of 0 and a "maximum permissible level" of 100. Each parameter is then given a relative weight out of a total weight of unity to reflect its importance to the use. Relative economic weights for all uses are derived by means of a damage coefficient. Segments of the index are combined where parameters are common to more than one use, and a bivariate use and parameter economic index is constructed. A set of subgroup bivariate indices by both use and individual parameters is also produced to allow for quality trend evaluation.

Landwehr, J.M., "A Statistical View of a Class of Water Quality Indices", Water Resources Research, Vol. 15, No. 2, Apr. 1979, pp. 460-468.

Water quality indices are treated as random variables; the class of quality "averaging" indices is examined and several statistics are derived. The probability density functions of the water quality constituents and the structure of the transform functions (rating curves) influence the ability of indices to measure policy performance.

Lee, C.D., Wang, S.B. and Kuo, C.L., "Benthic Macroinvertebrate and Fish as Biological Indicators of Water Quality, with Reference to Community Diversity Index", Water Pollution Control in Developing Countries, Proceedings of the International Conference, Held at Bangkok, Thailand, February 1978, 1978, Pergamon Press, Inc., New York, New York, pp. 233-238.

The community diversity index of benthic macroinvertebrates and fish can be an efficient and effective tool for quantifying the impact of wastewater pollution on stream environments. The full-scale evaluation of stream environments can be more accurately determined if indicator organism systems are applied to facilitate water quality assessments. Examples of proper indicator organisms are: Hydropsyche formosae, Neoporia sp., Eubrianax sp., Sigara sp., Acheilognathus himontegus, Rhinogobius similis, Pseudogobio brevirostris, Zacco temmincki, Caridina denticula, Macrobrachium asperulum, Macrobranchium japonicum, Thiara granifera, Gambusia patruelis, Tilapia mossambica, Eriocheir japonicus, and Ophidonais sp. In terms of biological assessment, almost all the upstream parts of 15 major rivers in western Taiwan have good water quality for aquatic life. However, half of the downstream parts of these rivers have been exposed to moderate or even heavy contamination.

Ott, W.R., Environmental Indices--Theory and Practice, 1978, Ann Arbor Science Publishers, Inc., Ann Arbor, Michigan.

This book systematically describes environmental index systems along with principles for their design, application and structure. Chapter I introduces the reader to environmental data, presenting simple communicative approaches such as environmental quality profiles. It also describes national monitoring activities. Chapter II presents a new conceptual framework that is designed to embrace nearly all existing environmental indices, allowing the behavior of different index structures to be compared and probed in detail. Chapter III concentrates on air pollution indices, using the conceptual framework introduced in Chapter II to analyze and compare published air pollution indices. Chapter III also gives a detailed summary of the historical evolution and scientific basis for the Pollutant Standards Index (PSI), which has been developed for uniform application throughout the United States. Computational aids (equations, tables and nomograms) for applying PSI to actual air quality data are included. Chapter IV covers water pollution indices, using the theoretical framework and concepts from Chapter II to examine currently used water pollution indices; it also presents design principles for an ideal water quality index and discusses a candidate index structure. In both Chapters III and IV, the current air and water index usage patterns in the United States are described in detail. Finally, Chapter V presents conceptual approaches, such as Quality of Life and environmental damage functions, that extend beyond the traditional fields of air and water pollution.

Polivannaya, M.F. and Sergeyeva, O.A., "Zooplankters as Bioindicators of Water Quality", Hydrobiological Journal, Vol. 14, No. 3, 1978, pp. 39-43.

The purpose of this study was to conduct a comparative evaluation of methods of biological analysis of water quality by using, as an example, zooplankton of the Dnieper River and the Kanev reservoir (USSR). The material consisted of samples collected in the river in 1969 and 1972, and in the reservoir in 1975. The results indicate that pollution leads to less varied species composition of the zooplankton. The first to suffer are crustaceans and among them, the nonpredator forms. Polyphagous insects (Asplanchnidae, Brachionus calyciflorus, Brachionus angularis and predator Cyclopoidae) tend to maintain themselves in the plankton. The data indicate an abrupt dip of the number of zooplankton dominants in river zones subjected to pollution. Age composition, replenishment and loss, fertility, morphometric characters of populations of leading species under normal conditions and during disturbances associated with pollution, may serve as supplementary indices of water quality that are calculated in terms of zooplankton.

Provencher, M. and Lamontagne, M.P., "A Method for Establishing a Water Quality Index for Different Uses (IQE)", July 1979, Environmental Protection Services, Montreal, Quebec, Canada.

This report presents the various stages in developing a water quality index (IQE), as well as its possible applications. The IQE is a mathematical method of determining the quality of a given stretch of water in terms of a given use, with a number rating of 0 to 100. Three special types of needs were taken into account--industrial, social and ecological. Three categories of water quality were used--chemical, biological and physical characteristics--and for each category of use, a set of weighted parameters is assigned. Six computer subroutines were used to input data in order to assess the characteristics of the quality evaluation function under different conditions with diferent combinations of parameters. Categories of water use, such as drinking, cooling, and industry, were compared to their respective percentage use of the total water supply. Data processing was found to be both rapid and simple. The Index, which can be calculated for 17 categories, has a fairly wide field of application. The method should be useful for several purposes--to prepare relative classifications of different bodies of water, and to determine changes in water quality in space and time scales; it could be a useful tool in carrying out technical studies of the aquatic environment, in establishing an indicator for incorporation into directories and data banks, and in providing information on water quality.

Reynolds, P.J., "Environmental Indicators in River Basin Management", Proceedings of International Symposium on Hydrological Characteristics of River Basins and the Effects of These Characteristics on Better Waste Management, Tokyo, Japan, Dec. 1-8, 1975, IAHS-AISH Pub. No. 117, 1975, International Association of Hydrological Scientists, Paris, France, pp. 557-569.

Results of methodology development for the application of a water quality index are demonstrated in a river basin site. Water quality indices are selected based on local expert opinion and measured against given water uses and water quality objectives. Relative economic weights for all uses are derived by means of a damage coefficient and are expressed in the form of an economic index. The combination of indices are used for deriving the proportionate share of treatment costs among the various users, and for establishing the general value of water uses within a river basin for more effective water management decisions.

Thomas, W.A., "Attitudes of Professionals in Water Management Toward the Use of Water Quality Indices", Journal of Environmental Management, Vol. 4, 1976, pp. 325-338.

Attitudes of professionals regarding the applicability of water quality indexes were surveyed by a questionnaire distributed to 1800 randomly selected subscribers to "Water and Sewage Works". Useful responses were received from 226 individuals, representing a broad range of professional training (engineering, chemistry, business, and science), employers (state, local and Federal government, industry, consulting firms, and university), and occupational designations (engineering, sales or administration, and science of teaching). Results indicated that water management professionals generally agreed that present methods for reporting water quality are inadequate, that politicians and the public do not really know what is meant by water pollution, and that a need exists for uniform and objective descriptors of water quality. However, they disagreed on the relative merits and limitations of numerical water quality indices, with some clear delineations according to professional training, type of employer, and occupational responsibility. The main disadvantages perceived were potential misinterpretation and misuse of an index, and loss of technical detail as a result of aggregating data.

Yu, J.K. and Fogel, M.M., "The Development of a Combined Water Quality Index", Water Resources Bulletin, Vol. 14, No. 5, Oct. 1978, pp. 1239-1250.

Use-oriented benefits and treatment costs analysis has been incorporated into a water quality index to show an economically optimized concentration for the treatment of the pollutants and the resulting water quality. This combined water quality index can be used in decision-making at the Federal and local government levels. Five major parameters, i.e., coliforms, nitrogen, phosphorus, suspended solids, and detergents, have been considered for the municipal wastewater. With each higher level of improvement, the treatment costs increase accordingly and the benefits associated with the reuse of this treated waste water will increase also, but not for the nutrient removal in agricultural use. The optimal concentration is determined when the marginal costs equal the marginal benefits. The combined water quality index is the combination of the maximum net benefits and the water quality index of the optimized residual concentrations. This water quality index is zero dollars for the Tucson region in this study. The possible reclaimed use of municipal wastewater is for agricultural irrigation and recreational lakes for the Tucson region.

APPENDIX I

WATER QUANTITY/QUALITY IMPACT PREDICTION AND ASSESSMENT

Abbott, J., "Guidelines for Calibration of STORM", Training Document No. 8, 1977, Hydrologic Engineering Center, U.S. Army Corps of Engineers, Davis, California.

This report provides specific information on calibration and application of the Storage, Treatment, Overflow Runoff Model (STORM). The STORM model is intended for use in simulation of the quantity and quality of storm water runoff. In particular, the report discusses procedures for collection of rainfall, runoff quantity and quality data. Procedures were recommended for management of the collected data. Recommendations were provided for use of the site-specific data in calibration of the model. The calibrated model can then be used for two important planning components of a storm water study. These are: (1) prediction of wet-weather pollutographs (mass loading curves) for use in a receiving water assessment model, these pollutographs can include both surface runoff and dry weather flow in combined system; and (2) preliminary sizing of storage and treatment facilities to satisfy desired criteria for control of storm water runoff. The model will analyze a matrix of combinations of storage and treatment rates. Results include frequency information on quantity and quality of washoff of pollutants and soil erosion, as well as frequency information of the quantity and quality of storage overflows.

Ahlgren, I., "A Dilution Model Applied to a System of Shallow Eutrophic Lakes After Diversion of Sewage Effluents", Archieve fur Hydrobiologie, Vol. 89, No. 1/2, June 1980, pp. 17-32.

A simple hydraulic dilution model was used to compare predicted nitrogen and phosphorus concentrations with those observed in a chain of four heavily eutrophied shallow lakes north of Stockholm, Sweden, after all sewage effluents were diverted in 1970. Total P loadings were reduced by 80-90 percent, and total N loadings by 50-90 percent. The model successfully predicted P concentrations using a sediment retention coefficient of zero, but nitrogen fixation and denitrification produced differences between calculated and observed N concentrations. The average annual amplitudes in P levels were correlated with wind fetch over lake surface divided by mean depth, thus showing that differences among the lakes may depend on wind-generated turbulence at the sediment surface. Nitrogen was the limiting factor in chlorophyll-a concentration. The deepest of the four lakes showed decreased chlorophyll contents and increased Secchi disc transparencies since 1974. The other lakes did not have this tendency; in fact, algal blooms have been more severe in recent years.

Ahmed, R. and Schiller, R.W., "A Methodology for Estimating the Loads and Impacts of Non-Point Sources on Lake and Stream Water Quality", Proceedings of a Technical Symposium on Non-Point Pollution Control--Tools and Techniques for the Future, Technical Publication 81-1, Jan. 1981, Interstate Commission on the Potomac River Basin, Rockville, Maryland, pp. 154-162.

A model for computing loading estimates from nonpoint sources in a watershed (CLENS) was used to quantify the phosphorus in 16 lakes in Connecticut and Massachusetts as part of the development of preliminary management plans. The model is simple and can be used to develop quantitative estimates on nonpoint sources of pollution and their impact on water bodies. It can also be used to develop cost-effective management planning. CLENS is not computer-based although it is amenable to the use of computers. CLENS considers the following nonpoint sources: washoff from urban areas; erosion from other areas; washoff from barnyards and feedlots; leachate from landfills; washoff from roads; leachate from septic systems; and wet and dry fallout. CLENS was recently applied to Lake Waramaug, the second largest lake in Connecticut. Data from this application are given. CLENS has several advantages over a sampling program, it provides long-term estimates, and it locates the specific sources and details the primary causes of the pollution.

Austin, T.A., Landers, R.Q. and Dougal, M.D., "Environmental Management of Multipurpose Reservoirs Subject to Fluctuating Flood Pools", Technical Completion Report No. ISWRRI-84, June 1978, Water Resources Research Institute, Iowa State University, Ames, Iowa.

Mathematical models were developed and used to simulate the effect of fluctuating water levels in multipurpose reservoirs in Iowa. The models are designed as management and operational tools to evaluate trade-offs between the environmental impacts in the flood pool upstream of the dam and the economic benefits downstream of the dam. The Saylorville Reservoir located on the Des Moines River was used for the study because it periodically inundates an environmentally sensitive and scenic area in the Ledges State Park, one of the area's most popular recreational parks. A dynamic programming optimization model was developed to select the optimal operating policy for the reservoir. The model is forward-looking and has only a single decision variable, the release rate from the reservoir. State incremental or discrete differential solution algorithms are not needed as the number of discrete states used is not excessive. The ten largest historical floods were input to determine the operating policy which best minimizes damages. Also used were a reservoir routing model and a vegetative succession model. The vegetative model simulated general vegetative responses to flooding but it does not give precise results for specific sites. The lower areas of Ledges State Park will be heavily impacted by high reservoir water levels and extensive cleanup will be needed after each inundation. Five vegetative zones were defined for degree of damage and various techniques for mitigation are recommended.

Austin, T.A., Riddle, W.F. and Landers, Jr., R.Q., "Mathematical Modeling of Vegetative Impacts from Fluctuating Flood Pools", Water Resources Bulletin, Vol. 15, No. 5, Oct. 1979, pp. 1265-1280.

A flood control reservoir protects valuable developments on the downstream floodplain by storing flood waters and releasing them at a rate that will reduce the downstream damage. The water surface level of the flood pool behind the dam can fluctuate considerably during the occurrence of a large magnitude flood, thus causing severe impacts on shoreline vegetation and water-based recreation facilities located in the flood pool. A mathematical simulation model describing shoreline vegetative succession in response to flooding was presented. Plant species were grouped into ecologically similar compartments. Differential equations describing compartment intrinsic growth, intraspecies competition, interspecies competition, and other growth-limiting factors were solved numerically. The model was used to evaluate the impacts of various operating policies on plant succession for a new reservoir in central Iowa.

Baca, R.G. et al., "A Generalized Water Quality Model for Eutrophic Lakes and Reservoirs", Nov. 1974, Battelle Pacific Northwest Laboratory, Richland, Washington.

A multisegment deep reservoir water quality simulation model was developed for the U.S. Environmental Protection Agency and applied to American Falls Reservoir, Idaho. A hydrothermal submodel accurately predicts vertical temperature profiles with little or no subjective effort, and requires only standard meteorological data to predict seasonal temperature variations. A water quality submodel simulates natural seasonal patterns of algal growth and death and nutrient cycling, and predicts DO-BOD dynamics in reservoirs and impoundments. Principal environmental variables are: (1) water flows; (2) temperature; (3) solar radiation; (4) dissolved oxygen; (5) total and benthic BOD; (6) phytoplankton; (7) zooplankton; (8) nitrogen and phosphorus; (9) toxic material; and (10) coliform bacteria. Data used in modeling American Falls Reservoir is included in this report, along with results of sensitivity analyses and model performance. Model results were found to be most sensitive to algal-sinking velocities, benthic BOD, and diffusion coefficients. Potential applications are: (1) assessing future trophic state of reservoirs and lakes as a function of nutrient inputs, waste loadings, and hydraulic and hydrological site characteristics; (2) waste allocation studies; (3) waste discharge effects; (4) reservoir operation management; (5) pre-impoundment analyses; and (6) nutrient diversion and lake recovery studies.

Baca, R.G. et al., "Water Quality Models for Municipal Water Supply Reservoirs, Part 2. Model Formulation, Calibration and Verification", Jan. 1977, Battelle Pacific Northwest Laboratory, Richland, Washington.

Formulations, calibration, and verification of a eutrophication model and a limnological model for predicting and simulating water quality changes in municipal water supply reservoirs of Adelaide, Australia, are described. These computer models apply to both shallow and deep lakes and reservoirs. The eutrophication model incorporates inflows and outflows, fluctuations of the thermocline, nutrient fixation and mineralization, and sediment-water interactions to simulate monthly changes of four eutrophication indicators: (1) soluble phosphorus; (2) total phosphorus; (3) chlorophyll-a; and (4) Secchi disc depth. The limnological model is based on dynamics of heat and mass transport, hydromechanics, and chemical and biological transformations. The model simulates daily vertical and horizontal variations of: (1) water flow and temperature; (2) phytoplankton and zooplankton biomass; (3) nitrogen and phosphorus forms; (4) BOD; (5) DO; (6) total dissolved solids; and (7) suspended sediments. The eutrophication model was verified with data from Lake Washington (Washington) for 1922-72, which showed its ability to predict changes in lake trophic state; the limnological model was tested with data from Mt. Bold Reservoir near Adelaide for 1973-75 with good results, except for suspended sediment for which data were insufficient. Three other volumes provide a summary of the project, a user's manual, and Mt. Bold Reservoir data acquisition and evaluation.

Baca, R.G. et al., "Water Quality Models for Municipal Water Supply Reservoirs. Part 3. User's Manual", Jan. 1977, Battelle Pacific Northwest Laboratory, Richland, Washington.

This manual gives detailed user's instructions for two mathematical models developed to predict and simulate eutrophication and other water quality changes in municipal water supply reservoirs in Adelaide, Australia. The eutrophication model and limnological model apply to both deep and shallow lakes and reservoirs. This manual provides detailed input instruction, an explanation of required input data, and samples of model input and output. The eutrophication models predict monthly average changes in lake trophic state in terms of soluble and total phosphorus, chlorophyll-a, and Secchi disc depth over many years (10 or more). The computer program of this model is written as one main program. The limnological model predicts daily horizontal and vertical flow and water quality patterns over short-term periods (less than 10 years), simulating water flow and temperature, phytoplankton and zooplankton biomass, nitrogen and phosphorus forms, BOD, DO, total dissolved solids, and suspended sediments. The computer program of this model consists of four main subprograms coupled by common input and output, and run in sequence.

Booth, R.S., "A Systems Analysis Model for Calculating Radionuclide Transport Between Receiving Waters and Bottom Sediments", Proceedings of the 18th Rochester International Conference on Environmental Toxicity, 1975, Oak Ridge National Laboratory, Oak Ridge, Tennessee.

The model reported upon has four variables: the receiving water; interstitial water intermingled with the bottom sediments; bottom sediment particles that undergo sorption-desorption reactions with the interstitial water; and bottom sediment particles that undergo only sorption reactions with the interstitial water. From the model two tables were generated; one of these gives equilibrium radionuclide concentrations in the receiving water when radionuclide transfers to bottom sediments are possible, divided by the receiving water concentrations when transfers to bottom sediments are ignored; the other lists ratios of equilibrium radionuclide concentrations in sediments, divided by their corresponding receiving water concentrations. Results indicated that the usual effect of a neglect of sediment interactions is an overestimate of the total potential dose to man from the radionuclides.

Bourne, R.G., Day, G.N. and Debo, T.N., "Water Quality Modeling Using Hydrocomp Simulation Programming (HSP)", Proceedings of the 26th Annual Hydraulics Division Specialty Conference on Verification of Mathematical and Physical Models in Hydraulic Engineering, 1978, American Society of Civil Engineers, New York, New York, pp. 358-362.

Hydrocomp Simulation Programming (HSP) is a mathematical model capable of continuously simulating the hydrologic and water quality responses of a watershed. Continuous simulation essentially refers to use of long-term precipitation records to generate long-term flows. From the flows, the dynamic processes controlling water quality can be simulated for extended periods. The basic philosophy undergirding the simulation of water quality (using HSP) is that hydrology is a major mechanism controlling critical situations, assimilative capacity, and many other water quality processes. From the long-term simulation of water quality, probabilities of a particular event can be determined. For example, the percent of time DO falls below 5 mg/l can be estimated. The model incorporates many of the basic physical, chemical, and biological processes controlling water quality and can include factors which impact it (e.g., point and nonpoint loads).

Brandstetter, A. et al., "Water Quality Models for Municipal Water Supply Reservoirs, Part I. Summary", Jan. 1977, Battelle Pacific Northwest Laboratory, Richland, Washington.

This volume summarizes a project to develop a eutrophication model and a limnological model to predict and simulate eutrophication and other water quality changes in the municipal water supply reservoirs of Adelaide, Australia, and to evaluate the effectiveness of lake restoration schemes. Three other volumes cover: (1) model formulation, calibration, and verification; (2) user's manual; and (3) Mt. Bold Reservoir data acquisition and evaluation. The detailed limnological model predicts daily changes of all important water quality phenomena, including thermal stratification, dissolved oxygen, nutrient cycling, and algal growth and decay, over several seasons (less than 10 years). The eutrophication model, which requires minimal data, can predict monthly changes in key trophic indicators over many years (10 or more). The two models together provide the information necessary for assessing detailed short-term water quality fluctuations and general long-term eutrophication trends resulting from alternative land use and lake management plans.

Brown, J.A.H. et al., "A Mathematical Model of the Hydrologic Regime of the Upper Nile Basin", Journal of Hydrology, Vol. 51, No. 1-4, May 1981, pp. 97-107.

The Upper Nile Basin Model was developed as part of the hydrometeorological survey of the catchments of Lakes Victoria, Kyoga, and Mobutyu Sese Seko (Albert), sponsored by the United Nations Development Program. The hydrology of the basin is dominated by the role of the lakes, which cover about 20 percent of the total surface area. The model comprises three separate models combined into an overall system, which is the Upper Nile Basin Model. A catchment model enables the runoff of the catchments contributing to the three lakes to be estimated from rainfall and evapotranspiration data. A lake model computes outflows and changes in storage for various inflows and operational rules. A channel model is used to predict the movement of water in the main river channels between the lakes. The Upper Nile Basin Model includes a total of 36 programs and a master program which controls their selection. The 36 programs can be called up automatically in any predetermined sequence. Beginning with only rainfall and evaporation data, the model is capable of simulating the behavior of the whole basin and of computing the outflows from the system downstream of Lake Mobuto Sese Seko. Provisions for preselection of programs for a particular study, inclusion of new programs, and easy replacement or revision of individual existing programs contribute to the flexibility of the model.

Carrigan, B., "Water Quality Modelling--Hydrological and Limnological Systems. Volume 3. July 1977-June 1979 (A Bibliography with Abstracts)",

1979, National Technical Information Service, U.S. Department of Commerce, Springfield, Virginia.

The abstracts contain information on models used to describe water quality, including models of the chemical, physical, biological and hydrological processes important to water quality. Studies are included on the modeling of eutrophication, nutrient removal, pollutant dispersion, stream flow, heat dissipation, limnological factors, aquifer water quality, and water runoff quality. This bibliography contains 200 abstracts.

Charlton, M.N., "Hypolimnion Oxygen Consumption in Lakes: Discussion of Productivity and Morphometry Effects", Canadian Journal of Fisheries and Aquatic Sciences, Vol. 37, No. 10, Oct. 1980, pp. 1531-1539.

Hypolimnion oxygen is a function of thickness and temperature, as well as biological productivity, according to a mathematical relationship developed in this paper. When the equation was applied to Lake Erie, it explained the previously observed anomaly that phytoplankton production was similar in areas of deep and shallow hypolimnion, but oxygen depletion was twice as rapid in the shallow section. Calculated and observed data were in agreement for Lakes Superior, Ontario, Michigan, Georgian Bay and East Erie.

Freedman, P.L., Canale, R.P. and Pendergast, J.F., "Modeling Storm Overflow Impacts on a Eutrophic Lake", Journal of Environmental Engineering Division, American Society of Civil Engineers, Vol. 106, No. EE2, Apr. 1980, pp. 335-349.

A mathematical model was developed to predict the transient impact of storm loads on phosphorus, fecal coliform, and dissolved oxygen concentrations in Onondage Lake, and to determine the need for control of storm overflows. Model simulations demonstrated that combined sewer and storm loads have a significant impact on lake fecal coliform but little effect on phosphorus, CBOD, NBOD, and dissolved oxygen in Onondage Lake. Observed variations in lake dissolved oxygen were caused by changes in chlorophyll-a, light, and wind. Control of storm loads is recommended to alleviate violations of fecal coliform water quality standards but not for improvement of dissolved oxygen concentrations. As a consequence of the modeling analysis, a limited control program for combined sewer overflows was signed which included only disinfection and removal of objectionable solids. Reductions in storm loads of nutrients and BOD would not provide any significant improvements in lake water quality and were not recommended.

French, R.H. and Krenkel, P.A., "Effectiveness of River Models", Water Science and Technology, Vol. 13, No. 3, 1981, pp. 99-113.

Modeling of river water quality is a practical tool in water resources planning and water quality management, but only if used correctly. Eight factors affecting the effectiveness of models are discussed. Some processes are not easy to model, e.g., erosion, eutrophication, and toxicity relationships. Dissolved oxygen, temperature, and dissolved solids are relatively easy to model. Indicator bacteria, sediment transport, algal growth, metal transport, nutrient ransport, and pesticide transport are intermediate in complexity. General case models are not always applicable to specific rivers. The model should be chosen to fit the river, not vice versa. Concentration on a critical time period may suffice to answer the study question. For example, low dissolved oxygen may only occur during the summer. Calibration and verification are necessary for good results. The two are independent, and different data sets should be used for each. Any drastic change in the river system may require recalibration and verification. Proper sampling and monitoring are important for a statistically reliable data base. The model results must be presented in nontechnical language, understandable by managers and planners. The model should be developed well in advance of actual need, so that calibration, verification, sampling, etc., can be designed and executed in an optimum fashion. The validity of the model's basic assumptions needs periodic reevaluation. A significant use of models is sensitivity analysis, showing the effect of variation of a given parameter on the output if all other factors are constant. This is illustrated with a study on the Willamette River.

Ford, D.E. and Stefan, H.G., "Thermal Predictions Using Integral Energy Model", Journal of the Hydraulics Division, American Society of Civil Engineers, Vol. 106, No. HY1, Jan. 1980, pp. 39-55.

A one-dimensional integral energy model (mixed-layer) was used to simulate the seasonal temperature cycle of three, morphometrically different, temperate lakes. In the model, turbulent kinetic energy supplied by wind shear was used to entrain denser water into the upper mixed layer by working against gravity. The model was calibrated with data from one lake for one year and verified against data from two other lakes and also against data from other years. Predictions of the onset of stratification, surface and hypolimnetic temperatures, mixed layer depths, and periods of turnover were all in agreement with data.

Hoopes, J.A. et al., "Selective Withdrawal and Heated Water Discharge: Influence on the Water Quality of Lakes and Reservoirs, Part II--Induced Mixing with Submerged, Heated Water Discharge", Technical Report No. WIS SRC 79-04, 1979, Water Resources Center, University of Wisconsin, Madison, Wisconsin.

Various methods have been proposed to improve the water quality of stratified lakes and reservoirs. Artificial destratification using pumps, air injection and water jets have been successfully employed. Hypolimnetic aeration has been used to remove chemical stratification, leaving the thermal stratification unchanged. This report presents a field and theoretical investigation of the induced, vertical mixing in a temperature-stratified impoundment resulting from the submerged discharge of heated water. This work shows that a subsurface, heated water discharge can induce vertical mixing of a stratified impoundment. The feasibility of and considerations in using this method of mixing along with relations and procedures for applying the results of this study to an impoundment are presented. General relations are developed for the rate of change of stability in a stratified impoundment due to a heated water discharge, atmospheric energy exchange and impoundment inflows and outflows, for the percent of impoundment mixed, and for the mixing time. A numerical model of an impoundment's temperature structure and changes thereto, resulting from a submerged, vertical, heated water discharge and/or atmospheric energy exchange is presented. The model is in good agreement with observations of the natural temperature distribution (no heated water discharge) in two lakes; model predictions show the changes in the "natural" temperature distribution for one lake induced by a submerged, heated water discharge.

Horst, T.J., "A Mathematical Model to Assess the Effects of Passage of Zooplankton on Their Respective Populations", Proceedings of the Clemson Workshop on Environmental Impacts of Pumped Storage Hydroelectric Operations, FWS/OBS-80-28, Apr. 1980, U.S. Fish and Wildlife Service, Washington, D.C., pp. 177-189.

A mathematical model is used to study the operational effects of the proposed Vastana pumped storage plant on the Lake Ivosjon, Sweden, ecosystem by assessing the effects of withdrawal and discharge of water on zooplankton. Bosmina coregoni was selected as representative of zooplankton with a relatively rapid development time, and Eudiaptomus graciloides was selected as the representative of long-lived zooplankton. A series of simulations was conducted to assess the effects of power plant operation on the populations of the zooplankters. Mortality values of 10%, 50%, and 100% were used for entrainment mortality. The percent reduction in the Bosmina population was 2% for the low entrainment mortality, and 8 to 12% for the assumption of total mortality of entrained organisms. The percent reduction in the Eudiaptomus population was 2% for the low entrainment mortality, and 9 to 16% for the assumption of total entrainment mortality. Except for the most extreme cases simulated, the predictions fall within the range of measured variation. Based on this study, the proposed Vastana plant is expected to have a tolerable effect on the lake ecosystem.

Huang, T., "Changes in Channel Geometry and Channel Capacity of Alluvial Streams Below Large Impoundment Structures", M.S. Thesis, 1979, Department of Civil Engineering, University of Kansas, Lawrence, Kansas.

Alluvial streams are known to form their own channels, and the characteristics of the channels is in a state of equilibrium with the bank material, and water and sediment discharge patterns. Construction of large impoundment structures on alluvial streams alters their water and sediment discharge patterns. Changes in channel geometry and channel capacity caused by large impoundment structures are examined. The theories of fluvial morphology and the theories of open channel flow are used for analyses. Both approaches lead to a similar conclusion that the stream below a dam tends to form a relatively narrower and deeper channel. U.S. Geological Survey streamflow measurement data from seven Kansas streams are used to test this conclusion. Hydraulically, the morphological changes of the stream channel are attributed to channel degradation and increased channel roughness. A channel degradation model is presented and applied on the Wakarusa River below Clinton Dam. The result shows a maximum degradation of 11 feet, which is probably too excessive. However, the result also indicates a consistent trend that has been predicted by the analyses. With carefully chosen values for the stream parameters, the model can be used satisfactorily to predict changes in channel geometry and channel capacity for a stream reach below a large impoundment structure.

Jorgensen, S.E., "Water Quality and Environmental Impact Model of the Upper Nile Basin", Water Supply and Management, Vol. 4, No. 3, 1980, pp. 147-153.

The HYDROMET project, which is an international survey effort on the Nile River Basin, has been given approval for expansion into a water quality model to give information on lakes, eutrophication, aquatic life, irrigation, industrial wastes and environmental impact. To meet all of these objectives accurately, more comprehensive biological data collection procedures have been adopted with storage of information into a data base, and submodels have been formulated which will feed data into the system. Areas of focus of the smaller models include algal growth, zooplankton, fisheries, nutrient exchange between sediment and water, mass balances of P, N, Si, O, and C, a bilharziosis (snail) model, DDT, Cu, and the water and salt balance. Flow charts for each submodel and calibration data are examined to simplify the understanding of the main model. The water quality model will be an important management tool in the Lake region as this area becomes more developed. The model will also help in the decision-making process and in the planning of environmental policy. An example of the application of the model to the decision-making process on a contrived scheme to build an ammonia fertilizer factory in the area is given. The model will require an intensive training program on its proper use, plus a system of dynamic feedback from applied situations to ensure its validity in comprehensive planning and water quality management.

Johanson, R.C. and Leytham, K.M., "Modeling Sediment Transport in Natural Channels", Watershed Research in Eastern North America, A Workshop to Compare Results, Volume II, February 28-March 3, 1977, Report No. NSF/RA-770255, 1977, Chesapeake Bay Center for Environmental Studies, Edgewater, Maryland, pp. 861-885.

This article discusses some of the problems which must be addressed when modeling the transport of sediment from the perspective of pollutant studies. The role played by colloidal material in conveying pollutants is stressed. A sediment transport model being developed by Hydrocomp, Incorporated, for the U.S. Environmental Protection Agency is outlined. The concepts and constructs which are used to represent the behavior of mixtures of cohensionless and cohesive sediments are presented. Some questions are posed which must be answered if significant progress is to be made in this field.

Karium, F., Croley, II, T.E. and Kennedy, J.F., "A Numerical Model for Computation of Sedimentation in Lakes and Reservoirs", Completion Report No. 105, 1979, Water Resources Research Institute, Iowa State University, Ames, Iowa.

A new computer-based numerical model, SEDRES, is developed for the calculation of amounts, rates, and spatial distributions of sediment in lakes and reservoirs. The principal components of SEDRES compute the following: sediment entrapment, distribution, and differential settling for three different size classes (clay, silt and sand); compaction of currently and all previously deposited sediments; correction to zero elevation for compaction; sediment slump correction due to compaction at zero elevation and at sediment-type interfaces; and alteration of elevation-area-capacity relations due to sedimentation. Inputs to the model are water inflows, reservoir operation levels, original reservoir elevation-area-capacity relation, sediment characteristics, type of sediment-entrapment, and sediment-distribution methods. The time-interval for simulation may be one week or any multiple thereof. SEDRES was applied to predict future sedimentation rates in the Coralville, the Red Rock and the Saylorville Lakes on the Des Moines River, Iowa. Model results were in good agreement with the observed data. Results from the 100-year sedimentation simulation of these three reservoirs are presented at 5-year intervals for the current and several alternative operation plans.

Lehmann, E.J., "Water Quality Modelling--Hydrological and Limnological Systems, Volume 2, 1975-June 1977 (A Bibliography with Abstracts)", June 1978, National Technical Information Service, U.S. Department of Commerce, Springfield, Virginia.

The abstracts contain information on models used to describe water quality, including models of the chemical, physical, biological, and hydrological processes important to water quality. Studies are included on the modeling of eutrophication, nutrient removal, pollutant dispersion, streamflow, heat dissipation, limnological factors, aquifer water quality, and water runoff quality. This bibliography contains 185 abstracts.

Lehmann, E.J., "Water Quality Modeling--Hydrological and Limnological Systems, Volume 3, July 1977-June 1978 (A Bibliography with Abstracts)", 1978, National Technical Information Service, U.S. Department of Commerce, Springfield, Virginia.

The abstracts contain information on models used to describe water quality, including models of the chemical, physical, biological, and hydrological processes important to water quality. Studies are included on the modelling of eutrophication, nutrient removal, pollutant dispersion, stream flow, heat dissipation, limnological factors, aquifer water quality, and water runoff quality. This bibliography contains 86 abstracts.

McCuen, R.H., Cook, D.E. and Powell, R.L., "Water Quality Projections: Preimpoundment Case Study", Water Resources Bulletin, Vol. 16, No. 1, Feb. 1980, pp. 79-85.

Recent regulations require impact statements for major water development projects, including reservoirs that will be used for water supply, recreation, and pollution control. A water quantity/quality model was developed and used for making water quality projections of a proposed reservoir in Montgomery County, Maryland. The study area was uncommon in that there was an extensive water quality data base. The results indicated that land use changes will have a significant effect on water quality, and that the proposed reservoir will improve the quality of the surface waters downstream from the reservoir. A major effect of land use changes is the increase in the variability of water quality.

Meinholz, T.L. et al., "Verification of the Water Quality Impacts of Combined Sewer Overflow", EPA-600/2-79-155, Dec. 1979, U.S. Environmental Protection Agency, Washington, D.C.

Water quality impacts in the Milwaukee River in terms of dissolved oxygen (DO) and fecal coliform concentrations following wet weather discharges were studied and modeled with a modified Harper's water quality model. Intensive field surveys and river dye studies show that river velocity is extremely slow in the lower reach near Lake Michigan, even during high flow periods. Due to this slowness, sediments accumulate in the lower river forming a significant sink for DO. Oxygen demand for these sediments, when disturbed, can be as high as 1000 gm/sqm-day or more than 100 times greater than the demand for undisturbed sediments. A rapid decline in DO is often observed following combined sewer overflows (CSO) that cannot be entirely attributed to the loadings from the CSO. The rapid DO decline is linked to the scouring of sediment oxygen demand by submerged CSO outfalls. Using regression analysis empirical equations were developed to predict CSO sediment scouring impacts on DO. Harper's water quality model was modified, calibrated, and verified for the simulation of fecal coliform and DO river conditions with various dry, wet, and CSO events. It is recommended that the feasibility of a periodic dredging program for the river be examined as a means of alleviating the DO problems.

Noble, R.D., "Analytical Prediction of Natural Temperatures in Rivers", Journal of the Environmental Engineering Division, American Society of Civil Engineers, Vol. 105, No. EE5, Oct. 1979, pp. 1014-1018.

This paper presents an analytical solution to the prediction of the natural temperature regime of a river. This solution includes transient, convective, and dispersive effects. In many instances, dispersive effects are negligible. A simplified analytical solution was also presented for this case. The ground water temperature of the water entering the headwaters was used as a boundary condition at the upstream end. The natural temperature regime of a river was defined as the temperature regime in the absence of any man-made alterations. The alterations can come about from channelization, impoundments, forest cover removal, irrigation, and industrial discharges. Knowledge of the natural temperature regime can be of great value when dealing with any man-made alteration so as to determine the effect of the alteration on the ecology of the river. A theoretical mathematical model was presented which can predict the axial temperature profile of an entire river basin given the meteorological and flow conditions present. The only simplification is a linearization of the heat source term. A simplified model also was presented which can be used when axial dispersion is negligible. Both models use the temperature of the ground water in the headwaters region as a boundary condition at the upstream end.

Onishi, Y., "Sediment-Contamination Transport Model", Journal of the Hydraulics Division, American Society of Civil Engineers, Vol. 107, No. HY9, Sept. 1981, pp. 1089-1107.

An unsteady two-dimensional element model, FETRA, was developed to simulate sediment and pollutant transport in rivers and estuaries. It includes sediment/contaminant interactions such as adsorption and desorption of contaminants; and transport, deposition, and resuspension of contaminated sediments. The model handles three sizes of sediment separately. The sediment transport submodel includes mechanisms of: (1) advection and diffusion/dispersion of sediments; (2) fall velocity and cohesiveness; (3) deposition on the river bed; (4) erosion from the river bed; and (5) sediment contributions from point/nonpoint sources and subsequent mixing. The migration of sediments and the pesticide kepone, deposited in the James River Estuary, Virginia, in the early 1970s, was simulated. During calibration, the model was adjusted until total sediment concentrations at maximum ebb, slack tide, and maximum flood for a net fresh water river discharge of 58.3 cu meters per sec at River Mile 76, were reproduced. The model was verified for a net fresh water discharge of 247 cu meters per sec. The computed dissolved kepone concentrations were 0.0048-0.0084 micrograms per liter, with a peak concentration of 0.017 micrograms per liter at River Mile 48 at the maximum flood tide. The model suggests that suspended sediments carry 12-53 percent of the total kepone being transported, while 47-88 percent is dissolved. Over a tidal cycle the average kepone concentration was 0.0076 micrograms per liter, indicating that 14.0 kg of kepone is transported toward Chesapeake Bay each year at a fresh water discharge of 58.3 cu meters per sec.

Orlob, G.T., "Mathematical Modeling of Surface Water Impoundments, Volume I and II", 1977, Resource Management Associates, Lafayette, California.

A review of the state-of-the-art of mathematical modeling of surface water impoundments was conducted. Models reviewed included one-dimensional models for simulation of temperature and water quality in stratified reservoirs, two-dimensional circulation and water quality in shallow lakes, two-dimensional stratified flow, circulation in multi-layer large lakes, and eutrophication and ecological responses in lake systems. Models for simulation/optimization of single reservoir and multiple reservoir systems were also reviewed, including LP, DP, explicit and implicit stochastic methods, and simulation techniques. Recommendations are made for the formation of a "national register" of software for water resource planning and management, with functions of facilitating technology transfer, standardizing documentation procedures, and disseminating information on mathematical models to potential users.

Rahman, M., "Temperature Structure in Large Bodies of Water: Analytical Investigation of Temperature Structure in Large Bodies of Stratified Water", Journal of Hydraulic Research, Vol. 17, No. 3, 1979, pp. 207-215.

This article presents a two-layer mathematical model for water temperature prediction in stratified reservoirs. The model includes the nonlinear effects of the physical properties of water, and the heat budget which accounts for vertical motion of the water body. A sensitivity analysis was performed with respect to the linear and nonlinear effects of the physical properties of water. Temperature profiles were determined by considering variable density, conductivity, and diffusivity, taking into account the influence of vertical motion of water, but neglecting the effects of horizontal currents. Solution of the differential equation was affected by a similarity technique and applied to numerical problems. An analysis of profiles calculated under various assumptions indicated the importance of the major parameters. It also showed that neglecting the variation of physical properties with respect to temperature leads to significant differences in temperature profiles.

Rosendahl, P.C. and Waite, T.D., "Transport Characteristics of Phosphorus in Channelized and Meandering Streams", Water Resources Bulletin, Vol. 14, No. 5, Oct. 1978, pp. 1227-1238.

Comparisons were made between rates of movement of orthophosphate in a canal and a meandering stream. The meander system had greater algal and macrophyte phosphate uptake rates, and lower plankton and sediment release rates, compared to the canal. Chemical precipitation and direct rainfall influences on orthophosphate movement were insignificant relative to other terms. The major source of phosphorus to both systems was from upland runoff. The impact of this source was greater on the meandering system due to the smaller channel volume. When secondary effects of meandering were considered, such as marsh inundation, the net orthophosphate movement within the meandering channel was less than that for the canal, due to the lower concentrations of phosphorus in marsh effluent waters. Field experiments were conducted to compare the longitudinal dispersion coefficient between a canal and meandering river system; the meandering stream had a dispersion coefficient over 17 times that measured for the canal. Rates of orthophosphate movement were combined into a single mass transport equation, and a numerical solution was obtained. Internal river and canal channel processes were overshadowed by external point source loadings.

Snodgrass, W.J. and Holloran, M.F., "Utilization of Oxygen Models in Environmental Impact Analysis", Proceedings of the 12th Canadian Symposium on Water Pollution Research, 1977, McMaster University, Hamilton, Ontario, Canada, pp. 135-156.

A vertical one-dimensional temperature-oxygen model for reservoirs is used to estimate zones of stress on the aquatic environment of a series of reservoirs in Nova Scotia. Application to cold climates has necessitated a few novel developments for the temperature model. The oxygen model, whose sinks are water column decay and sediment oxygen demand (SOD), is calibrated using under ice measurements of oxygen stocks, and laboratory and in situ measurements of a zero-order kinetic model for sediment oxygen demand. These extensive studies are complementary and indicate a winter SOD of 0.1 gm 02/m2/day, and a higher summer value. High epilimnetic temperatures coupled with the predicted anoxic zones in lower waters cause a major stress upon fisheries potential. This model provides a tool for determining the effects of different reservoir management strategies upon water quality and for selecting among these strategies.

Thomann, R.V., "An Analysis of PCB in Lake Ontario Using a Size-Dependent Food Chain Model", 1979, Perspectives in Lake Ecosystem Modeling, Manhattan College, Bronx, New York, pp. 293-320.

Considered is the development of models for simulating the distribution and dynamics of toxic substances within an ecosystem. In order to incorporate both bioaccumulation of toxic substances directly from the water and subsequent transfer up the food chain, a mass balance model is constructed that introduces organism size as an additional independent variable. The model represents an ecological continuum through size dependence; classical compartment analyses are therefore a special case of the continuous model. The principal factors that influence the total toxicant concentration in various regions of the food chain include excretion and uptake rates, the rate of decrease of biomass density with organism size and the food chain transfer velocity, a parameter reflecting average predation along the food chain. The model behaves linearly with respect to external mass loading of the toxicant, and hence can be used to estimate the input that can be allowed without exceeding given levels in various regions of trophic space. The analysis of some PCB data from Lake Ontario is used as an illustration of the theory. The introduction of organism size as an independent variable in the mass balance of a toxicant provides a generalized analysis framework; this permits the integrated use of diverse laboratory experiments on uptake and excretion as well as an interpretive framework for field data of toxicant concentrations.

U.S. Environmental Protection Agency, "Modeling Phosphorus Loading and Lake Response Under Uncertainty: A Manual and Compilation of Export Coefficients", EPA-405-80-011, June 1980, Washington, D.C.

A procedure is proposed that may be used to quantify the relationship between land use and lake trophic quality. The methodology is based on an input-output phosphorus lake model. When it is employed to predict the impact of projected land use changes, it is necessary to use phosphorus export coefficients extrapolated from other points in time and/or space. These coefficients represent the mass loading of phosphorus to a surface water body per year per unit of source. Nitrogen and phosphorus export coefficients from different land uses including forest, crops, pasture and grazing land, urban, septic tanks and absorption fields, and sewage treatment plants, are given. The impact of atmospheric inputs is noted. The criteria for selecting the appropriate coefficients are: accuracy; precision; representativeness; temporal extent of sampling; nutrient flux estimation; and concentration and flow data. The model includes an error estimation procedure, and is applicable to a fairly wide range of lake types. The limitations of the methodology involve uncertainties in the input information and their impact on prediction precision.

Uzzell, Jr., J.C. and Ozisik, M.N., "Three-Dimensional Temperature Model for Shallow Lakes", Journal of the Hydraulics Division, American Society of Civil Engineers, Vol. 104, No. HY12, Dec. 1978, pp. 1635-1645.

A pseudo three-dimensional time-dependent analytical model was developed for the prediction of temperature distribution in lakes resulting from thermal discharges. For the special case of lakes having a uniform depth, rectangular geometry, and a constant axial velocity, explicit analytical solutions were presented. Sample calculations were performed to demonstrate the effects of the time variation of the thermal loading, different types of boundary conditions, circulation velocities, and eddy conductivity coefficients on the surface temperature distribution along the lake as a function of time.

Vick, H.C. et al., "Preimpoundment Study: Cedar Creek Drainage Basin: Evans County Watershed: Evans, Tatnall, and Candler Counties, Georgia", EPA 904/9-77-006, Mar. 1977, U.S. Environmental Protection Agency, Athens, Georgia.

Twelve sampling stations on Cedar Creek near Claxton in southeastern Georgia were monitored from May, 1974 to January, 1975 to obtain baseline water quality data prior to constructing a multipurpose impoundment. The proposed impoundment will have a normal pool area of 387 acres, and will drain 29,658 acres. Agricultural and livestock production and natural conditions are the only pollution sources; potential pollution problems are foreseen in the E-5 and E-6 arms. Fecal coliform densities ranged from 130-5600/100 ml in May and 100-2200/100 ml in August. Salmonella were isolated at four of five stations sampled in May. Nitrogen and phosphorus concentrations were occasionally high. Postimpoundment water quality was predicted by the Hydrocomp Simulation Programming model; no major violations of Georgia water quality standards were forecast. The major conclusions were: (1) high fecal coliform densities plus salmonella isolations represent stormwater runoff under free-flowing stream conditions; following impoundment retention, time will decrease both parameters and should produce water quality suitable for body contact recreation; (2) eutrophication potential will depend on such nutrient source control factors as assimilation by swampy areas; and (3) attempts should be made to contain runoff from livestock feeding areas and to limit fertilizer use.

Ward, A.D., "A Verification Study on a Reservoir Sediment Deposition Model", Transactions of the American Society of Agricultural Engineers, Vol. 24, No. 2, Mar./Apr. 1981, pp. 340-352.

Seven years' data from monitoring two agricultural reservoirs in Missouri were used to verify the DEPOSITS model, a conceptual design model for predicting the sediment trapping performance of small impoundments. Both the model and its sedimentgraph default option gave good estimates of the reservoirs' performance during the 13 storm events in this study. The EPA overflow rate method gave poorer results.

Webster, J.R., Benfield, E.F. and Cairns, J., "Model Predictions of Effects of Impoundment of Particulate Organic Matter Transport in a River System", The Ecology of Regulated Streams, 1978, Plenum Publishing Corporation, New York, New York, pp. 339-364.

Three empirical equations relating elevation, stream width, and mean annual stream flow to distance were used in a model of the effects of impoundment on particulate organic matter (POM) dynamics in a river-reservoir ecosystem. The model considers the entire river system above an impoundment and for a distance downstream. Data needed to numerically parameterize the model were taken from the New River and Clayton Lake in North Carolina and Virginia. The river goes from first to sixth order and is impacted by impoundment, industrial and urban effluents, and agricultural runoff. The impoundment, a hydroelectric power dam, is located 290 km below the headwaters. Equations are presented for hydraulic parameters, reservoir parameters, biological parameters, and sedimentation and erosion. A 14-month simulation identified several areas where the model does not accurately reflect actual POM behavior, including: the seasonality of allochthonous inputs; autochthonous sources of POM; filter-feeder use of POM; and deposition and resuspension of POM. Simulation results did show that POM concentration was highest during leaf fall; during leaf fall POM concentration peaked a short distance downstream from the headwaters; benthic POM was greatest just after leaf fall; and there was a consistent downstream decrease in POM standing crop. Changes occurring within and below the impoundment were also discussed.

Woodward, F.E., Fitch, Jr., J.J. and Fontaine, R.A., "Modeling Heavy Metal Transport in River Systems", Apr. 1981, Land and Water Resources Center, University of Maine, Orono, Maine.

This study resulted in a mathematical model for prediction of the probable fate of heavy metals in rivers downstream from wastewater discharges. The developed model accounts for hydraulic mixing processes of rivers, and relies heavily on the relationship between suspended particulate matter transport and heavy metals transport. While heavy metals are subject to many reactions, depending on pH, pFe, pMn, dilution, and the physical characteristics of the river, an extensive literature review indicated that most metals are found adsorbed to suspended particulate matter surfaces. The developed model, hence, coupled mixing of a waste stream with river flow and suspended sediment load transport. Adsorption-desorption reactions were used as a basis for portioning metals among various particulate size ranges. The ultimate fate of the metals was determined primarily by the predicted transport/deposition of particulate matter. The sediment transport model of Einstein (1950), as modified by the U.S. Bureau of Reclamation, formed the basis for the developed heavy metal transport model.

Wycoff, R.L. and Singh, U.P., "Application of the Continuous Storm Water Pollution Simulation System (CSPSS): Philadelphia Case Study", Water Resources Bulletin, Vol. 16, No. 3, June 1980, pp. 463-470.

This paper describes the Continuous Stormwater Pollution Simulation System (CSPSS) as well as a site-specific application of CSPSS to the Philadelphia urban area and its receiving water, the Delaware Estuary. Conceptually, CSPSS simulates the quantity and quality of urban stormwater runoff, combined sewer overflow, municipal and industrial wastewater effluent, and upstream flow on a continuous basis for each time step in the simulation period. In addition, receiving water dissolved oxygen, suspended solids, and lead concentrations resulting from these pollutant sources may be simulated. However, only receiving water dissolved oxygen (DO) response is considered in this paper. The continuous DO receiving water response model was calibrated to existing conditions using observed data at Chester, Pennsylvania, located on the Delaware Estuary approximately 10 miles downstream from the study area. Average annual pollutant loads to the receiving water were estimated for all major sources, and receiving water quality improvements resulting from removal of various portions of these pollutant loads were estimated by application of the calibrated simulation model. It was found that the removal of oxygen-demanding pollutants from combined sewer overflow and urban stormwater runoff would result in relatively minor improvements in the overall dissolved oxygen resources of the Delaware Estuary; whereas removal of oxygen-demanding pollutants from wastewater treatment plant effluent would result in greater improvements. The results of this investigation can be used along with appropriate economic techniques to identify the most cost-effective mix of point and nonpoint source pollution control measures.

APPENDIX J

BIOLOGICAL IMPACT PREDICTION AND ASSESSMENT

Belyakova, O.V., "Model of the Seasonal Dynamics of an Ecosystem of a Shallow Lake", Water Resources (English Translation), Vol. 7, No. 5, Sept./Oct. 1980, pp. 450-457.

In order to understand and predict the reaction of a lake to varying external effects, it is first necessary to create a mathematical model of the seasonal dynamics of the components of the ecosystem. The development of a mathematical model of the seasonal dynamics of the biomass and production of the main trophic groups of aquatic organisms and of the concentrations of biogenic elements is discussed. The inputs of the system are the mean monthly values of absorbed solar radiation, the temperature of the water mass, inflow and outflow of the lake, and the amount of production of macrophytes. The outputs of the system include dissipation of heat energy during the vital activity of aquatic organisms, the biomass of chironomids that have emerged and flown away, and the consumption of zooplankton and benthos by fishes. The results obtained on the model were compared with data obtained by direct observation on a mesotrophic lake in northwestern U.S.S.R. This comparison showed that the model provided a satisfactory reproduction of both the overall level of development and seasonal dynamics of the main components of the lake ecosystems. This model can be used to evaluate the role of the main trophic links in the seasonal dynamics of the flow of material and energy in the lake, and to evaluate the possible responses of the system to quantitatively different regimes of the input variables. However, it is not suitable for evaluation of potential responses to large disturbances which would create the basis for a different energy regime.

Bovee, K.D., "The Determination, Assessment, and Design of 'In-Stream Value' Studies for the Northern Great Plains Region", Sept. 1974, Department of Geology, Montana University, Missoula, Montana.

This study proposes a methodology for the recommendation of minimum discharges for a warm water fishery. The method utilizes field measurements of critical stream areas and biological criteria determined from the use of indicator species. For large rivers, the indicator species was the paddle fish (Polyodon Spathula); for smaller rivers it was the sauger (Stizostedion Canadense); and for determining adequate fish habitat, the stonecat (Notorus Flavus) was used. Rearing flows are determined on the basis of stream productivity by analyzing macroinvertebrate habitats, and on the basis of fish habitat typing. A number of variables might require a greater amount of instream flows than the fishery. These variables include streamflow needs for riparian and other sub-irrigated vegetation, water quality parameters, anchor ice formation, and the relationship between discharge and sediment yield. Information concerning these variables is presently insufficient to determine whether a variable will override the requirement for the fishery. Further research is needed in these areas, and several investigative methods for conducting such research are proposed. An extensive literature review was conducted to determine the discharge requirements of a warm water fishery. Where exact hydrologic parameters were not measured directly, they were estimated. From this information it was determined which components of the stream community would be most seriously affected by reduced discharges. A method for recommending minimum discharges should not sacrifice reliability for expediency.

Bovee, K.D. and Cochnauer, T., "Development and Evaluation of Weighted-Criteria, Probability-of-Use Curves for Instream Flow Assessments: Fisheries", Report No. FWS/OBS-77/63, IFIP-3, Dec. 1977, U.S. Fish and Wildlife Service, Fort Collins, Colorado.

This information paper documents the methods and procedures used in the construction of probability criteria curves. Weighted criteria are used to assess the impacts of altered streamflow regimes on a stream habitat. They are developed primarily for those habitat parameters most closely related to stream hydraulics; depth, velocity, substrate, and temperature. Guidelines for data collection, analysis and curve development are discussed.

Brungs, W.A. and Jones, B.R., "Temperature Criteria for Freshwater Fish: Protocol and Procedures", EPA/600/3-77/061, May 1977, U.S. Environmental Protection Agency, Duluth, Minnesota.

Temperature criteria for freshwater fish are expressed as mean and maximum temperatures; means control functions such as embryogenesis, growth, maturation, and reproductivity; and maxima provide protection for all life stages against lethal conditions. These criteria for 34 fish species are based on numerous field and laboratory studies, and yet for some important species the data are still insufficient to develop all the necessary criteria. Fishery managers, power-plant designers, and regulatory agencies will find these criteria useful in their efforts to protect fishery resources.

Camougis, G., Environmental Biology for Engineers, 1981, McGraw-Hill Book Company, Inc., New York, New York.

This book is intended as a practical guide to environmental biology for engineers and other professionals concerned with environmental affairs and not trained primarily in the biological sciences. Three primary objectives are reflected in the format and organization of this book. First, the book presents a practical vocabulary in environmental biology. The text, the appendixes and glossary all define terms used frequently in environmental studies. Next, the book discusses environmental biology in the context of both environmental engineering and environmental legislation. In that sense it relates the theory and practice of environmental biology to "real-world" applications in engineering projects. Finally, the book provides environmental engineers, project engineers, environmental planners and project managers with practical guidelines on the application of biology to environmental assessments.

Casti, J. et al., "Lake Ecosystems: A Polyhedral Dynamics Representation", Ecological Modeling, Vol. 7, No. 3, Sept. 1979, pp. 223-237.

Polyhedral dynamics, a type of systems methodology based on topology, is presented as a technique for modeling the global structure and dynamic behavior of water quality problems in the aquatic ecosystem. The lake ecosystem, composed of organic and inorganic nutrients, light, phytoplankton, zooplankton, and fish is presented as a multidimensional graph using a technique called Q-analysis. Then a flow pattern is superimposed upon the Q-analysis to describe the dynamic behavior of the system. The utility of the method is demonstrated by looking at the effects of eutrophication. Q-analysis results correlate well with the traditional concept of a food chain. Also, the dynamic changes in the ecosystem caused by nutrient addition are seen to be a result of limits on matter and energy flow through the ecosystem imposed by the structure as it is defined by the Q and obstruction vectors. This insight cannot be easily reached by traditional differential equation representations. Consideration of the dynamic behavior gives a quantitative basis to the qualitative changes which an ecosystem undergoes as a result of increased nutrient loading. Thus the approach may be used to assess the effectiveness of alternative eutrophication control methods. However, the nonlinear nature of the pattern defined by the analysis makes a more quantitative analysis difficult.

Cowardin, L.M. et al., "Classification of Wetlands and Deepwater Habitats of the United States", FWS/OBS-79/31, Dec. 1979, U.S. Fish and Wildlife Service, Washington, D.C.

This report highlights a classification scheme to describe ecological taxa, arrange them in a system useful to resource managers, furnish units for mapping, and provide uniformity of concepts and terms. Wetlands are defined by plants (hydrophytes), soils (hydric soils), and frequency of flooding. Ecologically related areas of deep water, traditionally not considered wetlands, are included in the classification as deepwater habitats. Systems form the highest level of the classification hierarchy; five are defined--Marine, Estuarine, Riverine, Lacustrine, and Palustrine. Marine and Estuarine systems each have two subsystems, Subtidal and Intertidal; the Riverine system has four subsystems, Tidal, Lower, Perennial, Upper Perennial, and Intermittent; the Lacustrine has two, Littoral and Limnetic; and the Palustrine has no subsystem. With the subsystems, classes are based on substrate material and flooding regime, or on vegetative life form. The same classes may appear under one or more of the systems or subsystems. Six classes are based on substrate and flooding regime: (1) Rock Bottom with a substrate of bedrock, boulders, or stones; (2) Unconsolidated Bottom with a substrate of cobbles, gravel, sand, mud, or organic material; (3) Rocky Shore with the same substrate as Rock Bottom; (4) Unconsolidated Shore with the same substrate as Unconsolidated Bottom; (5) Streambed with any of the substrates; and (6) Reef with a substrate composed of the living and dead remains of invertebrates (corals, mollusks, or worms). The bottom classes, (1) and (2) above, are flooded all or most of the time and the shore classes, (3) and (4), are exposed most of the time. The class Streambed is restricted to channels of intermittent streams and tidal channels that are dewatered at low tide. The life form of the dominant vegetation defines the five classes based on vegetative form: (1) Aquatic Bed, dominated by plants that grow principally on or below the surface of the water; (2) Moss-Lichen Wetland, dominated by mosses or lichens; (3) Emergent Wetland, dominated by emergent herbaceous angiosperms; (4) Scrub-Shrub Wetland, dominated by shrubs or small trees; and (5) Forested Wetlands, dominated by large trees. The dominance type, which is named for the dominant plant or animal forms, is the lowest level of the classification hierarchy. Only examples are provided for this level; dominance types must be developed by individual users of the classification.

The structure of the classification allows it to be used at any of several hierarchical levels. Special data required for detailed application of the system are frequently unavailable, and thus data gathering may be prerequisite to classification. Development of rules by the user will be required for specific map scales. Dominance types and relationships of plant and animal communities to environmental characteristics must also be developed by users of the classification. Keys to the systems and classes are furnished as a guide, and numerous wetlands and deepwater habitats are illustrated and classified. The classification system is also compared with several other systems currently in use in the United States.

Elwood, J.W. and Eyman, L.D., "Test of a Model for Predicting the Body Burden of Trace Contaminants in Aquatic Consumers", Journal of the Fisheries Research Board of Canada, Vol. 33, 1976, pp. 1162-1166.

A model for predicting the accumulation and retention of trace contaminants obtained through food ingestion in aquatic consumers was tested for short-term exposure conditions. Model parameters were determined in a single-feeding experiment using bluegill (Lepomis macrochirus) and food labeled with 137-Cs over a 16-day period, and the predicted and measured body burden of the radionuclide were compared. The model realistically simulated the absorption of 137-Cs from the gastrointestional tract and its accumulation over the 16-day period. The average body burden of 137-Cs in bluegill was within 25 percent of the predicted body burden when the experiment was terminated. Apparent equilibrium of 137-Cs in bluegill by day 26 suggests that this two-compartment linear model does not apply to the long-term accumulation of cesium in fish. The model appears most applicable for predicting body burdens of trace contaminants under acute exposure conditions that simulate an accidental release.

Fieterse, A.J.H. and Toerien, D.F., "The Phosphorus-Chlorophyll Relationship in Roodeplaat Dam", Water SA (Pretoria), Vol. 4, No. 3, 1978, pp. 105-112.

The relationship between phosphate phosphorous (PO4-P) and chlorophyll-a was investigated to establish a model for the eutrophication of the Roodeplatt Dam. Many of the impoundments in South Africa have excessive algae and/or macrophyte growths, which result in poorer water quality. Surface samples were taken at various points in the dam and analyzed for inorganic nitrogen and phosphorous ions. Regression analysis of the averaged data indicated that algal growth was limited by the PO4-P concentration rather than by inorganic nitrogen. A reduction in the PO4-P concentration would then reduce algal growth and PO4-P concentration data is more applicable to South Africa conditions than Total P values. The significant relationship between chlorophyll-a and PO4 concentrations upholds this conclusion. A model for eutrophication was developed for the Roodeplatt Dam which may be applicable to similar bodies of water. Algal nuisance conditions can be expected to occur above a phosphate phosphorous concentration of 26 micrograms per liter. The eastern sections of the dam represented the upper threshold conditions for algal growth and phosphorous loading, probably because of loading via the Pienaars River. The trophic status and the intensity of eutrophication was accurately represented by the PO4-P parameter.

Hazel, C. et al., "Assessment of Effects of Altered Stream Flow Characteristics on Fish and Wildlife, Part B: California, Case Studies", FWS/OBS-76/34, Dec. 1976, U.S. Fish and Wildlife Service, Washington, D.C.

The results and conclusions are given for 47 case studies of California water projects that altered natural streamflow regimes and causally affected the fish and wildlife. Surveys were conducted on existing conditions below dams and diversions to assess the actual effects of the streamflow characteristics on fish and wildlife, and to evaluate the adequacy of the methodologies used to determine necessary flows.

O'Connor, D.J., Di Toro, D.M. and Thomann, R.V., "Phytoplankton Models and Eutrophication Problems", Ecological Modeling in a Resource Management Framework, 1975, Resources for the Future, Inc., Washington, D.C., pp. 149-209.

The primary purpose of this paper is to present the applications of a set of equations which describe the seasonal distribution of phytoplankton to the analysis of eutrophication problems in various locations throughout the country. A brief review of the theoretical structure of the analysis is presented with a qualitative description of the pertinent equations and a discussion of the general procedure of the verification process. Examples from various natural water systems are presented to demonstrate the utility of this type of analysis in evaluating alternative plans to restore or maintain appropriate levels of water quality. The systems considered are: (1) the fresh water segment of the San Joaquin River; (2) the estuarine regions of the Sacramento-San Joaquin delta; (3) the Potomac River; (4) Western Lake Erie; and (5) Lake Ontario. The individual studies cited herein contain complete bibliographies of the scientific and engineering literature which was used in the analysis of the general problem and the development of the equations. Specific applications included prediction of the effect of water diversions and increased nutrient discharges on phytoplankton populations wherein estimates of the effect of alternative policies regarding both diversions and required sewage treatment levels are considered.

Oglesby, R.T. and Schaffner, W.R., "Phosphorus Loadings to Lakes and Some of Their Responses. Part 2. Regression Models of Summer Phytoplankton Standing Crops, Winter Total P, and Transparency of New York Lakes with Known Phosphorus Loadings", Limnology and Oceanography, Vol. 23, No. 1, Jan. 1978, pp. 135-145.

Using a new method of calculating phosphorus loading to lakes based on a composite of several phosphorus forms rather than on total phosphorus, and as input to the summer mixed zone rather than surface area, this study examines responses of 16 lakes in central New York State to such loadings as an interrelated series of simple regression models. Regressions describing the dependence of summer phytoplankton standing crop and winter total phosphorus concen-trations on loading, and of standing crop on total phosphorus, are linear. Those characterizing water transparency as a function of standing crop and winter total phosphorus are parabolic. All regressions showed high correlation coefficients. The overall model comprising these regressions supports the hypothesis that lake phosphorus loading is an important factor controlling phytoplankton standing crop. The regressions are useful for trophic state description and in developing lake management strategies. The revised phosphorus loading calculation method incorporates the sum of soluble reactive and unreactive phosphorus, and phosphorus potentially desorbable from particulate matter. The lakes studied represent a wide range of morphology, hydrology, and phosphorus loading. The methodology presented is valid for all lakes with mean hydraulic retention times of 0.5 yr. It is noted that with very high loadings, phosphorus may no longer control phytoplankton production.

O'Neill, R.V., "Review of Compartmental Analysis in Ecosystem Science", CONF-780839-1, 1978, Oak Ridge National Laboratory, Oak Ridge, Tennessee.

The compartment model has a large number of applications in ecosystem science. An attempt is made to outline the problem areas and objectives for which this type of model has particular advantages. The areas identified are an adequate model of tracer movement through an undisturbed but nonequilibrium ecosystem; an adequate model of the movement of material in greater than tracer quantity through an ecosystem near steady state; a minimal model based on limited data; a tool for extrapolating past trends; a framework for the summarization of large data sets; and a theoretical tool for exploring and comparing limited aspects of ecosystem dynamics. The review is set in an historical perspective which helps explain why these models were adopted in ecology. References are also provided to literature which documents available mathematical techniques in an ecological context.

Ostrofsky, M.L. and Duthie, H.C., "An Approach to Modelling Productivity in Reservoirs", Proceedings: Congress in Denmark 1977, Part 3: Internationale Vereingung fur Theoretische und Angewandte Limnologie, Vol. 20, 1978, pp. 1562-1567.

Leaching forest litter is the key nutrient source in the Smallwood Reservoir in Labrador, Canada. New reservoirs pass through a stage of high biological activity, trophic upsurge, and then give way to a much reduced activity state, trophic depression, trophic upsurge, and trophic depression; this 1974-75 study tested a phosphorus dynamics hypothesis for predicting magnitude and duration of reservoir stages. Weekly sampling during the ice-free season was carried out on the reservoir and an adjacent natural lake, Mile 83 lake. Primary productivity rates were higher in the reservoir in both years, 1974-75, averaging 107-168 mg C/sq m/d, respectively. Phosphorus budgets were constructed with the atmosphere contributing 0.44 kg P/ha/yr and the land surface exporting 3.72 mg P/sq m/yr. Predicted and observed P concentrations in the lake agreed, but in Smallwood Reservoir the predicted P concentration (6.1 mg/cu m) was less than 50 percent of the observed value (12.6 mg/cu m). The most obvious important P source overlooked was the inundated vegetation and soils. Mathematical studies on the mass balance equation and leaching input were performed and concluded that an average P contribution of 15 mg P/sq m resulted from leaching. Phosphorus was graphed as a function of time and high P concentrations corresponded to trophic upsurge and decreasing P concentrations to that of trophic depression.

Scavia, D. and Robertson, A., Editors, Perspectives on Lake Ecosystem Modeling, 1979, Ann Arbor Science, Ann Arbor, Michigan.

Modeling should be an integral part of both basic research and applied management programs for aquatic ecosystem study. This book consists of papers which consider lake ecological model usage, possible model improvements, and new directions for development. Most of the papers included have been taken from a special symposium on ecological modeling at the 20th Conference on Great Lakes Research at the University of Michigan in 1977. The book is divided into four sections, each dealing with an individual area of aquatic ecosystem research and the role of models in that area. Section One, Improved Model Components, considers: scale in modeling large aquatic ecosystems; an experimental and modeling review of water column death and decomposition of phytoplankton; zooplankton grazing in simulation models--the role of vertical migration; and mathematical modeling of phosphorus dynamics. Section Two, Identification of Research Needs through Model Studies, considers: modifications to the model Cleaner; and the use of ecological lake models in information synthesis. Section Three, Models in Management, contains: predictive water quality models for the Great Lakes; empirical lake models for phosphorus; and a least-cost surveillance plan for water quality trend detection in Lake Michigan. Section Four, New Directions in Ecosystem Analysis, presents: preliminary insights into a three-dimensional ecological hydrodynamic model; study of ecosystem properties of Lake Ontario using an ecological model; and an analysis of PCB in Lake Ontario using a size-dependent food chain model.

Schnoor, J.L. and O'Connor, D.J., "A Steady State Eutrophication Model for Lakes", Water Research, Vol. 14, No. 11, Nov. 1980, pp. 1651-1665.

An alternate approach to eutrophication modeling is presented which simplifies assumptions of the kinetic and transport equations. Both steady and nonsteady state cases were studied for LBJ Lake, Texas (a short detention time reservoir) and Lake Ontario (a long detention time reservoir). Steady state solutions are also given for 81 of the phosphorus-limited lakes of the U.S. National Eutrophication Survey. Use of the model requires the estimation of the sedimentation, hydrolysis, autocatalytic growth, and death rate constants. The sedimentation rate constant determines the amount of phosphorus lost to the deep sediment, and the total phosphorus levels of the lake. The other three constants control the partitioning of nutrients among the various organic, inorganic, and phytoplankton fractions. In the lakes sampled in the northeast and north central United States, phytoplankton accounted for 10-40% of the total phosphorus, while dissolved phosphorus was 35-75%, and organic phosphorus 0-40%, of the total. The steady state model was recommended for use as a management tool when detailed analyses are not practical.

Stalnaker, C.B. and Arnette, J.L., "Methodologies for the Determination of Stream Resource Flow Requirements: An Assessment", FWS/OBS-76/03, Apr. 1976, U.S. Fish and Wildlife Service, Washington, D.C.

This report summarizes and evaluates techniques for determining instream flow requirements and assessing the effects of changing stream flows. The state-of-the-art summation is intended to provide a basis for further work in developing methodologies and defining research needs. Standard nomenclature is presented, together with an overview of hydrologic techniques, the calculation of essential hydraulic parameters, and determination of other quantitative relationships, including a summary of applicable modeling approaches. The report discusses methods for assessing instream flow needs for fish, terrestrial wildlife, and water quality. Problems of determining the impact of streamflow changes on recreational activities and aesthetic values are considered.

Taylor, M.H., "An Indicator-Prediction Model for Ecosystem Parameters of Water Quality", Technical Completion Report, Nov. 1979, Water Resources Center, University of Delaware, Newark, Delaware.

A mathematical model, based on fit to a rectangular hyperbola, has been developed for use in predicting the toxicity of chemical pollutants in fish. The model was tested by exposing Fundulus heteroclitus to a range of concentrations of napthalene between 0 and 30 mg/l. Observations of mortality, condition index, tissue changes, and serum cortisol and glucose levels were tested as indicators of toxic effect. Mortality data obtained in these experiments were used to calculate the dose for which appearance of the response would require an infinite exposure time. This was defined as the "safe level" for the indicator in question. Using mortality as the indicator, the model predicted the safe level to be 1.69 mg/l. This prediction appears to be valid since 1.73 mg/l was lethal at 15 days, and no mortality occurred in the fish exposed to 1.60 mg/l naphthalene for 30 days. Of the other toxicity indicators tested, only sensory cell necrosis was a more sensitive indicator of toxicity than mortality. Serum cortisol and glucose, and ischemia of brain and liver occurred only in animals which were near death. The condition index did not change in exposures as long as 30 days.

Thomas, J.M. et al., "Statistical Methods Used to Assess Biological Impact at Nuclear Power Plants", Journal of Environmental Management, Vol. 7, No. 3, Nov. 1978, pp. 269-290.

This paper illustrates and discusses a number of analytical procedures used to assess ecological impact studies conducted at three nuclear power stations. The advantages, disadvantages and limitations of each statistical procedure are discussed. It is suggested that use of one of these procedures, accompanied by a field design based on the same statistical model and field methods recommended by other studies, will be suitable for the detection of changes in biota. However, since litigation has been directed to the question of effects on future populations, a plan for conducting environmental monitoring studies is devised in which that question is addressed.

U.S. Fish and Wildlife Service, "Habitat Evaluation Procedures", Mar. 1979, Division of Ecological Services, Fort Collins, Colorado.

This document is a procedural manual designed to provide the user with a quantitative methodology for estimating and comparing resource development project impacts on fish and wildlife resources. The methodology is continuing to be refined and supplemented.

Veith, G.D., DeFoe, D.L. and Bergstedt, B.V., "Measuring and Estimating the Bio-Concentration Factor of Chemicals in Fish", Journal of the Fisheries Research Board of Canada, Vol. 36, 1979, pp. 1040-1048.

A method of estimating the bioconcentration factor of organic chemicals in fathead minnows (Pimephales promelas) is described. Water at 25°C was intermittently dosed with the chemical at a nontoxic concentration in a flow-through aquarium. Thirty minnows are placed in the aquarium, and composite samples of five fish are removed for analysis after 2, 4, 8, 16, 24, and 32 days of exposure. The bioconcentration process is summarized by using the first-order uptake model, and the steady-state bioconcentration factor is calculated from the 32-d exposure. A structure-activity correlation between the bioconcentration factor (BCF) and the n-octanol/water partition coefficient (P) of individual chemicals is summarized by the equation log BCF=0.85 log P - 0.70, which permits the estimation of the bioconcentration factor of chemicals to within 60 percent before laboratory testing. The facilities and resources for testing need be used only for those chemicals that are likely to result in substantial bioconcentration in organisms. The bioconcentration factors derived from tests of mixtures of chemicals are shown to be the same as those derived from tests with the chemicals individually.

Walters, R.A., "A Time- and Depth-Dependent Model for Physical, Chemical and Biological Cycles in Temperate Lakes", Ecological Modeling, Vol. 8, Jan. 1980, pp. 79-96.

A numerical model designed for studying the complex relationships that exist between chemical, physical, and biological processes which occur in deep stratified lakes of the temperate zone is described. Results of a mathematical model of the thermal stratification cycle of a deep lake are combined with a phytoplankton growth and nutrient concentration model to ensure consistency of the vertical eddy diffusion of algal cells and dissolved nutrients with the mixing processes that determine the lake's thermal stratification. Results of simulation trials were in good agreement with measured distributions from Lake Washington in Washington state. Turbulent mixing processes in the thermal model controlled the chlorophyll-a and distribution of nutrients. The thermal model utilizes a heat diffusion equation which is nonlinear and reflects the interaction of wind-induced turbulence and buoyancy gradients related to surface heating and cooling. Changes in surface heat are described by standard meteorological parameters, and both finite-difference and finite-element algorithms are used to solve the thermal model. A pair of coupled, nonlinear partial differential equations used to form the biological production model is solved by the finite differences approach and an iteration technique. The equations govern the distribution of chlorophyll-a and dissolved phosphorus.

Watanabe, M., "Modeling of the Eutrophication Process in Lakes and Reservoirs", Proceedings of the Baden Symposium on Modeling the Water Quality of the Hydrological Cycle, Publication No. 125, 1978, International Association of Hydrological Sciences, pp. 200-210.

A closed, energy-driven, matter-flow loop of nitrogen-phosphorus cycles was proposed as a fundamental structure for understanding and controlling the eutrophication process in aquatic environments. It consists of 11 storage variables of nitrogen and phosphorus, and 14 transformations which represent observable biochemical reactions and ecological interactions. The mathematical model was applied to a microcosm experiment, and the comparison between predicted versus measured values supported the validity of the matter-flow rate mechanism of phosphorus and nitrogen cycles, although some modifications for modeling of phytoplankton and zooplankton are necessary.

Williams, L.R. et al., "Relationships of Productivity and Problem Conditions to Ambient Nutrients: National Eutrophication Survey Findings for 418 Lakes", EPA 600/3-78-002, Jan. 1978, Environmental Monitoring and Support Laboratory, U.S. Environmental Protection Agency, Las Vegas, Nevada.

Data collected on 418 lakes, ponds and reservoirs east of the Mississippi River in 1972-73 for the National Eutrophication Survey were analyzed to determine correlations between chlorophyll-a (as an indicator of lake productivity), nutrients, and other water quality parameters. Conclusions include: (1) productivity, as measured by mean chlorophyll-a concentrations, is strongly related to ambient phosphorus levels, especially in lakes with hydraulic retention times greater than 14 days; (2) significant regional differences in chlorophyll-a response per unit total phosphorus exist, the reasons are being investigated; (3) classifying lakes on the basis of stratification, dominant vegetation, or fishery type results in no differences in chlorophyll-a response to phosphorus; (4) no algal blooms occur in lakes with mean total phosphorus concentrations less than 19 micrograms/liter; (5) nuisance aquatic weed growths generally occur at lower phosphorus levels than did algal blooms, and in many cases control of phosphorus inputs is unlikely to have much impact on macrophytes; (6) fishkills are generally unrelated to mean phosphorus or chlorophyll-a levels, or even to chronic low-oxygen conditions; and (7) many oxygen problems in southeastern lakes arise from establishment of trout fisheries in marginal habitats. Positive correlations were found between chlorophyll-a and phosphorus, Kjeldahl nitrogen, pH, and total alkalinity; negative correlations occurred with Secchi disc transparency and the nitrogen/phosphorus ratio.

Yahnke, J.W., "Water Quality of the Proposed Norden Reservoir, Nebraska, and Its Implications for Fishery Management", REC-ERC-81-8, May 1981, U.S. Bureau of Reclamation, Denver, Colorado.

After the filing of the Final Supplement to the Environmental Impact Statement on the O'Neill Unit, the U.S. Environmental Protection Agency expressed concern over the potential for excess productivity in Norden Reservoir, a major feature of the O'Neill Unit, and consequent hypolimnetic oxygen depletion in the reservoir if it should become thermally stratified for an extended period of time. A combination of nutrient loading models and a temperature simulation model were used to project the trophic state, productivity, and duration of stratification of the reservoir. The trophic state index indicated that the reservoir would be eutrophic; the estimated mean annual chlorophyll-a concentration indicated a high likelihood that algal blooms would occur. A classification function was applied to the reservoir to evaluate whether the blooms would consist of blue-green or nonblue-green algae. The results show a 98 percent probability that blue-green algal blooms will occur. Dissolved oxygen curves for the reservoir were then developed for the period immediately prior to the fall overturn. Curves were developed based on a reservoir with similar nutrient loadings and on theoretical curves. From these and the temperature model, oxygen concentrations in the reservoir discharge were estimated. The dissolved oxygen may be sufficient to support aquatic life but may fall below the Nebraska standard under certain circumstances. The reservoir and the river downstream are expected to support a cold-water fishery; most species now present in the river will be eliminated.

APPENDIX K

ESTUARINE IMPACT PREDICTION AND ASSESSMENT

Armstrong, N.E., "Effects of Altered Fresh Water Inflows on Estuarine Systems", Proceedings of the Gulf of Mexico Coastal Ecosystems Workshop, FWS/OBS-80/30, May 1980, U.S. Fish and Wildlife Service, Washington, D.C., pp. 17-31.

There are three critical elements in this framework for determining freshwater release needs for estuarine systems. The first is the determination of appropriate water quality levels needed to sustain the diversity and productivity of the bay system, particularly commercially important species. The second is the process or predictive basis by which the effects of spatial and temporal distributions of freshwater flows are determined. The third is the determination of the quality and scheduling of freshwater inflows. Studies have related inflows to biological changes in estuaries using the community approach, the population approach, and the individual approach. The method chosen for such studies should be a function of the data available for the analysis. By exercising mathematical models for the system being studied, freshwater inflows may be correlated with changes in water quality at any point. Using this information and estimates of freshwater inflows to the bay, the effects of changes in freshwater inflows can be determined.

Armstrong, N.E. and Wart, Jr., G.H., "Effects of Alternatives of Fresh Water Inflows into Madagorda Bay, Texas", Proceedings of the National Symposium on Fresh Water Inflows to Estuaries, FWS/OBS-81-04, Vol. II, Oct. 1981, U.S. Fish and Wildlife Service, Washington, D.C., pp. 179-196.

The Matagorda Bay, Texas, region is experiencing increased development on its periphery and in its drainage basin. The Matagorda Bay Project is a three-year study focusing primarily on the effects of altered freshwater inflows on the living resources of the bay. Freshwater inflows will be reduced because of increased withdrawals of water from the Colorado, Lavaca, and Navidad Rivers for agricultural, industrial, and municipal purposes. The reduced inflows will cause an increase in salinity, but model studies show that this will be less than 1 ppt. Nutrient concentrations may be increased or reduced depending on upstream practices. The biological effects are expected to be the result of loss of nursery areas in secondary and tertiary bays, and the possible loss of nutrients. A navigation and diversion project is planned for the Colorado River, in which water from the river will be diverted into Matagorda Bay, and a navigation channel will be maintained to the Gulf of Mexico. The major effect of the diversion channel will be to alter the circulation in Matagorda Bay and increase sedimentation. The increased sedimentation will eliminate some shallow bay habitat, but at the same time, build up new marsh habitat. These results are preliminary; biological and hydrological studies are continuing.

Bella, D.A. and Williamson, K.J., "Simulation of Sulfur Cycle in Estuarine Sediments", Journal of the Environmental Engineering Division, American Society of Civil Engineers, Vol. 106, No. EE1, Feb. 1980, pp. 125-143.

A mathematical model of estuarine sediment was developed using rate coefficients and field measurements. The model had a particular emphasis on the sulfur cycle and included specific chemical components of dissolved oxygen, soluble organic carbon, sulfates, free sulfides, total sulfides, sulfide capacity, sulfur, and pyrite. Different levels of sediment organics and turnover rates (RST) were mathematically imposed, and the subsequent levels of chemical components were determined after a 210-day period.

Browder, J.A. and Moore, D., "A New Approach to Determining the Quantitative Relationship Between Fishery Production and the Flow of Fresh Water to Estuaries", Proceedings of the National Symposium on Fresh Water Inflow to Estuaries, FWS/OBS-81-04, Vol. I, Oct. 1981, U.S. Fish and Wildlife Service, Washington, D.C., pp. 403-430.

Freshwater inputs to estuaries appear to enhance the production of marine organisms, because the highest standing stocks along shorelines are found in or near estuaries, which receive freshwater inputs. Activities of man that affect the quantity or timing of the flow of fresh water to estuaries include: dams for irrigation and power; diversions; canals in uplands; deforestation; clearcutting; grazing; road construction; and paving. The flow of fresh water into estuaries may influence fishery production, either directly or indirectly, in at least five ways: transport of nutrients; transport of detritus; transport and deposition of sediments; reduction of salinity; and mixing and transport of water masses. Diverse estuarine systems found along the gulf coast from Florida to Texas have similar problems as a result of changes in the quantity and seasonal patterns of the freshwater flow they receive. In a conceptual model, direct river inputs are sediments, nutrients, and fresh water. Sediments build and maintain tidal wetlands, counteracting both natural and anthropogenic processes that destroy wetlands. Nutrients in river water stimulate productivity of wetland vegetation. The physical force of runoff flushes decaying vegetation into tidal creeks and open waters, where it is processed by microorganisms into food for benthic animals, which are fed on by juvenile fish and invertebrates. Nutrients released to open water stimulate productivity of phytoplankton and seagrasses. Freshwater inputs establish chemical potential energy that sometimes drives the mixing of the estuary.

Chu, W.S. and Yeh, W.W., "Two-Dimensional Tidally Averaged Estuarine Model", Journal of the Hydraulics Division, American Society of Civil Engineers, Vol. 106, No. HY4, Apr. 1980, pp. 501-518.

Two types of models have been used extensively in the studies of estuaries: (1) physical hydraulic models, and (2) numerical models. This study deals with vertically averaged two-dimensional numerical models for estuarine hydrodynamics and salinity. The governing equations are the hydrodynamics equations coupled with the transport equation. The effect of the density gradient was included in the model which requires simultaneous solutions for all variables. The tidally averaged solutions and the associated numerical scheme were verified by solutions obtained from a corresponding transient model where long term integration was performed to reach a dynamic steady-state condition. The proposed model can be utilized to: (1) obtain dynamic steady-state solutions directly without performing long term integrations over time; and (2) serve as an efficient forward solution scheme for parameter identification, since parameters imbedded in the governing equations, such as the roughness coefficients, are essentially time-in-variant. The tidally averaged model was derived under the assumption that amplitudes of the transient variables are small in magnitude.

Green, K.A., "A Conceptual Ecological Model for Chesapeake Bay", FWS/OBS-78/69, Sept. 1978, U.S. Fish and Wildlife Service, Washington, D.C.

A conceptual model for the Chesapeake Bay ecosystem (wetlands, tributaries, and bay proper) has been developed as an interrelated series of diagrams showing carbon as a nutrient. Information was based on an analysis of local literature and discussions with scientists who are studying the Bay. The ecological functions that produce the resources of commercial and recreational fisheries, habitat for migratory birds and other wildlife, waste disposal, and aesthetic water quality are indicated. Physical (light, turbidity, mixing, transport, sedimentation) and chemical (sediment-water interaction, presence of pollutants) aspects of the environment modify the rates of biological processes (primary production, nutrient regeneration, and larval survival). A detailed ecosystem model combining the wetlands, plankton, seagrasses, other benthos, and fish trophic dynamics submodels shows the importance of material transfer and interactions between subsystems.

Jennings, M.E., "Characterization of Fresh Water Inflow Modification to Estuaries Resulting from River Basin Development", Proceedings of the National Symposium on Fresh Water Inflow to Estuaries, FWS/OBS-81-04, Vol. II, Oct. 1981, U.S. Fish and Wildlife Service, Washington, D.C., pp. 375-384.

There are three approaches for determining quantity and modification of freshwater inflow to estuaries: continental-scale assessment, statistical studies of long-term records at the mouth of a river, and river basin modeling. Continental-scale assessments are of limited use in determining modifications of freshwater flows. The statistical approach is illustrated using long-term streamflow records for the Brazos River, Texas, and the Santee River, South Carolina. For both rivers, the influence of upstream impoundments is obvious, and for the Brazos River, the 9-year moving averages show the dominance of the long-term regional climatic effect on the flow pattern. River basin modeling, illustrated using data from the lower Santee River Basin, is believed to be the best approach for general field studies. The Large River Basin Transport Model is being developed, which will make long-term (20-50 yr) simulations. The model was used with the Santee River data to optimize power production through a study of reservoir operating rules. The observed and computed flows were in good agreement for flow in the river and discharges used in power production.

Klein, C.J. et al., "Assessment Methodologies for Fresh Water Inflows to Chesapeake Bay", Proceedings of the National Symposium on Fresh Water Inflow to Estuaries, FWS/OBS-81-004, Vol. I, Oct. 1981, U.S. Fish and Wildlife Service, Washington, D.C., pp. 185-199.

The EPA Chesapeake Bay program is conducting an in-depth study to assess the principle factors having adverse impacts on the water quality of the bay. The program focuses on the point and nonpoint sources of pollution, including nutrients and toxic chemicals that are associated with various land use practices. Water quality data will be evaluated through use of stochastic and deterministic models. Field data collected on specific land uses from five test basins will be the basis for research to verify nonpoint source runoff rates. The field data will be used to calibrate and verify mathematical models in the test basins including nonpoint source loading, stream transport and estuarine processes. In particular, the estuarine models will simulate the impacts of nutrients on water quality. Fall line water quality data will serve as an independent data set to compare the point and nonpoint source projections associated with various land use activities. Mathematical models will be employed on a bay-wide scale to generate nonpoint source loadings basin-wide and to assess the impact from those loading on the tidal bay for the present (1980) and future (2000) conditions. Several growth scenarios that include consumptive freshwater use will be evaluated for their impact on water quality in the bay.

Lauria, D.T. and O'Melia, C.R., "Nutrient Models for Engineering Management of Pamlico Estuary, North Carolina", Report No. 146, July 1980, Water Resources Research Institute, University of North Carolina, Raleigh, North Carolina.

The objectives of this research were: to develop predictive models for nutrients and associated water quality parameters in the Pamlico Estuary for use in managing that aquatic system; to calibrate and verify these models using available data; to use the models to simulate or predict water quality in the Pamlico Estuary under different nutrient loadings; and to use the models to evaluate the significance of selected physical, chemical and biological processes in the estuary. Two steady-state, one-dimensional models have been developed and verified. One model, designed for winter conditions, considers phosphorus as the nutrient limiting algal growth during that season. The second model, developed for summer seasons, is based on nitrogen as the limiting nutrient. Simulations using these models indicate that past industrial discharges of phosphorus have had significant effects on water quality in the estuary. Research indicates nitrogen incorporated into algal blooms during a winter season is detained in estuary sediments until the following summer.

Linton, T.L. and Appan, S.G., "A Dynamic Methodology for Characterizing and Monitoring Estuarine Ecosystems", Proceedings of the National Symposium on Fresh Water Inflow to Estuaries, FWS/OBS-81-04, Vol. II, Oct. 1981, U.S. Fish and Wildlife Service, Washington, D.C., pp. 448-462.

A methodology is proposed that utilizes extant data from disparate sources for the purpose of determining or predicting transitory and/or permanent effects that natural and/or man-made changes will cause in an estuarine ecosystem. Data on estuaries in Alabama, Louisiana and Texas were used to construct a computer file, from which the natural variability of the estuary may be defined. Temperature and salinity changes were recorded for one year, and substrate type and polychaete species were determined for the following estuaries: Timbalier; Lavaca; Aransas; Copano; and Mobile. The Lavaca estuary characteristics were used to select monitoring stations, and the principal months for seasonal monitoring. The spatial, temporal, and species-level variability for polychaetes, and seasonal variability of temperature and salinity were determined. Data from the monitoring program that fall within these ranges would indicate that activities are not detrimental to the estuary. This method provides a direct way to measure actual impacts, can be updated as more data becomes available, provides for prediction of impacts in similar estuaries, can aid in the design of monitoring programs, and can detect the cumulative effects of many activities.

Najarian, T.O. and Harleman, D.R.F., "A Real Time Model of Nitrogen-Cycle Dynamics in an Estuarine System", Progress in Water Technology, Vol. 8, No. 4-5, 1977, pp. 323-345.

A computer analysis of the response of an estuary to the input of waste water treatment effluent reveals the importance of coupling tidal transparent dynamics with biochemical and ecological processes. A real-time model obtains better results than a tidal average model because it realistically simulates ocean boundary conditions. The real-time model explains effluent transport upstream of the point of injection due to flow reversal in the estuary caused by tides. An analysis of simple batch systems reveals the structured closed loop nitrogen-cycle has limited cycle characteristics; system response depends on whether or not it is open to matter flow; system variations occur with changes in parameters which govern the rate of transformation processes. Three objectives were established in formulating the study: (1) presenting a developed model for aerobic nitrogen limited aquatic systems; (2) predicting the estuarine response to future increases in waste water treatment facilities; and (3) coupling transport dynamics with nitrogen cycle dynamics. The basic principle of mass conservation of nitrogen was applied to various storage forms of the nutrient throughout its cycle. This closed loop nitrogen cycle model is important for estuarine phytoplankton dynamics. Seven storage variables and twelve biological and ecological transformations of nitrogen from one storage form to another were used in through-flow and unsteady-flow runs. Transformation rates were functions of nutrient concentrations and available energy levels.

Neu, H.J., "Man-Made Storage of Water Resources--A Liability to the Ocean Environment, Part II", Marine Pollution Bulletin, Vol. 13, No. 2, Feb. 1982, pp. 44-47.

Regulating natural seasonal runoff for power production may produce serious damage to marine life. Biological activities in large ocean currents such as the Gulf Stream depend on a seasonal nutrient supply, which in turn is regulated by the seasonal fresh water runoff. Modifications of these patterns, which have evolved over thousands of years, may be responsible for declines in fish catches, as seen in the late sixties and early seventies off the North Atlantic Coast. Other areas where marine life has been disturbed by water resources development are the Nile Delta, the Black and Azov Seas, and the Grand Banks of Newfoundland. Over 300 water storage projects have been built in Canada during the last 25 years. This has changed the balance of water storage in this region from the natural ratio of 4:1 (summer:winter) to 3:2. Possibilities for achieving more natural seasonal flows are power production without storage (run of the river stations), closed-circuit twin lake systems in which water is pumped into the upper lake during periods of low power demand and excess available energy, and determining the minimum spring flow necessary to maintain primary production in estuaries.

Onishi, Y. and Wise, S.E., "Mathematical Modeling of Sediment and Contaminant Transport in the James River Estuary", Proceedings of the 26th

Annual Hydraulics Division Specialty Conference on Verification of Mathematical and Physical Models in Hydraulic Engineering, 1978, American Society of Civil Engineers, New York, New York, pp. 303-310.

Various contaminants (e.g., pesticides, heavy metals, nutrients, and radionuclides) discharged to surface waters are dispersed by currents, surface waves, and turbulent mixing. These contaminants are also adsorbed from solution onto sediment; thus, otherwise dilute contaminants are concentrated. This process may create a significant pathway to man. For example, contaminated sediments may be deposited onto river and ocean beds. Contaminated bed sediments in turn may become a long-term source of pollution through desorption and resuspension. In contrast, sorption by sediment can be an important mechanism for reducing the area of influence of these contaminants by reducing concentrations of dissolved constituents. To identify the effects of sediment transport on contaminant migration, the finite element sediment and contaminant model (FETRA) was applied to the James River estuary. The modeling procedure for FETRA involved simulating the transport of sediments (organic and inorganic materials) within the water body. The results were then utilized to simulate dissolved and particulate contaminants by taking into account interaction with the sediments. Finally, changes in river bed conditions were calculated, including: (1) river bottom elevation change, (2) distributions of the ratios of each bed sediment component in the bed, and (3) distribution of contaminants in the river bed.

Ozturk, Y.F., "Mathematical Modeling of Dissolved Oxygen in Mixed Estuaries", Journal of the Environmental Engineering Division, American Society of Civil Engineers, Vol. 105, No. EE5, Oct. 1979, pp. 883-904.

One of the basic difficulties in modeling DO in a given estuary is the uncertainties involved in the BOD and reaeration rates. The oxidation of both carbonaceous and nitrogenous substances simultaneously and their representation by first-order kinetics may be justified in estuaries. The carbonaceous and nitrogenous BOD rate coefficients were determined from the standard BOD tests of estuary water samples by considering the two stages of BOD separately and by adjusting the time origin of the nitrification state. The available aeration rate coefficient equations developed for nontidal systems and applied in modeling DO in tidal estuaries did not fully represent the tidal estuaries. A new rearation rate coefficient equation was developed for tidal systems. The proposed equation was related to the dispersion coefficient and was a function of flow depth and the four-thirds power of tidal velocity.

Radford, P.J. and Joint, I.R., "The Application of an Ecosystem Model to the Bristol Channel and Severn Estuary", Water Pollution Control, Vol. 79, No. 2, 1980, pp. 244-254.

The formulation and application of the GEM-BASE model, an ecosystem model describing the Bristol Channel and Severn Estuary, are presented. Research in the channel, data from the literature, and laboratory experiments were used to derive the model. The model showed that the riverine input of nutrients was not sufficient to support the observed primary production rate and that local nutrient recycling must occur. Nutrient regeneration studies in the Carmarthen Bay have commenced as a result of model information. GEM-BASE will also be utilized to answer applied problems and to predict the consequences of building a tidal barrage on the channel ecosystem. Expansions of the model could be used to evaluate the effects of pollutants on the ecosystem where data on effects already exist. The hydrodynamic subsystem would simulate the distribution of any conservative pollutant and, if the relationship is known between a pollutant's concentration and its biological effect, the effect on the entire ecosystem could be simulated. Data collected from 1977 to 1981 on monitoring cruises in the British Channel will be used to validate the verified version of GEM-BASE.

Shea, G.B. et al., "Aspects of Impact Assessment of Low Fresh Water Inflows to Chesapeake Bay", Proceedings of the National Symposium on Fresh Water Inflows to Estuaries, FWS/OBS-81-04, Vol. I, Oct. 1981, U.S. Fish and Wildlife Service, Washington, D.C., pp. 128-148.

Modification of the Chesapeake Bay hydrologic environment has occurred over the past several decades, and is expected to continue at least until the end of the century. An attempt to assess the impact of low freshwater inflows (due to drought and consumptive losses) upon the Chesapeake Bay biota is currently underway. Some of the tools being used in this assessment are: distribution, tolerance, and life history studies of selected estuarine species; hydraulic modeling of salinity and circulation regimes; and computer simulation of representative segments of the ecosystem. Critical life history stages, habitat and food requirements, and tolerance to physical stress have been used to select representative study species for evaluation of effects of reduced freshwater flows. Information on community structure and trophic relationships has been used to develop a conceptual model of major energy flows within the Chesapeake Bay ecosystem. A computer simulation model will be used to provide insight into the effects of low flow, and the propagation of these effects throughout the ecosystem. The results of this study will aid managers in planning consumptive use patterns of freshwater flows into the estuary.

APPENDIX L

GROUND WATER, NOISE, CULTURAL, VISUAL AND SOCIO-ECONOMIC IMPACT PREDICTION AND ASSESSMENT

Altshul, D.A., "Guidelines: The Use of Cultural Resource Information in Water Resource Environmental Impact Reports", MS Thesis, 1980, Department of Hydrology and Water Resources, University of Arizona, Tucson, Arizona.

Techniques and guidelines for use by public officials when conducting cultural resource analysis on Federally funded water projects were developed. The more important Acts and Executive Orders which deal directly with cultural resources and environmental impact analysis are summarized. The guidelines developed are divided into seven steps: (1) contact with the State Historic Preservation Office; (2) contact with state historical societies, museums, universities, and other recognized institutions; (3) contact with a qualified agency or consulting firm; (4) preparation of a list of prehistoric and historic sites and finds in the area; (5) incorporation of a report from the consultants into the initial environmental assessment; (6) modify plans incorporating the assessment into the cultural resource section of the final environmental report; and (7) assess the need for the recovery of significant data. Cultural resource management techniques and the guidelines should be incorporated into the earliest stages of project planning for large scale projects. Use of the guidelines for both large and small scale projects is discussed and illustrated using hypothetical case studies. Three case studies are given: (1) Buckhorn-Mesa Watershed Plan; (2) Lower Queen Creek Watershed Plan; and (3) a plan for a Demonstration Recharge Project in the Salt River Valley. Use of definite procedural guidelines will provide a measure of standardization and consistency in environmental reports.

Carlson, J.E. and Sargent, M.J., "A Dynamic Regional Impact Analysis of Federal Expenditures of a Water and Related Land Resource Project--The Boise Project of Idaho, Part IV: A Social Impact Analysis of Federal Expenditures on a Water Related Resource Project: Boise Project, Social Subproject", Technical Completion Report, Mar. 1979, Water Resources Research Institute, Idaho University, Kimberly, Idaho.

Social changes resulting from the Boise Project, an irrigation and power project in southwestern Idaho are analyzed. The Project was built by the U.S. Bureau of Reclamation between 1910 and 1956 and is now managed for irrigation, power, recreation, and flood control. Both spatial and temporal impacts on Ada and Canyon Counties and changes from 1940 through 1970 are analyzed. A comparison is made of the quality of life "with" the project and "without" the project. Quality of life indicators used include education, housing, and neighborhood, formal achievement, health (mental and physical), law enforcement, accessibility, and recreation. The most significant social impact has been on population numbers, with the project adding 38,000 to 22,000 residents from 1940 to 1970. Farm population has declined but not as sharply as it would have without the project. The project has apparently had little impact on education. Income has somewhat increased with fewer families classified as poverty level, however, the percent employed is little changed. Housing has been unaffected except for a quality improvement in 1940. Health has also been unaffected. Violent crimes and accessibility were slightly increased by the project. Water-based recreation was greatly increased by the project. Overall, the apparent social impacts of the Boise Project have not been major.

Chang, S. and Beard, L.R., "Social Impact Studies: Belton and Stillhouse Hollow Reservoirs", Technical Report No. CRWR-164, June 1979, University of Texas, Austin, Texas.

The social effects resulting from the implementation of reservoir water resources projects are examined. As examples, the social impacts resulting from the Belton and Stillhouse Hollow Reservoirs, Bell County, Texas, are documented and used to develop principles for evaluating projects in an integrated manner. Beneficial and adverse social effects are evaluated in the areas of: (1) personal effects; (2) community and institutional effects; (3) regional socio-economic effects; (4) national and emergency preparedness effects; and (5) aggregate social effects. Impact information was gathered from public data such as the U.S. Census, from expert informants, and from personal interviews and questionnaires. Questionnaires and results from 11 land owner interviews are presented. Belton Dam, costing $13.8 million, was constructed by the U.S. Army Corps of Engineers beginning in 1949 and completed in 1954. The reservoir has 13.6 miles of shoreline and is 3 by 26 miles in area. Stillhouse Dam was begun by the Corps in 1962 and completed in 1968. Its total cost was $20.1 million. The reservoir has 58 miles of shoreline. Overall impact analysis shows that social impacts are only occasionally related directly to economic effects and the most important impacts are those in the areas of health, welfare, and quality of living. Also, most of these impacts could be created in an alternative manner not involving reservoir construction. The quantitative value of social impacts should be limited by the cost of providing these alternative equivalent impacts.

Coughlin, R.E. et al., "Assessing Aesthetic Attributes in Planning Water Resource Projects", Environmental Impact Assessment Review, Vol. 3, No. 4, 1982, pp. 406-416.

The authors recommended a four-phase procedure for evaluating the effects of water resource projects on aesthetic resources: (1) defining aesthetic resources; (2) inventorying aesthetic resources, to determine the future of the landscape, both in the absence of a project and with each of the project alternatives; (3) assessing the effects of the project from significant viewpoints; and (4) appraising the effects of project-related changes on the aesthetic resource. The method, an evaluative approach that relies on surveys of public judgments of aesthetic qualities, allows alternatives to be compared and provides information that might guide strategies to mitigate undesired aesthetic effects.

Daneke, G.A. and Priscoli, J.D., "Social Assessment and Resource Policy: Lessons from Water Planning", Natural Resources Journal, Vol. 19, No. 2, Apr. 1979, pp. 359-375.

Water resource development species have advanced the state-of-the-art in social well-being and quality-of-life accounting systems. As social and environmental assessments become widely utilized, the importance of presently held distinctions between land, air and water planning, as well as natural versus built systems, will diminish. Various forms of social accounting have emerged as alternatives to traditional resource assessment strategies. Social assessments can identify a range of value concerns not often expressed in the standard economic account. A variety of distinct life-quality accounting methodologies have also emerged. Some of the more widely used techniques are: (1) social profiling; (2) institutional analysis; (3) community assessment; (4) construction impact analysis; (5) mitigation design; and (6) survey research. However, there is no guarantee that the results of life-quality accounts will be given full weight in the decision-making process; some contend that they are too amorphous and subjective to warrant practical acceptability. These social indicators expand understanding of the distribution of costs and benefits. Social information will earn public confidence if it is used to display and help adjudicate conflicting claims, to design socially useful projects and programs which produce minimal social disruptions, and to explore means of enhancing the general quality of life.

Dickens, Jr., R.S. and Hill, C.E., editors, Cultural Resources--Planning and Management, 1978, Westview Press, Boulder, Colorado.

Cultural resource management is a new and vital field that has come about as a result of intensified Federal efforts to identify, evaluate, and manage cultural resources as an element of the environment. This volume--an edited collection of 16 of the papers given at a symposium on cultural resources planning and management--offers a wide range of perspectives from academia, private industry, and governmental agencies. It includes discussions of how a resource is defined, legal and government requirements, the agency-contractor relationship, and improved methods for implementing cultural resource studies and increasing public involvement. An important feature is the expansion of the concept "cultural resource"--which traditionally has referred to historical sites and archaeological remains--to include living people and their differing cultural heritages.

Eckhardt, W.T., "Cultural Resource Inventory of Areas Affected by Reject Stream Replacement Projects", July 1979, Westec Services, Inc., San Diego, California.

Approximately 14,990 acres of East Mesa and Imperial Valley, California, were culturally inventoried in compliance with Federal regulations for areas affected by Federal projects. The Yuma Desalting Plant in Arizona, a Federal project, will reject approximately 42,000 ac-ft of brine per year which it is required to replace with fresh water from other sources. Four alternative sources are: (1) lining a section of the All-American canal; (2) desalting Alamo River water; (3) desalting geothermal fluid; and (4) extracting ground water. The alternatives are evaluated by the areas and cultural resources affected. From the field survey 95 cultural resource sites were identified. Three archaeological sites and two districts were nominated for inclusion to the National Register of Historic Places. Recommended mitigation measures for the protection of these historic sites include restricted public knowledge of exact site location, careful construction of roads and other areas so that archaeological sites are avoided, and subsequent archaeological clearance before future development. Also, if it is impossible to design alternative construction to avoid archaeological sites, they must be salvaged prior to any project activity and a reasonable amount of time must be allowed for this. The inventory is designed to be a systematic base for future planning and cultural resource management.

Felleman, J.P., "Coastal Landforms and Scenic Analysis: A Review", Proceedings, The First Annual Conference of the Coastal Society, Nov. 1975, Arlington, Virginia, State University of New York, College of Environmental Science and Forestry, Syracuse, New York, pp. 203-217.

Scenic quality is related to man's perception of natural and built forms. A review is made of three visually related landform description approaches: numerical, geometric, and geomorphic. Diversity and complexity of coastal features are examined. Desirable analysis approaches are found to be sensitive to varying scales, offshore, beach, bluff and upland elements. A visual assessment approach ideally is suitable for both area-wide activity allocation planning and local site design decisions. It is necessary to establish a multi-tiered framework which aggregates characteristic groupings of similar features at the macro scale, and utilizes individual landforms or sets of landforms at the local scale. The latter would be applicable to analysis of actual planning and design relating to landscape scenes.

Fletcher, J.L. and Busnel, R.G., Effects of Noise on Wildlife, 1978, Academic Press, New York, New York.

This book brings together data from many disciplines to give, for perhaps the first time, a comprehensive picture of the effects of noise on wildlife--what is known, what needs to be investigated, and how these data are relevant to protecting wildlife and preserving endangered species. Topics covered include: effects of high voltage power transmission noise on wildlife; effects of sound on endocrine function; effects of sound on the fetus; effects of sonic booms on reproduction; effects of noise on hearing in fish; effects of noise on insects; snowmobile noise effects on wildlife; acoustics and habitat; and quantification of acoustic dose in studying the effects of noise. The book also addresses the problem of coordination between the scientific community and those who formulate and enforce wildlife protection legislation.

Harper, D.B., "Focusing on Visual Quality of the Coastal Zone", Proceedings, The First Annual Conference of the Coastal Society, Nov. 1975, Arlington, Virginia, State University of New York, College of Environmental Science and Forestry, Syracuse, New York, pp. 218-224.

Consideration of aesthetic values on an equal basis with ecologic, economic, and other values is mandated for planning decisions in the coastal zone. This research program in New York seeks to provide use-oriented methods for visual quality protection and control along the state's coastline.

Harvey, E.J. and Emmett, L.F., "Hydrology and Model Study of the Proposed Prosperity Reservoir, Center Creek Basin, Southwestern Missouri", Geological Survey Water Resources Investigation 80-7, June 1980, U.S. Geological Survey, Rolla, Missouri.

A reservoir has been proposed on Center Creek, Jasper County, southwestern Missouri. Ground water levels in the limestone uplands adjacent to the reservoir will rise when the impoundment is completed. The site is a few miles upstream from the Oronogo-Duenweg belt in the Tri-State zinc district. Grove Creek joins Center Creek downstream from the reservoir separating it from the mining belt. A model study indicates water-level rises varying from about 20 feet near the reservoir to 0.5 to 1.0 foot in the southern part of the Grove Creek drainage basin. A significant rise in the water table adjacent to the reservoir could increase mine-water discharge if Grove Creek is not an effective drain. However, it is probable that Grove Creek is an effective drain, and the higher ground water levels in the reservoir area will increase ground water discharge to Grove Creek, and in turn, Center Creek. The increase in ground water discharge to Grove Creek will have the beneficial effect of diluting mine-water discharge from the Oronogo-Duenweg belt during periods of low flow.

Hitchcock, H., "Analytical Review of Research Reports on Social Impacts of Water Resources Development Projects", IWR Contract Report 77-3, Mar. 1977, Program of Policy Studies in Science and Technology, George Washington University, Washington, D.C.

The purpose of this review is to aid water resources planners in identifying and evaluating the impacts of water resources development project actions. The review summarizes existing research on social impacts, identifies patterns and gaps in coverage, and suggests further areas of research. The review concentrates on case studies discussing social impacts of specific projects, thus providing the planner with a substantial foundation for evaluating impacts. There are three levels of summary; the most specific level is the individual study summaries; the next level provides summaries of study characteristics and impacts; and the final and most general level summary discusses patterns formed by characteristics and impacts presented earlier. The review indicates that certain areas of weakness are apparent in the study of social research. Construction phase impacts were found to be virtually ignored; economic impacts in the distributional and opportunity categories have received less attention than the sociological areas of community cohesion and service delivery; little has been done to measure intra-community conflict; and the relationship between water resources development projects and the provision of educational and cultural opportunities has also received very little attention. The study finds that these gaps are not unbridgeable and that more emphasis should be placed on identifying the full range of social impacts deriving from project actions.

Hoffman, W.L., "A Socio-Economic Feasibility Study of the Proposed Rochester Dam", Technical Assistance Report, Nov. 1977, U.S. Department of Commerce, Washington, D.C.

This study examines the desirability of construction of a multipurpose dam on the Green River in Butler, Kentucky. The purpose of the proposed Rochester Dam is to create a wide range of economic and social benefits for the area. The proposed project would flood 52,000 land acres, requiring the purchase of 71,000 acres in all. An estimated 240 families would have to be relocated, with some 50,500 acres of crop and pasturelands being inundated. The proposed maximum pool elevation of 420 feet would have no effect on the subterranean streams in Mammoth Caves, but the author believes further study should be undertaken to insure the integrity and well-being of this natural resource. The area is rural, with low median income, and suffers from out-migration and under-employment, with limited industrial and commercial potential. Costs and benefits of the project are presented. The findings indicate that although agricultural crop losses would occur--estimated at over $300 million over a 50 year period--and a loss of a sense of community by part of the population would result from resettlement, the overall benefits justify construction of the dam. These benefits include revitalization of the area's economy through construction employment; purchases of goods and commodities by the federal government and local construction workers, which would spur employment and purchase of services; recreation benefits, providing both recreation resources and increased employment; increases in land values and the construction of vacation homes; opening of the area to river navigation; providing industrial opportunities and benefits; and hydroelectric energy production.

Kessler, F.M. et al., "Construction-Site Noise Control Cost-Benefit Estimating Procedures", CERL-IR-N-36, Jan. 1978, U.S. Army Construction Engineering Research Laboratory, Champaign, Illinois.

This report aids the U.S. Army Corps of Engineers construction cost estimator in determining the level of noise generated at construction sites, in comparing this level with Corps of Engineers criteria, and in estimating costs to a contractor of reducing the noise.

King, T.F., "The Archaeological Survey: Methods and Uses", 1978, Heritage Conservation and Recreation Service, U.S. Department of the Interior, Washington, D.C.

This report summarizes various methodologies available for the conduction of archaeological surveys.

Leatherberry, E.C., "River Amenity Evaluation: A Review and Commentary", Water Resources Bulletin, Vol. 15, No. 5, Oct. 1979, pp. 1281-1292.

River amenity evaluation refers to the assessment of features, or conditions, in riparian environments that may provide recreational, or preservational and esthetic values. This report reviews methods of evaluating river amenity values and divides them under three main headings: River Recreation Potential Evaluation, River Esthetic Evaluation, and River Preservation Evaluation. Under the first heading the different methods attempt to evaluate recreation potential of rivers by assessing the utility of physical characteristics and other environmental effects for recreation activities. These methods did not adequately represent the dynamic nature of the resource and the potential users. The two methods for evaluating river esthetics had many shortcomings, e.g., they were dependent on the evaluator's personal judgment and photographs were not considered. The evaluation of river esthetics needs to consider what effect people and their impacts have on the esthetic quality of the resource. Under the heading of River Preservation Evaluation the methods reviewed are designed to assess the suitability of rivers or river segments for classification in state or federal river preservation programs. The passage of the National Wild and Scenic River Act (PL 90-542) in 1968 instigated the classification of rivers for preservation and recreation purposes. In order to improve the methods they should: (1) be capable of identifying and predicting change and its impact on amenity values; (2) incorporate concepts and techniques that measure the dimensions of people's feelings about specific environmental features; and (3) place more emphasis on the quality of the experience to be provided.

Michalson, E.L., "An Attempt to Quantify the Esthetics of Wild and Scenic Rivers in Idaho", Proceedings: River Recreation Management and Research Symposium, Jan. 24-27, 1977, Minneapolis, Minnesota, U.S. Forest Service General Technical Report No. NC-28, U.S. Forest Service, Department of Agriculture, St. Paul, Minnesota, pp. 320-328.

Described is the procedure used to estimate demand for outdoor recreation on rivers and in the development of a Likert-Type Scale to distribute the net resource values estimated in the demand analysis according to perceptions that users indicated as being important to the wild and scenic river experience. Outdoor recreation demand models for three study areas on the Salmon River provide the basis for valuing outdoor recreation and hence for valuing esthetics. Searched was a way to allocate value to the esthetic portions of the whole outdoor recreation experience. To establish this relation, the methodology determines the amount of "consumer surplus" involved in the outdoor recreation experience and then estimates how much of it is related to esthetics. Consumer surplus is defined as the difference between total utility and the market value of a good or service. It is concluded that this technique needs more study and research to determine its consistency and the stability of the distributions that have been developed.

Munter, J.A. and Anderson, M.P., "The Use of Ground Water Flow Models for Estimating Lake Seepage Rates", Ground Water, Vol. 19, No. 6, Nov./Dec. 1981, pp. 608-616.

Two- and three-dimensional ground water flow models were developed to estimate seepage rates for Bass Lake and Nepco Lake, Wisconsin. Bass Lake, ground water controlled with no surface water inlet or outlet, was described by a two-dimensional profile model. Results showed that the ratio of horizontal to vertical hydraulic conductivity of the aquifer around the lake is related to both the magnitude of vertical hydraulic gradients near the lake and the distribution of seepage from the lake as a function of distance from the shore. The presence of fine-grained sediments in the littoral area can have a significant effect on lake seepage rates, but the presence or absence of fine sediments in the deep parts of the lake is relatively unimportant. Nepco Lake, created by construction of an earthen dam, has an anomalous ground water flow system. A three-dimensional model was far superior to a two-dimensional model for simulating the flow system around this lake. It showed that much of the lake bottom loses water. The three-dimensional model produces a much lower water table slope at a point where the influence of two creeks, receiving ground water discharge, is noticeable.

Nelson, T.L., Warnick, C.C. and Potratz, C.J., "A Dynamic Regional Impact Analysis of Federal Expenditures of a Water and Related Land Resource Project--The Boise Project of Idaho, Part III: Economic Scenario of the Boise Region 'Without' a Federal Irrigation Project, Economics Subproject", Technical Completion Report, Mar. 1979, Water Resources Research Institute, University of Idaho, Moscow, Idaho.

Two models were used to estimate the economic conditions in Idaho and the Boise region if the Federally funded Boise Irrigation Project had never been built. A hydrologic model of the natural unregulated flows of the Boise and Payette Rivers were used to estimate the gross crop output that would have occurred without the Boise Project. Based on data from the hydrologic model an interregional trade flow model was also developed. Most of the precipitation for the two watersheds occurs as snow resulting in a heavy spring runoff. Flows then decline as summer progresses, with August flows often being only 15 percent of June flows. Without the storage water provided by the Boise Project there would be little water for irrigation during the critical summer season. From the hydrologic model, average annual irrigation diversions available from both rivers without the Project was 705,000 acre-feet. This diversion would have irrigated approximately 261,000 acres which in turn would produce $8,600,000 in crops by 1973. A comparison of these "without" results to actual "with" data shows that in 1972 while the "without" acreage represents 77% of the actual "with" acreage, the "without" diversion is only 36% of the "with" diversion. Further, the "without" income represents only 18% of the "with" diversion. Trade flow model results show that area output "without" the Project would have been $1.6 billion compared to $1.8 billion "with" the project as of 1970. The pattern for income is similar. The relative success of the project is justified on the basis of economic efficiency.

Nieman, T.J., "Assessing the Visual Quality of the Coastal Zone", Proceedings: The First Annual Conference of the Coastal Society, Nov. 1975, Arlington, Virginia, State University of New York, College of Environmental Science and Forestry, Syracuse, New York, pp. 247-251.

The visual quality of the coastal zone is an important aspect of coastal management. However, mechanisms for objectively analyzing visual resources in relation to the perceptions and attitudes of coastal users is not well developed. The problem is further complicated by the diverse nature of the groups utilizing various coastal resources.

Pickering, J.A. and Andrews, R.A., "An Economic and Environmental Evaluation of Alternative Land Development Around Lakes", Water Resources Bulletin, Vol. 15, No. 4, Aug. 1979, pp. 1039-1049.

Reported herein is a study which made an evaluation of alternative land developments around New Hampshire lakes. Alternative development patterns, evaluated by their impacts on the lake area environment and area economy, included residential patterns, commercial patterns, and combinations of these two types. Phosphorus loading of the lake water was used as a proxy variable for changes in the lake water supply. Commercial developments yielded the highest revenues to the town and the local area. It also attracted the most lake users to the area as well as contributing the largest phosphorus loading in the lake waters. Residential developments, although contributing high revenues to the businessmen in the area, yielded less net income to the town. Phosphorus loading levels from residential developments were much lower than lake phosphorus loading by commercial developments.

Ricci, P.F., Laessig, R.E. and Glaser, E.R., "The Preoperational Effects of a Water-Resources Project on Property Prices", Water Resources Bulletin, Vol.

14, No. 3, June 1978, pp. 524-531.

Using time series analysis, residential properties nearby a park-lake system, Nockamixon State Park at the fringe of the Philadelphia metropolitan area, have been studied to detect possible changes in the land market during the period between announcement and completion of the impounding reservoir, 1962 to 1973. The analysis identifies a period of increases in the values of properties which were concomitant with the announcement of the park-lake. This is followed by a second period characterized by lower prices, which ends in 1970, as construction activities take place. During construction, from 1970 to 1973, property values have been found to increase rapidly.

Shapiro, M., Luecks, D.F. and Kuhner, J., "Assessment of the Environmental Infrastructure Required by Large Public and Private Investments", Journal of Environmental Management, Vol. 7, No. 2, Sept. 1978, pp. 157-176.

Large capital projects, such as highways and barge canals, may induce significant residential, commercial, and industrial growth. This paper discusses an approach for evaluating the infrastructure requirements of such secondary development. Issues considered include the level of detail and time horizon used in making projections, service area delineation, facility selection and staging, and the impacts of local and state government actions. The approach can be used to compute residual fluxes and to evaluate economic impacts such as changes in tax rates and user charges. The methodology is applied to assess certain of the secondary impacts of the proposed Cross-Florida Barge Canal. A county-level aggregate economic forecasting model has been developed to assist in projecting the major economic and demographic trends in the canal region.

Sloane, B.A. and Dickinson, T.E., "Computer Modeling for the Lake Tahoe Basin: Impacts of Extreme Land Use Policies on Key Environmental Variables", Journal of Environmental Systems, Vol. 9, No. 1, 1979, pp. 39-56.

A socio-economic computer simulation model of population, growth and land-use in the Lake Tahoe Basin utilizes the best available statistical data and incorporates views of regional planners and interested citizens. The model compares broad long-range social and economic implications of potential governmental policies. Major variables or submodels are available jobs; locally employed seasonal residents; housing demand; housing supply; housing market reconciliation; basin area year-round residents employment status; business units; year-round residents' age structure; tourist visitor days; tourist amenities; and special factors. Computer simulation runs were made of seven extreme land-use policy packages which included the following policies: (1) construction and subdivision moratorium; (2) housing occupancy limitation; (3) removal of most campground and commercial limitations; (4) increase in number of residential units per acre; (5) additional acres zoned residential; (6) doubling of gambling casino facilities; and (7) new light industry. Each policy package was simulated using three different model versions to accommodate its high sensitivity to assumptions about tourism. Results show that the number of acres developed and the related lake water clarity reduction would be greater with upzoning policies such as density variances than with simulatory ones such as new industry promotion.

APPENDIX M

METHODOLOGIES FOR TRADE-OFF ANALYSIS AND DECISION-MAKING

Anderson, B.F., "Cascaded Tradeoffs: A Multiple-Objective, Multiple Publics Methods for Alternatives Evaluation in Water Resources Planning", Aug. 1981, U.S. Bureau of Reclamation, Denver, Colorado.

Cascaded Tradeoffs is a method for arriving at an overall ranking of planning alternatives on the basis of public values. The key feature of the method is that it provides for making tradeoffs across both issue dimensions and publics. The method shows not only how to deal with tradeoffs where tradeoffs must be made, but also how to identify or create situations where tradeoffs do not need to be made. In addition, it provides for examination of the extent to which the overall ranking of the alternatives is sensitive to the addition of mitigation measures and to uncertainty in the data.

Ahmed, S., Husseiny, A.A. and Cho, H.Y., "Formal Methodology for Acceptability Analysis of Alternate Sites for Nuclear Power Stations", Nuclear Engineering Design, Vol. 51, No. 3, Feb. 1979, pp. 361-388.

A formal methodology is developed for the selection of the best sites from among alternate suitable sites for a nuclear power station. The method is based on reducing the various variables affecting the decision to a single function that provides a metric for the level of site acceptability. The function accommodates for well known site selection criteria as well as other factors; such as public reactions to certain choices. The method is applied to the selection of a site from three acceptable alternate sites for Wolf Creek nuclear power station, Kansas.

Baram, R. and Webster, R.D., "Interactive Environmental Impact Computer System (EICS) User Manual", CERL-TR-N-80, Sept. 1979, U.S. Army Construction Engineering Research Laboratory, Champaign, Illinois.

This report describes the Environmental Impact Computer System (EICS) and provides instructions for obtaining and using output for the current interactive version of the system. It is recommended that the instructions be used to obtain the most efficient use of the system.

Bohm, P. and Henry, C., "Cost-Benefit Analysis and Environmental Effects", Ambio, Vol. 8, No. 1, 1979, pp. 18-24.

Cost-benefit analysis (CBA) is used in three case studies to illustrate its value in reducing a problem with multi-dimensions to one with fewer dimensions, thus making the problem more manageable. CBA also can insure that no particular aspect is unduly favored or ignored and that uncertainties, indivisible concerns, risk premiums, irreversibility, and option value are taken into account. Because CBA examines relevant questions and gives appropriate weight to both the institutional framework and to distribution of benefits and costs among various affected publics, it helps environmental considerations in political decision-making. The problems considered are: (1) whether or not an oil refinery should be constructed in an unpolluted fishing and marine life zone in Roadstead Bay of Brest, France; (2) whether a highway connecting the suburbs of Paris should be allowed to cross the beautiful forests of Malmaison, Versailles, etc.; and (3) whether or not unique Hell's Canyon, in the United States should be used for a hydroelectric dam. These case studies show difficulties which arise where decisions are made without analysis, or with analysis but without an overall perspective.

Brown, C.A., Quinn, R.J. and Hammond, K.R., "Scaling Impacts of Alternative Plans", June 1980, Center for Research on Judgment and Policy, University of Colorado, Boulder, Colorado.

This report describes procedures for improving measurement of the effects of alternative water development plans--particularly environmental and social impacts. Basic issues in measurement are related to the task of comparing alternative plans. Four types of measurement scales are evaluated in terms of their usefulness in the planning process. The concepts of reliability and validity are discussed in detail as is the concept of a measurement standard. Measurements required for water resources planning are divided into four types: Type I--objective application of a measurement standard; Type II--subjective application of a measurement standard; Type III--subjective application of general scientific principles; Type IV--subjective application of personal values. Procedures are presented for improving measurement types II, III, and IV, focusing on: (1) disaggregating multidimensional factors (e.g., social well-being); (2) defining factors and subfactors; (3) specifying relations between factors and subfactors; (4) integrating subfactor impacts; and (5) defining higher order factors.

Bryant, J.W., "Modelling for Natural Resource Utilization Analysis", Journal Operations Research Society, Vol. 29, No. 7, July 1978, pp. 667-676.

A methodology is described which can be used to assess the environmental impact of human activities. The approach is based on a purpose-built computer modelling language which may be used to trace the flows of resources generated by any process of interest. Models developed using this approach may be used by planners to appraise environmental effects of their decisions and to explore strategies for energy or other resource savings.

Budge, A.L., "Environmental Input to Water Resources Selection", Water Science and Technology, Vol. 13, No. 6, 1981, pp. 39-46.

This paper outlines the method that has been used, in conjunction with engineering, economic and hydrological criteria, in water resources evaluations. It consists of the identification of environmental impacts through the use of a matrix, the collection and collation of information on each impact and the comparison of options, using a combination of ordinal rankings in preference to cost-benefit techniques. The methodology is illustrated with reference to water resource developments, but is applicable to any study involving a large number of development options.

Burnham, J.B., Nealey, S.M. and Maynard, W.S., "Method for Integrating Societal and Technical Judgments in Environmental Decision Making", Nuclear Technology, Vol. 25, No. 4, Apr. 1975, pp. 675-681.

A methodology was developed for environmental decision-making that combines societal and technical judgments. Eight factors that characterized the major economic and environmental impacts of nuclear power plant sitings were identified. These factors were used to construct "mini-environmental impact statements" for six siting alternatives. The impact statements formed the core of a survey questionnaire administered to three groups of respondents. Data analysis produced estimates of the relative importance of each factor. A procedure is described for using these estimates of importance as weighting factors to be applied to techno-economic scores. These latter scores would be generated by technical experts and would represent the actual or anticipated impact of a plant siting upon the eight factors.

Davos, C.A., "A Priority-Tradeoff-Scanning Approach to Evaluation in Environmental Management", Journal of Environmental Management, Vol. 5, No. 3, 1977, pp. 259-273.

A priority-tradeoff-scanning (PTS) approach is proposed to help satisfy three objectives in environmental management evaluation: (1) to record the impact of each option on all goals and interest groups; (2) to scan all feasible goal-priority tradeoffs for each interest group; and (3) to scan priority tradeoffs that are actually acceptable to those groups. On the decision-making evaluation level, the objectives of the PTS approach are: (1) to identify decision choices which will maximize consensus on goal priorities; (2) to identify choices which will maximize the satisfaction of individual interests of competing groups; and (3) to scan priority tradeoffs for maximizing achievement of goals, satisfying the aspirations of planners, and maximizing consensus. This paper is an approach to evaluation, rather than an evaluation methodology, in providing a framework for interpreting and synthesizing all information inputs pertinent to evaluation, instead of generating particular information inputs. The paper also focuses on evaluation of options, not actions. In PTS, pertinent information for each of the three evaluation objectives is synthesized in the form of the following matrices: (1) goals-achievement matrix (GAM); (2) goal-priority-tradeoff matrix (GPTM); and (3) interest-priority-tradeoff matrix (IPTM). Each of these matrices is discussed and explained.

Duckstein, L. et al., "Practical Use of Decision Theory to Assess Uncertainties About Actions Affecting the Environment", Completion Report, Feb. 1977, Department of Systems and Industrial Engineering, Arizona University, Tucson, Arizona.

The determination of the environmental impact of man's action has many uncertain components. A conceptual framework has been developed to assess the effect of uncertainty on environmental impact statements. A multiobjective system framework utilizing Bayesian decision theory was developed as a result of the research. As a result, salient features of an environmental impact statement that properly considers uncertainty are given. Decision theory can be used to advantage in applying this methodology to assess the effect of natural and informational uncertainty. Even the complexity of computational procedures involving a Bayes analysis can be overcome by use of approximations, or by simulation. Multiobjective decision-making models are necessary to analyze the noncommensurate objectives found in studying environmental impact. Current procedures used to assess environmental impact do not take proper cognizance of uncertainty and the problems of tradeoff among multiple objectives in the analysis and evaluation of actions affecting the environment. Some specific and legal concepts such as proof and model validation must be thought of in terms of probability, if optimal decisions are to be made affecting the future environment. The methodology developed by this project shows how uncertain information and scientific models can be used to advantage in planning the future.

ESSA Environmental and Social Systems Analysts, Ltd., "Review and Evaluation of Adaptive Environmental Assessment and Management", Oct. 1982, Environment Canada, Vancouver, British Columbia.

Adaptive Environmental Assessment and Management (AEAM) is a collection of concepts, techniques, and procedures intended for the design of creative resource management and policy alternatives. Development of AEAM was initiated in the early 1970s, and since that time, the approach has been applied to over 60 projects. These applications have provided exposure to a wide variety of institutions, problems, and disciplines resulting in varying degrees of success. This report documents a critical review of AEAM that consisted of three stages: (1) a workshop which brought together practitioners, users, and senior policy designers; (2) case study evaluation by both practitioners and users; and (3) a synoptic analysis of AEAM procedures, literature and case studies.

French, P.N. et al., "Water Resources Planning Using Computer Graphics", Journal of the Water Resources Planning and Management Division, American Society of Civil Engineers, Vol. 106, No. WR1, Mar. 1980, pp. 21-42.

Computer simulation and optimization models that are used to assist in multipurpose, multiobjective water resource planning often suffer from the lack of an efficient data input system and the lack of an easy, yet comprehensive, means of interpreting and communicating the results of model studies to others. These deficiencies may be minimized with the help of computer graphic input and display methods. Interactive computer graphics was applied to four planning problems, which included the prediction and management of water quality, multi-reservoir simulation for water supply, multiobjective analyses for reservoir sizing, cost and yields, and flood management. Tablet digitizing routines were frequently used to input spatial and other data, while the graphical output was accomplished by vector display methods. Visual feedback was obtained at all stages of the procedures. As the cost of computer memory declines, the use of the graphics input and display devices is expected to increase.

Herzog, Jr., H.W. "Environmental Assessment of Future Production-Related Technological Change: 1970-2000 (An Input-Output Approach)", Technological Forecasting, Vol. 5, No. 1, 1973, pp. 75-90.

This paper presents a technique, or model, to systematically assess the environmental impact of specific technological changes forecast to occur over this and the next two decades. The core of the model is a dynamic technical coefficient matrix of a large input-output model. The technological change considered is that which affects the coefficients of this matrix and thus the distribution of material inputs over time into the various sectors of the U.S. economy. An environmental assessment of this production-related technological change is achieved through a submodel that registers production residuals on an industry basis for 14 waste categories.

Hill, D., "A Modeling Approach to Evaluate Tidal Wetlands", Transactions 41st North American Wildlife and Natural Resource Conference, March 21-25, 1976, Washington, D.C., Wildlife Management Institute, Washington, D.C., pp. 105-117.

A resource allocation model of the linear programming type was devised to provide an economic evaluation of wetlands. Exercise of the model shows the values that are inputed to wetlands by alternative resource management decisions. The model consists of a set of equations that define the biological, physical, chemical, and economic boundaries on the possible uses of a hypothetical bay. With these equations, the combination of uses is determined that maximizes the value of the objective function, the components of which are annual net income (in dollars), the environmental value, and the social value. Finding the maximum dollar value (or, equivalently, the minimum cost) of this complex of uses identifies those that give the salt marsh its greatest monetary value. The annual dollar value of the salt marsh is determined to the extent to which it contributes to net income or reduces the need for costly alternatives. No distinction is made in this model between income that results from using the wetland and that which results from destroying it; the typical year considered is assumed to extend into the indefinite future.

Hill, D., "A Resource Allocation Model for the Evaluation of Alternatives in Section 208 Planning Considering Environmental, Social and Economic Effects", Proceedings of the Conference on Environmental Modeling and Simulation, April 19-22, 1976, Cincinnati, Ohio, EPA 600/9-76-016, July 1976, U.S. Environmental Protection Agency, Washington, D.C., pp. 401-406.

Modeling for 208 planning should be designed to facilitate participation by planners and representatives of the affected public. Intangibles and incommensurables must be considered, and the ultimate need for value judgments to assess the importance of environmental, social, and economic effects must be accommodated without obscuring the factual analysis. Ideally, population groups that are affected differently should be accounted for separately. Model building should therefore proceed in successive stages of greater precision. An initial qualitative analysis leads to a conceptual model that identifies the differential impacts of the alternatives, making only the judgment that the impacts are beneficial or detrimental. With land use decisions among the alternatives for satisfying water quality goals, a resource allocation model is useful to account for other costs and benefits. This can be solved for the minimum dollar cost mix of activities as a datum; alternative plans can then be generated by assigning additional importance to intangible and incommensurable values, with the plan selection informed by knowledge of its incremental dollar cost.

Hodgins, D.B., Wisner, P.E. and McBean, E.A., "A Simulation Model for Screening a System of Reservoirs for Environmental Impact", Canadian Journal of Civil Engineering, Vol. 4, No. 1, Mar. 1977, pp. 1-9.

Analysis of the potential impact of development alternatives is becoming increasingly complex with the imposition of more and more constraints, many of which are environmental. Restricted planning budgets require that a rapid identification be made of the most promising alternatives to avoid unnecessary expenditures. A screening model to quickly isolate the most promising alternatives is thus becoming considerably more important. Described is a computer model that satisfies these concerns in applications involving a series of reservoirs. The model, with both hydrologic and water quality components, simulates and thereby indicates probable changes in downstream flows, reservoir surface fluctuations, and temperature and dissolved oxygen changes in the reservoirs and streams. Through easily adjusted operating policies, reservoir sizes, etc., the model can rapidly determine the potential impact of alternative developments. This information is then available to biologists, wildlife, forestry, and social disciplines as an aid in the determination of environmental impact assessments. A case study application of the model that reflects eastern Canadian conditions is described.

Keeney, R.L., "Preference Models of Environmental Impact", IIASA-RM-76-4, Jan. 1976, International Institute for Applied Systems Analysis, Laxenburg, Austria.

Environmental problems usually involve multiple conflicting objectives, large uncertainties concerning the possible environmental impact, and several individuals or groups whose preferences are very different, but yet very important in choosing an alternative. If one wishes to influence the decision-making process using analysis, the above issues should be addressed. One critical aspect, which is usually conducted informally, involves weighting the advantages and disadvantages of the possible impacts of the various alternatives by each of the interested parties. The basic ideas of multiattribute utility are presented and placed in the context of a decision-making framework. The manner in which it aids the analysis of environmental issues is discussed. Several problems having environmental components in which multiattribute utility was used are surveyed. These include siting nuclear power facilities, developing fisheries, controlling a forest pest, examining energy policy, and transporting hazardous materials.

Kemp, W.H. and Boynton, W.R., "Integrating Scientific Data into Environmental Planning and Impact Analysis, General Methodology and a Case Study", The Environmental Impact of Freshwater Wetland Alterations on Coastal Estuaries, Conference held at Savannah, Georgia on June 23, 1976, Florida University, Gainesville, Florida, pp. 61-86.

Seven basic steps in environmental evaluations are outlined. These guidelines are pertinent to any environmental evaluation issue, including land use and water management interactions between upstream (freshwater) and downstream (estuarine) ecosystems: (1) conceptualization of models describing a problem; (2) selection of alternative plans to be presented for evaluation; (3) definition of the objective function of a plan in terms of the net energy channeled into work processes in the region, where work includes both work of man and nature; (4) collection and/or measurement of data descriptive of the systems under evaluation; (5) development of mathematical models based on conceptual models; (6) conversion of model outputs and other pertinent data into a common measure unit; (7) selection of the final plan on the basis of comparison of calculated costs and benefits to the objective function for all the plans under consideration. Following the seven outlined steps, the impact of a proposed dam over the Apalachicola River, Florida was evaluated.

Loran, B., "Quantitative Assessment of Environmental Impact", Journal of Environmental Systems, Vol. 5, No. 4, 1975, pp. 247-256.

The history of environmental impact and the current requirements, guidelines, and coverage of environmental impact statements are briefly reviewed. A procedure for the assessment of environmental impact is presented. The individual components of the environment induced by a proposed action and of the natural environment of the project area are listed, and each impact generated by their interaction is assigned value ratings. The ratings are assembled in an impact-incidence matrix, which is then reordered using a data analysis technique, the bond energy algorithm. The new matrix obtained displays interrelated clusters of high-valued ratings, corresponding to critical environmental areas. The impacts comprising each area are grouped according to similarity of action and effect. The environmental impact from the construction of an additional wastewater treatment plant in a resort area is assessed to illustrate the effectiveness of the procedure.

Lincoln, D.R. and Rubin, E.S., "Cross-Media Environmental Impacts of Coal-Fired Power Plants: An Approach Using Multi-Attribute Utility Theory", IEEE Transactions for Systematic Management of Cybernetics, Vol. SMC-9, No. 5, May 1979, pp. 285-289.

The types and rates of pollutant emissions from a coal-fired power plant depend on plant design, coal characteristics, and environmental control policy. By linking a human preference model, developed using multi-attribute utility theory, and an environmental emissions model for a coal-fired power plant, a methodology which can be applied to identify optimal control strategies reflecting specific pollutant trade-off decisions by environmental decision-makers is described.

Meyers, C.D., "Energetics: Systems Analysis with Application to Water Resources Planning and Decision Making", IWR Contract Report 77-6, Dec. 1977, U.S. Army Engineer Institute for Water Resources, Fort Belvoir, Virginia.

Energetics is evaluated as a possible analytical tool for use in water resources planning and decision-making. Basic scientific concepts which comprise the comprehensive subject of energetics are explained through the use of easily understandable correlaries. Examined are the potential which these concepts have in water resources work as well as the limitations which curtail their present usability and acceptance. It is concluded that energetics has promise as an analytical tool for assessing environmental quality in general, and water resources in specific. So far, no other proposed analytic technique has the promise energetics has for total environmental assessment which includes quantification of values of natural systems. However, energetics is a data-limited tool, its major drawback.

Motayed, A.K., "Alternative Evaluation of Power Plant Sites", Journal of the Energy Division, American Society of Civil Engineers, Vol. 106, No. EY2, Oct. 1980, pp. 229-234.

Selection of sites for power plants requires a detailed evaluation of a large number of alternative and feasible sites. Site evaluation techniques in practice essentially compare and evaluate the large number of alternative sites with respect to a number of environmental, engineering, socio-economic, and other site characteristics. Elements of such evaluation techniques are mostly subjective. A method using the "weighting-scaling" technique used in the assessment of water resources projects is presented which allows some of this subjectivity to be eliminated. A hypothetical example is presented wherein 10 sites are ranked for acceptability using the proposed methodology.

Olenik, S.C., "A Hierarchical Multiobjective Method for Water Resources Planning", MS Thesis, 1978, School of Engineering, Case Western Reserve University, Cleveland, Ohio.

Hierarchical multiobjective optimization was used in two approaches for the possible combination of the TECHCOM methodology and the Surrogate Worth Trade-off (SWT) methodology for use by decision-makers in water resources planning. TECHCOM has a hierarchical structure (goals, subgoals, social indicators, and action variables) that preserves information at all levels so that alternatives can be analyzed without the loss of lower level information. The SWT method, which solves the standard vector optimization problem where objectives are noncommensurable, can find a preferred solution by finding Pareto-optimal solutions and tradeoffs, and using systematic decision-maker input. All that is needed to generate a preferred solution is decision-maker preference input. Two sample integrations of TECHCOM and SWT were tested on a sample problem using the Maumee River Basin Level-B land resources model. One alternative, the mathematical integration alternative, integrates the SWT method within the information structure of the TECHCOM goal and subgoal hierarchy. The second alternative, the operational integration alternative, features an iterative process using TECHCOM and SWT in succession. Some of the requirements for use of the combined methodology are given. Usefulness of the combined methodology depends on the complexity of the hierarchy submodels. A major advantage of TECHCOM-SWT is that it eliminates the need for a priori preference information.

Peterson, J.H., Clinton, C.A. and Chambers, E., "A Field Test of Environmental Impact Assessment in the Tensas Basin", Proceedings of the 14th Annual Mississippi Water Resources Conference, Sept. 1979, Mississippi State University, Mississippi State, Mississippi, pp. 27-32.

A field test of Water Resources Assessment Methodology (WRAM) procedures was conducted at the Tensas River Basin project in northeastern Louisiana. WRAM is a procedure for assessing and evaluating impacts in the areas of economic development, environmental quality, regional development, and social well-being. The Tensas River Basin has a flooding problem and the project calls for extensive channel improvements along 99 miles of the river. Five alternative plans and a "no action" alternative are being considered. WRAM requires the formation of an interdisciplinary team representing the areas of ecology, economics, engineering, and sociology/anthropology. A preliminary list of mandatory and critical variables is assembled and the variables are weighted. Data are then collected and analyzed for impact projection and scaling. Finally a summary table is constructed for each account listing each variable and its weight and impact for each alternative. It is concluded that the Tensas study was a useful test of the WRAM procedures. The use of the interdisciplinary team proved to be quite successful in that it allowed division of labor, and a system of checks and balances, and duplication of effort was minimized.

Rubinstein, S. and Horn, R.L., "Risk Analysis in Environmental Studies. I. Risk Analysis Methodology: A Statistical Approach; II. Data Management for Environmental Studies", CONF-780316-8, Mar. 1978, Atomics International Division, Rockwell Hanford Operations, U.S. Department of Energy, Richland, Washington.

A new approach to risk analysis as applied to environmental studies is proposed in this paper. Advantages of this methodology are pointed out. It utilizes modern statistical methods in order to treat risk analysis as an integral part of the study design.

Schrender, G.F., Rustagi, K.P. and Bare, B.B., "A Computerized System for Wild Land Use Planning and Environmental Impact Assessment", Computers and Operations Research, Vol. 3, No. 2/3, Aug. 1976, pp. 217-228.

A computerized system useful for examining the physical, economic, and environmental consequences of alternative wild land use decisions is described. The system consists of a set of simulation models linked to a geographic data base by an information storage and retrieval subsystem. The simulation models cover forest production, timber harvesting, recreation, fish-wildlife-insect dynamics, and atmospheric and hydrologic processes. System inputs consist of land-use and management decision alternatives. The consequences of these system inputs can be evaluated at varying scales of spatial and temporal resolutions in terms of goods, services, and environmental impacts. Uses to date suggest that the system is a valuable aid to land-use planners and forest management decision-makers.

Schwind, P.J., "Environmental Impacts of Land Use Change", Journal of Environmental Systems, Vol. 6, No. 2, 1977, pp. 125-145.

Better measurement and analysis of environmental impacts are prerequisites to better land use planning. Developed is an empirical matrix method for expressing the environmental impacts of land uses in a form compatible with economic evaluation techniques such as cost-benefit analysis. Impacts are first measured in standardized physical units, then converted to monetary value, modified by the locationally variable effects of several land characteristics, and finally summed into composite impact cost estimates. The presentation emphasizes the preparation of the kind of data required to operate the method. In a case study example from Hawaii, the method is applied to calculate the environmental impact costs of three alternative land uses at each of three proposed development sites. The methodology combines the strengths of three kinds of techniques. In Step One, a "unit price" is applied to each environmental impact's average rate of occurrence in physical units per acre, to give the average impact cost per acre by land use. This step is accomplished as Matrix A (Cost per Impact Unit by Impacts) and is multiplied by Matrix B (Impacts by Land Use) to give Matrix C (Impact

Costs by Land Use). In Step Two, a set of locational weighting factors is produced to adjust the average impacts for the effects of the land characteristics at each location. In Step Three, the final matrix of composite impact cost for each land use at each location is the result of multiplying the products of Step One and Step Two. The problems considered include peak discharge, ground water recharge loss, and ground water consumption.

Seaver, D.A., "Application and Evaluation of Decision Analysis in Water Resources Planning", Dec. 1979, Office of Water Research and Technology, U.S. Department of Interior, Washington, D.C.

Decision analysis is a systematic way of structuring the considerations that enter into a planning or decision-making problem. It facilitates the planning or decision-making process primarily by enabling attention to be directed selectively at different components in a complex problem, and by properly integrating the various components. As such, decision analysis may offer a comprehensive, consistent framework for water resources planning that fills some existing methodological and practical gaps in current planning approaches. The work reported here was undertaken to explore the applicability of decision analysis to water resources planning. The approach was to use decision analysis in two ongoing problems, working closely with the responsible planning authorities in each case. These applications, and an evaluation of the usefulness of decision analysis, are reported here.

Sellers, J. and North, R.M., "A Viable Methodology to Implement the Principles and Standards", Water Resources Bulletin, Vol. 15, No. 1, Feb. 1979, pp. 167-181.

The "principles and standards for planning water and related land resources" were made effective October 25, 1973. The document was noticeably deficient in suggestions for the necessary implementing procedures to insure its success. Current implementing procedures are based on an incorrect premise of maximizing a single objective subject to nonquantified constraints. A successful implementation of multiple objective planning requires optimizing simultaneously several competitive goals. A system of goal programming has been developed and applied to decision-making situations as a test of its usefulness in planning for multiple objective water resources projects. The result is a project planning process which can be replicated for adjustment in expected resource supplies or demands to provide a tradeoff matrix between economic and environmental objectives, as well as traditional functional purposes. This procedure, tested on the Cross Florida Barge Canal, is an integrated analysis of economic and environmental values which may be as effective in implementing multiple objective planning as the "Green Book" was in developing the now inappproprite benefit-cost analysis.

Sicherman, A., "General Methodology and Computer Tool for Environmental Impact Assessment with Two Case Study Examples", Proceedings of the International Conference of the Cybernetics Society, Tokyo and Kyoto, Japan, Nov. 1978, IEEE, Vol. 1, 1978, New York, New York, pp. 638-642.

An environmental impact assessment in its broadest interpretation requires the documenting and accounting for multiple conflicting objectives such as cost, protection of various components of the environment, and economic benefits to society in selecting from alternative courses of action. Multiattribute utility theory is a methodology that addresses the difficult issues present in environmental impact assessments in a systematic and theoretically sound manner. It provides techniques for defining the problem alternatives and the preferences of different interest groups in quantitative terms. These techniques allow the alternatives to be compared in a consistent fashion. The models used in the methodology can be implemented on the computer and thereby provide a general tool for analyzing environmental impact assessment problems. This paper describes two applications that illustrate the use of the methodology and the computer tool. The first is a transmission line corridor analysis done for the U.S. Bureau of Reclamation. The second is an analysis of potential sites for oil terminals on the island of Kodiak done for the Alaska Department of Community and Regional Affairs.

Sondheim, M.W., "A Comprehensive Methodology for Assessing Environmental Impact", Journal of Environmental Management, Vol. 6, No. 1, Jan. 1978, pp. 27-42.

A methodology for assessing environmental impact is developed and tested. Advantages of this technique over other methods include: the ability to evaluate simultaneously a large number of project alternatives; the capability of incorporating directly a very broad definition of "environment" in the assessment process; the segregation of the subjective components of the study; the possibility of including direct public participation in the assessment process; the use of interval or ratio rating schemes instead of ordinal ones; and the examination of specific potential impacts in the way(s) deemed most suitable. The methodology was devised in response to a problem involving whether or not a dam should be constructed at a given site; however, it should be noted that the methodology is applicable to a wide variety of situations.

Tamblyn, T.A. and Cederborg, E.A., "Environmental Assessment Matrix as a Site-Selection Tool--A Case Study", Nuclear Technology, Vol. 25, No. 4, Apr. 1975, pp. 598-606.

Since passage of the National Environmental Policy Act of 1969, power plant siting procedures have undergone extensive change. The environmental assessment matrix was developed for use as an active tool in an ongoing nuclear power plant siting study. Its use is not intended to eliminate engineering judgment and ingenuity from the plant siting process, but rather to document the procedures used and conclusions drawn. When used in an iterative manner during a site-selection study, the environmental assessment matrix provides valuable insight into a complex evaluation problem, documentation of the logic used, and a graphic display that can be used for presentation at open meetings.

Whitlatch, Jr., E.E. "Systematic Approaches to Environmental Impact Assessment: An Evaluation", Water Resources Bulletin, Vol. 12, No. 1, Feb. 1976, pp. 123-137.

Assessment methods and procedures to fulfill a range of desirable characteristics are developed and offered as specific criteria against which currently proposed environmental assessment procedures should be judged. Such procedures as checklists, matrices, networks (stepped matrices), overlays, linear vector analyses, and nonlinear evaluation systems are presented and judged in light of the proposed criteria. It is concluded that the use of checklists does not constitute a viable assessment procedure. Use of the matrix or stepped matrix techniques in conjunction with the linear vector or nonlinear evaluation systems, with the latter modified slightly, is seen as an approach to environmental assessment that will achieve most of the established criteria.

Yapijakas, C. and Molof, A.H., "A Comprehensive Methodology for Project Appraisal and Environmental Protection in Multinational River Basin Development", Water Science and Technology, Vol. 13, No. 7, 1981, pp. 425-436.

A decision-making method for evaluating alternatives for multinational river basin development is presented. This method is intended as a tool, not as a panacea. Factors considered are benefit-cost ratio, capital outlay, environmental and social impacts, and manageability/technology level. Each government creates a coordinating group, a weighting panel, and a rating panel, each composed of specialists and experts. The coordinating group lists project alternatives, defines the factors in terms of its government's policies, and chooses the rating and weighting panels. The weighting panel establishes a weighting scheme for each of the aspects. Each member of the rating panel judges the alternatives in their field of expertise by any method--model construction, experiments, point assignment procedures, etc. Final selection of an alternative is done by one of two methods: ranking the alternatives from all countries in a single matrix on an equal weight basis, or producing a single list from the combined project ratings from each country.

Yorke, T.H., "Impact Assessment of Water Resource Development Activities: A Dual Matrix Approach", FWS/OBS-78/82, Sept. 1978, U.S. Fish and Wildlife Service, Kearneysville, West Virginia.

A dual-matrix system for reviewing and evaluating the impact of water development projects on fish and wildlife resources is discussed. The system will consist of generalized matrices; state-of-the-art literature reviews and synthesis for evaluating impacts, alternatives, and mitigation methodology; and a computer model for quantifying impacts. The generalized matrix presented in this report consists of summary statements of the impact of common water development projects on selected physical and chemical characteristics of streams.

APPENDIX N

PUBLIC PARTICIPATION IN WATER RESOURCES PLANNING

Albert, H.E., editor, "Education of Water Resources Planners and Managers for Effective Public Participation", Report No. 71, Feb. 1978, Water Resources Research Institute, Clemson University, Clemson, South Carolina.

Water resources development should provide measures and facilities which are responsive to the long-run needs and evolving preferences of the public. An important means of achieving this goal is effective participation in water resources planning and management. In order to achieve this, key representatives of the diverse interest groups and of the public must be sufficiently familiar with both the obvious and the subtle dimensions of public participation, and must be aware of objectives as well as alternative approaches to the process. Participants at this symposium included representatives from Federal, state, and local government agencies, and from private interest groups. The conclusions reached by this panel were: (1) the public must be divided into publics and having the right publics involved in the planning stage is crucial to the development stage; (2) the right publics must be involved as early as the pre-planning stage, although all of the publics who eventually will be involved need not be at the same time; (3) there are a variety of techniques for public involvement and no single technique is adequate for all projects. A case study was analyzed by leading participants from both the public and private sector, and this substantiated the accuracy of the above conclusions. In addition to this symposium, an extensive bibliography was compiled and annotated, and is included.

Arnett, W.E. and Johnson, S., "Dams and People: Geographic Impact Area Analysis", Research Report No. 97, Sept. 1976, Kentucky Water Resources Research Institute, University of Kentucky, Lexington, Kentucky.

An attempt is made to determine the efficacy of using geographic impact areas as analytical subgroups for the assessment of the impact of multipurpose reservoir projects on target communities. The impact areas utilized are: the lake area; the below-the-dam area; the urban area; and the adjacent area. Each area is described in detail and each is analyzed for differences in knowledge, previous experience, and perception of impact on community and family. Data originated from structured and open-ended interviews in Johnson County, Kentucky. Information was collected during two field efforts, in February, 1974, and in August, 1974. Frequency of response and content analysis are the chief analytical devices. Descriptions of the life styles of each region indicated significant differences exist between impact areas. In addition, findings concerning the key variables of knowledge, previous experience, and perception of impact support the efficacy of impact area analysis. Different impact areas represent different orientations to reservoir projects. These differences must be considered for a better understanding of the social impact of such reservoir projects.

Brown, R.J., "Public Opinion and Sociology of Water Resource Development (A Bibliography with Abstracts)", NTIS/PS-79/0515/1WP, June 1979, National Technical Information Service, U.S. Department of Commerce, Springfield, Virginia.

The citations give insight into the feelings of citizens and officials toward water resource programs involving water quality management, reservoirs, irrigation, and drinking water. The effects of water resource development on the social conditions of a region or a community are also presented. This bibliography contains 210 abstracts.

Bultena, G.L., Rogers, D.L. and Conner, K.A., "Characteristics and Correlates of Public Knowledge About a Water Resource Development Issue", OWRTB-020-IA(9), 1975, Iowa State University, Ames, Iowa.

Public knowledge about an environmental (reservoir) project proposed by the U.S. Army Corps of Engineers was examined. Study objectives were: (1) to determine how knowledgeable local citizens were about the proposed reservoir, for the purpose of assessing their ability to meaningfully respond to the Corps' proposal; and (2) to test the importance of several variables which were hypothesized as affecting the knowledge levels of individual respondents. The variables which were hypothesized as affecting knowledge of environmental issues included: social class; age; personal attitudes; perceptions of likely impacts of a reservoir; and amount of social interaction and group involvement of persons regarding community issues. Although the project proposal had become a prominent local issue, a relatively low level of public knowledge about the reservoir existed. As hypothesized, the best informed persons were those who were: (1) younger in age; (2) of higher socio-economic status; (3) more strongly opinionated on environmental issues; (4) more convinced of the efficacy of citizen actions in public affairs; (5) more conscious of likely community impacts of a reservoir; and (6) most often convinced they were benefactors, rather than beneficiaries, of the project. The strongest explanatory variable for such knowledge was extensive previous political involvement in activities designed to influence agency decision-making. Respondents who were receiving greatest personal benefits from a reservoir were no better informed about the project than those who were anticipating few, if any, benefits. Relatively few respondents perceived themselves deriving personal benefits from the reservoir.

Dinius, S.H., "Public Perceptions in Water Quality Evaluation", Water Resources Bulletin, Vol. 17, No. 1, Feb. 1981, pp. 116-121.

A Visual Perception Test, consisting of photographic slides of water sites, was designed to examine laymen's water quality perceptions. The slides were taken at five water sites where the level of visual pollution was artificially altered by the investigator. Analysis of variance indicated that the water sites were evaluated differently for each of five pollution levels. Increases in water discoloration and the quality of litter were viewed as increases in the level of pollution. Laymen not only evaluated visually polluted sites lower for uses such as picnicking, but they also evaluated the quality of the actual water lower. Stepwise multiple regression indicated that a combination of water color, scenic beauty appreciation, quality of the surrounding environment and industry as a pollution source explained 73 percent of the variance in predicting Overall Pollution. Application of factor analysis simplified the variables to an Overall Pollution Factor and a Boating Use Factor.

Edgmon, T.D., "A Systems Resource Approach to Citizen Participation: the Case of the Corps of Engineers", Water Resources Bulletin, Vol. 15, No. 5, Oct. 1979, pp. 1341-1352.

Citizen participation is identified as a process for decentralizing administrative decisions, making agencies "responsive" to grass roots or local needs and values. But while citizen participation has the potential to facilitate an alteration in administrative decisions, it is not sufficient in itself to affect this change. Alterations in administrative decisions require the alteration of agency goals. The goal process is the product of coalition formation and bargaining. Therefore, citizen participation is effective only if it allows for the participation of organized groups with linkages to policy areas where other resources necessary for the maintenance of the organization are allocated. In the Corps of Engineers Urban Studies Program, guidelines defined a citizen participation process going beyond the traditional public hearings format; procedures were to be established for developing public interaction in the planning process. This planning process was observed in three Corps offices: Atlanta; Pittsburgh-Wheeling; and Albuquerque. In Atlanta, substantive citizen-initiated issues find a way into the planning process, while in the other two study areas they do not. These outcomes can be attributed to Corps organization and the regional institutional environment of each of the offices. The study of these three offices illustrates the necessity of the citizen participation approach which assumes that in order to be able to exercise influence in goal setting, a group must be able to influence the flow of resources into the organization.

Ertel, M.O. and Koch, S.G., "Public Participation in Water Resources Planning: A Case Study and Literature Review", Publication No. 89, July 1977, Water Resources Research Center, University of Massachusetts, Amherst, Massachusetts.

Results are presented of the concluding phase of a four-year project that has analyzed and evaluated the public participation programs of three Level B Planning Studies conducted by the New England River Basins Commission. In this phase, the members of the Citizen Advisory Group to one of those studies, the Connecticut River Basin Program, were surveyed to determine what modifications in attitudes toward the advisory process and the planning objectives had occurred since a similar survey was conducted in the first phase of the project. The first section includes the findings of this survey plus a discussion of other developments related to the activities of this advisory group. The second section is a review and assessment of significant literature on the theory and practice of public participation in water resources planning.

Ertel, M.O., "Identifying and Meeting Training Needs for Public Participation Responsibilities in Water Resources Planning", Publication No. 107, 1979, Water Resources Research Center, University of Massachusetts, Amherst, Massachusetts.

Legislative mandates are increasingly assigning to water resources planning agencies the responsibility for conducting public participation programs. Few planning programs, however, have the resources to employ professionals with specialized training in the types of skills that are crucial to successful public participation programming, and planners themselves are often required to perform this function in addition to other duties. This project has documented the extent of this situation through a survey of planners in Coastal Zone Management and "208" programs in New England. The survey also determined these practitioners' educational and experiential preparation for performing public participation functions, as well as their own perceptions of the relative importance of those functions and the adequacy with which they were carried out. The research indicates that prior experience is more directly related to perceived adequacy than either academic or experimental background and; therefore, concludes that all planners should receive, as part of their professional preparation, specific training that will give them a viable substitute for this experience before they begin their professional careers. Recommendations for such a curriculum, stressing the importance of the development of public communications skills, are provided.

Fusco, S.M., "Public Participation in Environmental Statements", Journal of the Water Resources Planning and Management Division, American Society of Civil Engineers, Vol. 106, No. 1, Mar. 1980, pp. 123-135.

The author's involvement in several controversial projects since the passage of NEPA has resulted in the development of an effective methodology for public participation in the environmental assessment process. The methodology consists of seven elements. At the outset, a Methodology Document, written in lay terms, is distributed to interested parties. It describes the approach and critical issues of the proposed impact assessment. Comment is received on the Methodology Document and the approach is modified. During the process, the public is kept informed. The attitude of the general public is evaluated via surveys serving as a yardstick comparing those attitudes of the citizens who are actively involved in the process. The final element is the incorporation of public opinion into the development of an environmental impact matrix. The use of public opinion, via a procedure providing quantifiable relationships, overcomes the major weakness of impact matrices, that is, the inability to incorporate a system of subjective values.

Lake, L.M., editor, Environmental Mediation, 1980, Westview Press, Boulder, Colorado.

This book suggests that mediation is a feasible--and significant--alternative to litigation in solving environmental disputes. Included are nine chapters dealing with principles and case studies related to environmental mediation.

Lehmann, E.J., "Public Opinion and Sociology of Water Resource Development (A Bibliography with Abstracts)", NTIS/PS-78/0437/OWP, May 1978, National Technical Information Service, U.S. Department of Commerce, Springfield, Virginia.

The citations give insight into the feelings of citizens and officials toward resource programs involving water quality management, reservoirs, irrigation, and drinking water. The effects of water resource development on the social conditions of a region or a community are also presented. This bibliography contains 192 abstracts.

Ortolano, L. and Wagner, T.P., "Field Evaluation of Some Public Involvement Techniques", Water Resources Bulletin, Vol. 13, No. 6, Dec. 1977, pp. 1131-1139.

Public involvement in water resources planning is receiving much current attention, and there is a need to examine systematically how different public involvement techniques work in practice. The following techniques were among those used to involve the public in a recent Corps of Engineers' study of flooding on San Pedro Creek in Pacifica, California: a public workshop; citizen information bulletins (CIBs); and questionnaires. Interviews were held with 75 study participants to evaluate the effectiveness of these techniques. The interviews indicated that various study participants felt positively about the particular workshop format employed and about the use of a communications specialist to train workshop leaders. There were mixed reactions to the effectiveness of the CIBs and questionnaires. Although many individual citizens felt that the CIBs and questionnaires were useful, some of the Corps planners felt that CIBs and questionnaires would only be "cost-effective" on large studies and/or where the questionnaire response rate was high.

Ostrom, A.R., "A Review of Conflict Resolution Models in Water Resources Management", Workshop on the Vistula and Tisza River Basins, Feb. 11-13, 1975, International Institute for Applied Systems Analysis, Apr. 1976, Laxenburg, Austria, pp. 95-105.

Reviewed briefly are modeling techniques used to study problems of conflict resolution; these models ignore the underlying preferences of the decision-maker(s) and the subjective information available from relevant interest groups. Metagame analysis, however, allows one to deal with relative preferences for outcomes of a single issue where there are competing interests; but this method, as well as those reviewed, does not handle uncertainty rigorously. The decision analytic approach, however, is an improvement over metagame analysis because it allows the modeler to consider a greater amount of subjective information: preferences, risk aversion, and tradeoffs between objectives and uncertainty can all be treated in a rigorous fashion. Herein, metagame analysis and decision analysis are discussed and compared. Both methods are applied to a problem of finding efficient municipal wastewater treatment alternatives along a river. In addition, for the metagame approach, a problem of development on an international river basin where two countries have opposing interests is examined.

Potter, H.R. and Norville, H.J., "Perceptions of Effective Public Participation in Water Resources Decision Making and Their Relationship to Levels of Participation", OWRT-A-043-IND (1), Jan. 1979, Water Resources Research Center, Purdue University, Lafayette, Indiana.

This report focuses on how citizen participants perceive the effectiveness of their participation in natural resources decision-making, comparing very moderately and slightly active participants.

Potter, H.R., Grossman, G.M. and Taylor, A.K., "Participation in Water Resources Planning: Leader and Nonleader Comparisons", Technical Report No. 107, 1980, Water Resources Research Center, Purdue University, West Lafayette, Indiana.

The purposes of this research are to provide descriptive comparisons of community leaders and the general public on: (1) how they perceive environmental problems; (2) how they view opportunities for citizen participation; (3) the extent to which they participate in decision-making; and (4) to compare alternative models of community decision-making. Data are from personal interviews with 46 leaders and a sample of 139 residents of Tippecanoe County, Indiana. The differences between leaders and the public were not generally great in how they see environmental problems, however, leaders who see problems are more concerned, although they are more likely to think pollution will improve in the near future. The public is more likely to think not enough is being done. Both groups support citizen participation, however, leaders participate much more than the public. Nevertheless, both evaluate the effectiveness of techniques for political participation similarly. Two models of community decision-making, knowledge-participation and socio-economic models, produce similar, mixed results in explaining environmental attitude. A model that combines the variables of both of these models does somewhat better, but still varies depending on the indicator of the dependent variable. Political participation, leadership and education each have independent effects in the regression model, where R is statistically significant. Greater participation and more education leads to perceiving more environmental problems. Problems of measuring environmental attitude are discussed.

Sargent, H., Fishbowl Management: A Participative Approach to Systematic Management, 1978, John Wiley and Sons, Inc., New York, New York.

This book presents a common sense view of progressive management, with many specific recommendations for participation in decision-making by all levels of management and for decision-making that is under the scrutiny of everyone within an organization. It suggests that top level managers should be personally involved in studies, planning, alternatives evaluation and reviews in the decision-making process. Five management functions defined and stressed by the author include: decision-making; communication of decisions; follow-up on decisions; organizing; and motivating.

Schimpeler, C.C., Gay, M. and Roark, A.L., "Public Participation in Water Quality Management Planning", Handbook of Water Quality Management Planning, 1977, Van Nostrand Reinhold Company, New York, New York, pp. 336-372.

Urban planning and engineering must consider public participation in the effective selection of programs to be implemented. Public participation herein is defined as an open process in which the community's rights to be informed and to interact with the governing body are exercised through the communication of a cross section of affected citizens with appointed and elected officials on policies under consideration. This chapter discusses the requirements for public participation in the planning process, methods available for public participation and alternatives evaluation, and the applicability of these methods to various planning studies. Specific technical aspects of each structural component of a citizen participation program are considered. Of particular importance are analytical procedures used to determine the relative utilities of various criteria applied to water quality planning and how the utilities will be used in the total cost effectiveness evaluation. The chapter concludes with a discussion of two specific water quality management planning studies in which these procedures have been applied: a 201 facilities planning study; and a 303(e) river basin planning study.

Shanley, R.A., "Attitudes and Interactions of Citizen Advisory Groups and Governmental Officials in the Water Resources Planning Process", Publication No. 78, Aug. 1976, Massachusetts Water Resources Research Center, University of Massachusetts, Amherst, Massachusetts.

This case study of citizen advisory committees (CAC's) examined patterns of communication and recruitment in the context of political theory. It is concluded that these patterns placed these programs clearly within the pluralist approach to citizen participation and democracy. The CAC's analyzed were: the Citizens Advisory Committee of the proposed Mount Holyoke Park in Western Massachusetts (working initially with the U.S. National Park Service and later with the Massachusetts Department of Natural Resources); the Citizens Advisory Committee of the Charles River Study in eastern Massachusetts (working with the U.S. Army Corps of Engineers); and the Citizens Advisory Committee in the Long Island Sound Regional Study (under the auspices of the New England River Basin Commission). The CAC role in the planning process depended, in part, on the object's organizational structure. Responses to survey questions revealed that: lead agency officials differed significantly in their perceptions of the uses of CAC's; CAC members also had differing perceptions of their roles; officials might be willing to grant only slight incremental changes in CAC powers in planning; and some officials had elitist perspectives toward citizen participation, while a much smaller number had more fully developed democratic perspectives. These projects revealed that there were either serious problems of communication or coordination (usually both) which hampered the fullest impact of citizen advisory groups in the planning process.

Silberman, E., "Public Participation in Water Resources Development", Journal of the Water Resources Planning and Management Division, American Society of Civil Engineers, Vol. 103, No. WR1, May 1977, pp. 111-123.

The objectives of a public participation program should be to assure that planners and public have similar concepts of what the problems are and that the proposed solutions are perceived as solutions by both planners and public. Elements of a public participation program are identifying the public, reaching the public, and determining what the public wishes are. These elements are analyzed by recounting the experience of the Bassett Creek Flood Control Commission in establishing a flood control plan for an urbanized and urbanizing area in the Minneapolis metropolitan area. The Commission is composed of nine volunteer commissioners, one appointed by each of the nine cities in the basin. The Commission (with the aid of its consulting engineers) was very effective in melding bureaucratic and public input into an acceptable plan where a previous plan prepared under the auspices of a bureaucratic organization was practically shouted down.

APPENDIX O

IMPACT MITIGATION MEASURES

Anton, W.F. and Bunnell, J.L., "Environmental Protection Guidelines for Construction Projects", Journal of American Water Works Association, Vol. 68, No. 12, Dec. 1976, pp. 643-646.

Guidelines for minimizing the environmental impact of construction projects were issued by the East Bay Municipal Utility District (EBMUD) of Oakland, California for use by inspectors and design engineers. The guidelines cover: (1) noise abatement; (2) traffic controls and detours; (3) elimination of safety hazards; (4) protection of vegetation; (5) runoff and erosion controls; (6) dust control; (7) air quality control; (8) visual/aesthetic enhancement; (9) limitations on service interruptions; (10) historical and archaeological protection; (11) paving repair; (12) construction wastes disposal; and (13) completion of the project. For runoff and erosion control, reseeding, fertilizing, and mulching (where specified) are recommended to ensure recovery of grass cover; on steep slopes hemp mesh or other approved methods of mulching, seeding, and fertilizing should be used to minimize erosion until natural ground cover is established. Temporary drainage channels and structures should be provided to minimize erosion and prevent pond formation, and water used in washing down construction areas of flushing pipes must be diverted into storm sewers or natural channels. Detention ponds, silt dams, and settling basins should be used to lessen soil erosion and sedimentation. Solid wastes generated by the project should be recycled or reused where possible, or disposed of properly.

Darnell, R.M., "Minimization of Construction Impacts on Wetlands: Dredge and Fill, Dams, Dikes, and Channelization", Proceedings of the National Wetland Protection Symposium, June 6-8, 1977, Reston, Virginia, 1978, Texas A and M University, College Station, Texas, pp. 29-36.

Environmental impact may be minimized at many levels: through the establishment of national priorities; passage of appropriate legislation; adoption and implementation of managerial guidelines; evaluation of preconstruction alternatives; on-site activities and post-construction recovery; and application of the judicial process in the interpretation of environmental policy. General and operational principles of impact minimization are discussed along with specific recommendations for minimizing the impact of dredging and filling, damming, diking, and channelization activities.

Diener, R.A., "Man-induced Modifications in Estuaries of the Northern Gulf of Mexico: Their Impacts on Fishery Resources and Measures of Mitigation", Proceedings of the Mitigation Symposium: A National Workshop on Mitigating Losses of Fish and Wildlife Habitats, Technical Report RM-65, 1979, U.S. Fish and Wildlife Service, Washington, D.C., pp. 115-120.

The commercial and sport fisheries of the Gulf of Mexico are heavily dependent on the estuaries of the northern Gulf coast. A large variety of man-induced modifications in these estuaries are threatening these resources and the quality of the supporting habitat. This report summarizes potential impacts on the habitat and resource for each type of modification and lists mitigative measures which may be taken to offset these impacts.

Gangstad, E.O., Weed Control Methods for River Basin Management, 1978, CRC Press, West Palm Beach, Florida.

Aquatic plants continue to create problems associated with navigation, flood control, agriculture, irrigation and drainage, land values, wildlife and fisheries conservation, and water resource supply. This volume provides a scientifically documented treatise of the known facts as they apply to the control of aquatic weeds in river basins and their allied waterways, and the impacts of the control methods with particular emphasis on alligator weed and water hyacinth.

Mulla, M.S., Majori, G. and Arata, A.A., "Impact of Biological and Chemical Mosquito Control Agents on Non-Target Biota in Aquatic Ecosystems", Residue Reviews, Vol. 71, 1979, pp. 121-173.

In this review the ecological and environmental impact of biological and chemical control agents employed for the management and suppression of pest and mosquito vectors in aquatic habitats is analyzed. Most, if not all, available published information has been scrutinized with regard to the most widely used control agents in aquatic ecosystems for mosquito control. Among the biological control agents, larvivorous and predaceous fish are the most commonly utilized agents for mosquito control at the present time. The general notion that all biocontrol agents are specific and have little or no effect on the nontarget biota are dispelled on the basis of available research data. From the evidence prevailing in the literature, some chemical control agents are quite innocuous while others pose a great potential hazard to nontarget biota. Efforts are and should be made to screen, evaluate, and develop more specific, less hazardous, and less toxic agents, formulations, techniques, and control strategies for the management of disease vectors and pestiferous anthropods.

Ripken, J.F., Killen, J.M. and Gulliver, J.S., "Methods for Separation of Sediment from Storm Water at Construction Sites", EPA-600/2-77-033, 1977, U.S. Environmental Protection Agency, Washington, D.C.

The nature and amount of solids that may be transported by runoff at construction sites are discussed, and control methods are reviewed and evaluated. Simple sieves are usually not effective under most treatment conditions. Microstrainers should be considered if the site effluent contains solids above 25 to 30 microns in size. If a settling basin is used, a high rate gravity tube will not materially improve the effluent quality of a properly designed basin, but it can reduce the size of the required basins. A hydrocyclone is useful for removing fines, and centrifuges can produce effluent of a good quality but their costs are high. Stationary filters are effective for fines but are not useful at high flow rates. Rotary vacuum filters provide excellent effluent, but must be attended and are expensive. At present, electrophoresis is not practical, and the upflow rapid sand filter is suggested for future study. This study is based on a review of published and unpublished technical literature.

Therrien, D., "Environmental Corrective Programs at the La Grande Complex", Canadian Water Resources Journal, Vol. 7, No. 2, 1982, pp. 147-162.

Increased concern about the quality of the environment has made it necessary for developers of natural resources to provide for protection of the environment and to respect the desires of the people affected. In developing the La Grande Complex, which comprises a series of dams, reservoirs, and canals to the east of James Bay in Canada, the Societe d'energie de la Baie James has had to find original solutions to protect the integrity of the James Bay territory, avoid disruption of the traditional way of life practiced by the Amerindian populations in the area, and at the same time achieve the desired water resources management and hydroelectric power development goals. The first reservoir in La Grande Complex, LG 2, served as a trial area for the project. A variety of approaches were used to provide for protection of the biological quality of the environment and to allow use of the water bodies for both traditional and recreational activities. Other measures were designed to create a visually attractive environment near those sites likely to be visited. These corrective programs have been in operation at the LG 2 site since 1977. Although they are still being evaluated, information to date indicates that they have had a beneficial effect on the ecology of the area, minimizing the environmental harm caused by the project. Lessons learned at this site can be used to improve the development of future hydroelectric projects.

Tourbier, J.T. and Westmacott, R., "Water Resources Protection Measures in Land Development--A Handbook-Revised Addition", OWRT TT/81-5, Aug. 1980, Office of Water Research and Technology, U.S. Department of Interior, Washington, D.C.

This handbook contains descriptions of measures in urban development to prevent, reduce or ameliorate potential problems that would otherwise adversely affect water resources. These problems consist of runoff increases and decreases in infiltration, and a greater degree of erosion and sedimentation, flooding, runoff pollution and discharge of sewage effluent. Issues have been analyzed individually in the Christina River Basin and, where possible, quantified. Measures are presented in groups and related directly to the problems listed above. Each group is preceded by a flow chart that relates measures and can aid in the selection of alternative techniques. Each measure is described and site characteristics to which it is applicable identified. The application, advantages and disadvantages, design criteria and outline specifications, cost guidelines and maintenance, and legal implementation of each measure are individually covered. Illustrated case studies give examples of measures which have been constructed in the field.

U.S. Environmental Protection Agency, Proceedings of the Workshop on Aquatic Weeds, Control and Its Environmental Consequences, EPA-600/9-81-010, Feb. 1981, Washington, D.C.

Participants at the workshop reviewed the state-of-the-art of chemical, biological, mechanical, and integrated control of aquatic weeds, and discussed problems and how the EPA can assist in working toward their solutions. For each control method, the method is reviewed by the group leader and followed by an outline of important points and recommendations by the participants. Chemical control is the dominant control method, but information is needed in five areas: an inventory of weed problems; lack of basic data on aquatic weeds; maintenance level for aquatic weeds; effectiveness and safety of 2,4-D; and controlled-release formulations. Biological controls have shown some success. Research is needed in the areas of host specificity, cost, and potential for altering ecosystems when insects are the controlling organisms; lack of information on plant pathogens, cost of their production, and necessity of repeated treatments when plant pathogens are used; and the use of grass carp and hybrid carp when fish are the controlling organisms. The cost of harvesting and disposal of weeds after mechanical control is often prohibitive, but the method allows rapid removal of plants. Problem areas include: effects on dispersal of weeds; time of harvesting; concepts of aquatic plant control and aquatic plant management; disposal of harvested weeds; harvesting of weeds as a method of nutrient removal; and utilization of harvested weeds. Integrated control, using two or more of the above methods can achieve more precise vegetation management. Possible combinations include: herbicide treatment followed by stocking with fish; or pathogen application; mechanical harvesting followed by fish or pathogens; treatment with insects followed by pathogens; and mechanical or chemical control followed by competitive plants.

Walter, M.F., Steenhuis, T.S. and Haith, D.A., "Nonpoint Source Pollution Control by Soil and Water Conservation Practices", Transactions of the American Society of Agricultural Engineers, Vol. 22, No. 4, July-Aug. 1979, pp. 834-840.

There has been a tendency to equate best management practices, as defined in water quality legislation, with soil and water conservation practices. The effectiveness of soil and water conservation practices at controlling potential pollutants other than sediments depends on the characteristics of pollutants. Pollutants were categorized in groups having distinctly different soil adsorption properties which were related to the effect of soil and water conservation practices on water and soil movement. The effects of all the management practices on loss of pollutants were summarized in table form. The purpose behind categorization of pollutants was to allow a logical means of determining which if any management practices would effectively control a given pollutant. Problems occurred with this categorization when chemicals changed their adsorption behavior as they degraded or were transformed.

Whalen, N.A., "Nonpoint Source Control Guidance, Hydrologic Modifications", Feb. 1977, U.S. Environmental Protection Agency, Washington, D.C.

Man's land-disturbing hydrologic modifications activities is one of the main nonpoint source pollution categories and, management practices presented in this document apply in general to all types of hydrologic modifications. Control of nonpoint source pollution should be considered during the planning stages of a project in order to ensure that the most effective application of measures is achieved during the project implementation period. An adequately developed plan should involve preventing sediment losses; reducing peak surface runoff; and preventing the generation, accumulation, and runoff of oils, wastewaters, mineral salts, pesticides, fertilizers, solids, and organic materials from the site area.

Whisler, F.D. et al., "Agricultural Management Practices to Effect Reductions in Runoff and Sediment Production", Oct. 1979, Water Resources Research

Institute, Mississippi State University, Starkville, Mississippi.

This study reports on the soil erosion rates for three commonly occurring soil management systems on bottomland in soybean production, on the nutrient composition of runoff from these systems, on herbicide losses from these systems, on the relationship between sediment particle size and herbicide concentrations and the effect of different soil management systems on sediment particle (aggregate) sizes. A rainfall simulator was used on plots with soybean management systems of disc, chisel and plant (conventional); planting in old soybean stubble (minimum tillage); and broadcast seeding into standing wheat (double cropping). Fallow plots were used to evaluate the management systems effectiveness. Rainfall was applied in the fall of 1977 after soybean harvesting and again in the spring of 1978 after soybean planting. Soil loss was reduced, in order of decreasing effectiveness, by double cropping, minimum tillage, conventional, and fallow. From sediment nutrient chemical analysis the percent of fines or clay increased as the amount of sediment lost by a particular management system decreased. Greatest nutrient losses were associated with the sediment phase of the runoff. Herbicide losses were greatest for the more water soluble herbicides. The herbicides appeared to be selectively absorbed to either the large sediment particles (probably aggregates greater than 2 mm) or the fine particles (probably silt and clay less than 0.063 mm).

APPENDIX P

RELATED ISSUES AND INFORMATION (ACID RAIN, POST-EIS AUDITS, AND GENERAL REFERENCES)

Ciliberti, Jr., V.A., "Libby Dam Project: Ex-Post Facto Analysis of Selected Environmental Impacts, Mitigation Commitments, Recreation Usage, and Hydroelectric Power Production", Report No. 106, 1980, Water Resources Research Center, Montana State University, Bozeman, Montana.

Water impoundment projects generally impact the river below the dam. Frequently, the construction contract provides for mitigation of adverse impacts. The Libby Project, authorized primarily for hydroelectric power generation and flood control and to provide navigation and recreation as secondary benefits, caused the inundation of a big-game winter range, the relocation of a railroad and a highway in addition to damming a river. The Army Corps of Engineers in cooperation with the Montana Department of Fish and Game and others engaged in a number of wildlife mitigation projects. The effects of the Libby Project on the Kootenai River and the degree of realization of the wildlife mitigation were evaluated. Big-game mitigation was largely unsuccessful due to the increased frequency and magnitude of the river stage fluctuations. Nitrate and phosphate concentrations in the Kootenai River have been reduced. As a result of the channel alterations in Fisher River and Wolf Creek, substantial amounts of suspended sediment continue to be discharged into the Kootenai River. Recreation usage at the project site has been considerably below predictions but appears to be increasing. For water-years 1977 and 1978, electric power generation has been about 73% of the initial projection. Generator use has averaged 47% with four turbines. The installation of four additional turbines at Libby Dam will reduce current average turbine use to 21%. Water diverted out of the Kootenai River in Canada will further reduce the turbine load factor.

Golden, J. et al., Environmental Impact Data Book, 1979, Ann Arbor Science Publishers, Inc., Ann Arbor, Michigan.

This book serves as a data reference, to supplement other sources used regularly in the preparation of environmental impact assessments and statements (EIS). It assumes the user is already involved in the environmental impact process and knows the format and data required by law. The intent of the book is to provide needed data from one composite source. Information is presented, usually in tabular format, to aid in quantification of impacts. Substantive area chapters are included on air quality, water resources, noise, physical resources, ecosystems, toxic chemicals, cultural resources, energy, and transportation.

Hendrey, G.R. and Barvenik, F.W., "Impacts of Acid Precipitation on Decomposition and Plant Communities in Lakes", CONF-7805164-1, 1978, Brookhaven National Laboratory, Upton, New York.

Over the past few decades the acidity of lakes and rivers has been increasing in several areas of the world. In southern Norway, western Sweden, the Canadian Shield, and the northeastern United States, acidification of fresh waters has become a major environmental problem. It has been clearly established that acid precipitation is the cause of decreasing pH levels in waters of the affected areas. Over most all of northern Europe, southern Scandinavia, and essentially all of the United States east of the Mississippi River, the mean annual hydrogen ion concentrations in precipitation, expressed as pH, are below 5.0. Furthermore, the regions affected by very acid precipitation (pH less than

4.5) are rapidly expanding. The effects of acid precipitation on aquatic ecosystems depend primarily on the geology of the region, but also on the elevation (orographic precipitation). The regions now known to be the most seriously affected are mountainous districts of southern Scandinavia and the northeastern United States which are located hundreds of kilometers from SO_2 and NO_x emissions sources. Regions of the United States which are potentially sensitive to acid inputs because of their geology and surface water alkalinity are also found in the western United States. Local acid precipitation problems are now known on the West Coast, and some lakes in the Puget Sound region appear to be acidified; aquatic flora and fauna at all ecosystem levels are greatly impoverished in acidified freshwaters. The numbers of species are reduced and changes in the biomass of some groups of plants and animals have been observed. Decomposition of leaf litter and other organic substrates is hampered, nutrient recycling appears to be retarded, and nitrification inhibited at pH levels frequently observed in acid-stressed waters.

Marcus, L.G., "Methodology for Post-EIS (Environmental Impact Statement) Monitoring", Circ. No. 782, 1979, U.S. Geological Survey, Washington, D.C.

A methodology for monitoring the impacts predicted in environmental impact statements (EISs) was developed using the EIS on phosphate development in southeastern Idaho as a case study. A monitoring system based on this methodology: coordinates a comprehensive, intergovernmental monitoring effort; documents the major impacts that result, thereby improving the accuracy of impact predictions in future EISs; helps agencies control impacts by warning them when critical impact leels are reached and by providing feedback on the success of mitigating measures; and limits monitoring data to the essential information that agencies need to carry out their regulatory and environmental protection responsibilities. The methodology is presented as flow charts accompanied by tables that describe the objectives, tasks, and products for each work element in the flow chart.

Martin, R.G., Prosser, N.S. and Radonski, G.C., "Adequacy and Predictive Value of Fish and Wildlife Planning Recommendations at Corps of Engineers Reservoir Projects", Dec. 1983, Sport Fishing Institute, Washington, D.C.

This report summarizes the information collected, analyzed and published in a series of 20 individual case-history reports which were designed to evaluate the adequacy and predictive efficacy of fish and wildlife planning at U.S. Army Corps of Engineers (CE) reservoir projects. Total annual angling man-day use at the 20 projects was estimated at some 2,316,437 days including 1,375,095 man-days in the reservoirs and 941,342 man-days in project tailwaters. Hunting man-day use estimates obtained from the 13 projects with usable quantitative data totaled 135,284 man-days, including 79,895 man-days for small game (59 percent), 18,968 man-days for big game (14 percent) and 36,421 for waterfowl (27 percent). The median post-project increase in angling man-days at project reservoirs was some nine times above without-the-project predicted levels expected from the free flowing streams they replaced. Post-project angling man-day use in project tailwaters exhibited a four-fold median increase. Comparison of with-the-project angling man-day use predictions contained in the U.S. Fish and Wildlife Service (FWS) planning reports with post-project angler surveys indicated a strong tendency toward over-estimation of post-project man-day use within the reservoir proper and for under-estimation of angling man-day use in project tailwaters.

Total post-project hunting man-day use estimates were higher than without-the-project predictions contained in the FWS project planning reports at 11 of the 19 projects (58 percent) with sufficient data to permit comparison. Total hunting man-day use was higher than predicted by the FWS with the projects in place at 14 of the 18 projects (78 percent). The FWS planning reports particularly over-estimated the adverse impacts of project construction on big game and small game resources and post-hunting man-day use. Hunting man-day use, the principal parameter employed by the FWS and CE for evaluating project impacts on terrestrial wildlife, provided only a partial reflection of actual impacts of project construction on wildlife resources. A substantial portion of the increase noted for post-project hunting man-day use appeared to be attributable to improved hunter access to project lands as a result of public ownership which tended to mask actual adverse project impacts to wildlife habitat. Strategies involving habitat quality and/or wildlife density assessment would provide a more equitable assessment of project impacts. The currently used Habitat Evaluation Procedures (HEP) represents a significant improvement over past practices and, with further refinement, should expedite more appropriate habitat evaluation procedures in the future.

Ortolano, L., Environmental Planning and Decision Making, 1984, John Wiley and Sons, New York, New York.

This book reflects the multidisciplinary nature of environmental planning as a field. Environmental planning is made up of contributions from several academic disciplines including biology, engineering, geography, geology, and landscape architecture among others. Part 3 of the book addresses environmental impact assessment and includes chapters on environmental impact statements and government decision-making, approaches to forecasting environmental impacts, and methods and processes for evaluating environmental impacts. Part 5 relates to techniques for assessing impacts and includes chapters on biological considerations in planning, simulating and evaluating visual qualities of the environment, elements of noise impact assessment, estimating air quality impacts, and assessing impacts on water resources.

PADC Environmental Impact Assessment and Planning Unit, "Post-development Audits to Test the Effectiveness of Environmental Impact Prediction Methods and Techniques", 1983, University of Aberdeen, Aberdeen, Scotland.

Environmental Impact Assessment (EIA) attempts to identify, predict and assess the likely environmental consequences of proposed actions so that these consequences are taken into consideration along with technical and economic factors in the decision-making process. Whilst the application of EIA procedures, methods and techniques has spread from the U.S. to other parts of the world including many developing countries, only recently has research interest been directed towards evaluating the utility of the EIA concept. This research project on four case studies has focused on the question of the accuracy of environmental impact predictions. In the four case studies a total of 791 environmental impact predictions were identified; 19 predictions were so worded as to make them untestable; 184 could not be audited because of project design changes; 407 due to a nonfulfillment of the appropriate environmental conditions; and 87 because of a shortage of information concerning particular predicted impacts. Only 94 predictions were capable of being audited. For 17 predictions it proved impossible to determine whether or not the prediction was accurate. Of the remaining 77 predictions, 44 were considered accurate and 33 inaccurate. One of the four case studies involved a reservoir project.

Rau, J.G. and Wooten, D.C., Environmental Impact Analysis Handbook, 1980, McGraw-Hill Book Company, New York, New York.

This reference handbook is designed to provide environmental planners, analysts, and decision-makers with specific techniques and tools that can be used to assess and predict the environmental impact of projects such as residential, commercial and industrial developments, new communities, urban renewal or redevelopment, park and recreational facility development, dams and flood control projects, wastewater and sewage treatment plants, new airport construction, and power generating stations. Chapter 1 is designed to set the stage for the entire handbook by providing the legislative and legal background, a discussion of the general topics to be found in an environmental impact statement, and an identification of the typical kinds of environmental impacts for various types of projects and activities. Chapter 2 is concerned with tools and techniques for assessing the socioeconomic-related environmental impacts, such as the demand for public services, revenues and costs to local governmental agencies, employment, and population growth. Chapter 3 presents a detailed discussion of air quality impacts and the procedures to be used in measuring and predicting these types of impacts. Chapter 4 provides information on the effects of noise on the environment and how these effects can be measured and mitigated, if necessary. Chapter 5 deals with a relatively new impact area of concern, namely, the effects of energy--both demand and consumption. In this chapter the basic techniques for measurement and prediction of energy requirements by project type are presented. Chapter 6 focuses on water quality considerations in environmental impact assessments and the general procedures that are used in measuring and predicting water quality impacts. Chapter 7 addresses the approach to assessing the impacts on vegetation and wildlife (i.e., flora and fauna) as the result of land use activities. Finally, Chapter 8 presents a discussion of the various types of techniques used to "put all the pieces together" in the sense of being able to simultaneously consider both the positive and negative environmental impacts of projects with the desire to obtain an overall conclusion as to the "total impact" of the project.

Reuss, J.O., "Simulation of Soil Nutrient Losses Resulting from Rainfall Acidity", Ecological Modeling, Vol. 11, No. 1, Oct. 1980, pp. 15-38.

A simulation model is described which provides a quantitative system using basic soil chemistry to predict the most likely effect of rainfall acidity on the leaching of cations from noncalcareous soils. The model uses the relationships between lime potential and base saturation, the equilibrium between CO2 partial pressure and H+ and HCO3- in solution, the apparent solubility product of Al(OH)3, the equilibrium of cations and anions in solution, the Langmuir isotherm description of sulfate adsorption, and mass balance considerations. It predicts the distribution of ions between the solution of sorbed or exchangeable phases. Thus, ionic composition of leachate can be computed. The initial model results indicate that base removal due to rainfall acidity from actual soil systems can be investigated experimentally by increasing the acidity of applied water to levels in excess of that normally found in rain. This shortens the necessary experimental time.

Sport Fishing Institute, "Evaluation of Planning for Fish and Wildlife: Lake Sharpe Reservoir Project", Oct. 1976, U.S. Department of the Army, Office of the Chief of Engineers, Washington, D.C.

This study is an evaluation of the benefits and effects on fish and wildlife from the construction of the Lake Sharpe Reservoir Project on the Missouri River southeast of Pierre, South Dakota. The 55,800 acre lake, created by the construction of Big Bend Dam, in the middle of Crow Creek Indian Reservation, was delayed by negotiations with the Crow Indians for years after project approval in 1944. After construction in 1964, studies were initiated of project impact on waterfowl, big game, upland game, and fishery resources in light of predicted data before construction, actual data occurrences after construction, and the relevance and usefulness of the inputs into the planning process itself. Findings of impact on waterfowl resources were that predictions of greater goose kills by hunters were inaccurate, that duck kills increased, while both species became more evenly distributed over the area following impoundment. Project impacts on big game such as white-tailed deer, mule deer, and antelope indicated a sharp rise (more than 190 percent) in hunter kills from 1964 to 1973. Harvest of upland game such as pheasant dropped by one-half, while harvest of grouse increased 64 percent. Significant impacts on fish were an initial decline in rainbow trout due to temperature rise in the water, and a decline in carp. The report recommends establishment of additional wildlife management areas adjacent to the lake in order to increase waterfowl and upland game and offset losses due to harvesting by hunters.

Sport Fishing Institute, "Evaluation of Planning for Fish and Wildlife at Corps of Engineers Reservoirs--Ice Harbor Lock and Dam Project, Washington", Nov. 1977, Washington, D.C.

Fish and wildlife resources associated with the Ice Harbor Lock and Dam project have been the object of active planning at various times over a period of nearly 20 years. Original treatment of project-related impacts was based on limited data, which proved to be insufficient to enable adequate compensation recommendations. Over the years, the perceived importance of fish and wildlife resources has been upgraded dramatically. This evolution in the perception of fish and wildlife values is manifest in the recorded documentation of the Ice Harbor Lock and Dam project. Early project-related planning recommendations to protect the project-associated fish and wildlife resources were presented in the Fish and Wildlife Coordination report of 1959. This particular report anticipated the severity of certain losses, with reasonable precision, such as those incurred by the big game population and a severe loss of goose nesting. On the other hand, other losses, particularly upland game, were substantially underestimated. Actions recommended to offset or lessen fish and wildlife losses were inadequate, being limited to provisions for fish passage facilities and acqusition of a 97 ha (240 ac) tract and associated 81 ha (200 ac) embayment referred to as the River Mile 25 (R.M. 25) tract. Considerably more detailed and comprehensive recommendations were presented by the affected federal fish and wildlife agencies (FWS and NMFS) in a special compensation report released in 1972. These currently proposed compensation actions relate to fish and wildlife resources associated with all four lock-and-dam projects constructed by the CE on the lower Snake River, including Ice Harbor. While vastly improved over the 1959 document, the current report contains significant errors pertaining to project-related hunting predictions and to the actions recommended to compensate for the lost resident fishery.

Sport Fishing Institute, "Evaluation of Planning for Fish and Wildlife at Corps of Engineers Reservoirs--Keystone Lake Project, Oklahoma", Feb. 1979a, Washington, D.C.

Keystone Lake was built in the mid-1960's near Tulsa, Oklahoma. In all, 6,274 ha (15,504 ac) of project lands are open to public hunting. The severe terrestrial wildlife losses anticipated by the U.S. Fish and Wildlife Service (FWS) did not occur. Hunting effort currently supported by the project is greater than the FWS estimated hunting levels for resident terrestrial game species predicted for the area without-the-project. The post-impoundment hunting effort estimated for big game species was seven times higher than the level predicted assuming implementation of a mitigation plan. The optimistic 1961 FWS report prediction for waterfowl hunting use failed to materialize. FWS predictions for upland game hunter-day use were more accurate, although proving to be somewhat lower than estimated pre-impoundment occurrences. FWS prediction of angler-day use in both the lake and tailwater were substantially exceeded by post-impoundment estimates derived from creel surveys. The combined post-impoundment lake and tailwater angler-day use estimates was almost three times greater than predicted. Total annual post-impoundment angler-day usage estimated by creel surveys was over 16 times greater than without-the-project projections made by the FWS.

Sport Fishing Institute, "Evaluation of Planning for Fish and Wildlife at Corps of Engineers Reservoirs--Okatibbee Lake Project, Mississippi", Feb. 1979b, Washington, D.C.

The Okatibbee Lake Project is located on Okatibbee Creek near Meridian, Mississippi. At summer seasonal pool the lake covers 1,538 ha (3,800 ac). Lands acquired in fee surrounding the summer pool totals 2,895 ha (7,155 ac). Approximately 2,024 ha (5,000 ac) have been licensed to the Mississippi Game and Fish Commission (MGFC) for the purpose of wildlife management. A smaller project was proposed earlier with flood control and water supply as the only primary project purposes. Acquisition in fee was planned to include only the area expected to be inundated by the five-year flood pool, approximately 2,428 ha (6,000 ac). The CE's survey report was submitted to Congress in January, 1962. The 1962 report of the U.S. Fish and Wildlife Service (FWS) recommended acquisition and state management of approximately 1,133 ha (2,800 ac) located above the five-year flood pool, and additional water storage to permit flow releases for dilution of downstream pollution, as had already been recommended by the Public Health Service. The CE redesigned the Okatibbee project to include a seasonal (summer) water quality pool of 1.6 x $10^7 m^3$ (12,900 ac ft) and the acquisition in fee of 4,433 ha (10,955 ac) of land and water.

The small game hunting projection of 2,500 hunter-days was almost identical to the current use level of 2,473 hunter-days. The post-construction deer population on project lands is equal to the estimated pre-construction herd. As a result of special hunting restrictions, big game hunting (75 hunter-days) is considerably below the level expected to have been provided by the present time (425 hunter-days). Waterfowl hunting effort (356 hunter-days) is slightly less than one-half the predicted level (790 hunter-days). Development of the project for waterfowl has been handicapped by the inability to construct sub-impoundment habitat. Waterfowl production is probably double the project-associated harvest. Current angling use of Okatibbee Lake is estimated at 40,600 annual trips, compared with the predicted 60,000 angler-trips. The tailrace fishery supports approximately double the 7,000 angler-trips predicted prior to project construction. Striped bass have been successfully introuced in Okatibbee Lake. Although Okatibbee Lake releases are not believed to have resulted in direct fish kills, the discharge is low in dissolved oxygen during periods of lake stratification; chemical constituents of the outflow are believed responsible for oxygen sags observed downstream.

Sport Fishing Institute, "Evaluation of Planning for Fish and Wildlife at Corps of Engineers Reservoirs--Dworshak Reservoir Project, Idaho", Feb. 1981a, Washington, D.C.

The Dworshak project, located on the North Fork of the Clearwater River, in Clearwater County, Idaho, was authorized in Public Law 87-874, approved October 23, 1962. Navigation, power and recreation are contributors to the project purposes. Construction began on the 86.3 km (53.6 mi) long lake early in 1963, and the dam was closed on September 27, 1971. The project includes a 6,644 ha (16,417 ac) lake (at full pool) and 13,161 ha (32,521 ac) of fee lands located above the normal full pool. The project-associated wildlife habitat loss of 6,071 ha (15,000 ac) was expected to create serious losses to wildlife populations. Mitigation planning emphasized winter browse development primarily to benefit elk. Elk losses have not been as severe as the U.S. Fish and Wildlife Service's (FWS's) predicted loss of 2,700 animals. More elk are attracted to Smith Ridge during the spring calving season than during the winter and the majority of wintering elk on Smith Ridge and the "hard-core" lands are from the Little North Fork Clearwater Basin and not, as previously suspected, from the North Fork Clearwater drainage. Studies have not documented major migration problems. Some 1,000 white-tailed deer were eliminated by the project. Moose and mountain goats, as expected, were not harmed by the project. Losses of ruffed grouse, upland game and furbearers were expected but never quantified. The requested production of 300,000 resident fish has been possible at the Dworshak Hatchery. Since 1970, resident fish reared at the hatchery have been stocked into Dworshak Reservoir. Use and harvest at the reservoir greatly exceeded expectations expressed in 1962. Angler effort averaged 35,000 angler trips per year between 1973-76 or 5.4 times higher than the projected average life-time use of the lake. Harvest averaged 123,860 fish which was 9.5 times higher than the project-life prediction.

Sport Fishing Institute, "Evaluation of Planning for Fish and Wildlife at Corps of Engineers Reservoirs--Beltzville Reservoir Project, Pennsylvania", Apr. 1981b, Washington, D.C.

Beltzville Dam is located in Carbon County, Pennsylvania, on Pohopoco Creek, approximately 8.4 km (5.2 mi) above its confluence with the Lehigh River. The primary project purposes for the 383 ha (947 ac) lake are flood control and water supply. Most of the fish-and-wildlife-related planning for the Beltzville project occurred over the three-year period, 1961-1964. Total angler-use of the project was predicted to average 33,000 man-days annually, including insignificant use of the tailrace. It appears likely, if tailwater-use could be added to lake-use (25,285 angler trips), that the FWS's prediction of 33,000 angler-trips would be reasonably accurate. Estimated total harvest during 1979-1980 was less than 48 fish weighing 10.4 kg per ha (20 fish weighing 9 lbs per ac).

Sport Fishing Institute, "Evaluation of Planning for Fish and Wildlife at Corps of Engineers Reservoirs--Beaver Reservoir Project, Arkansas", Sept. 1981c, Washington, D.C.

Beaver Dam is located in the Ozark Highlands of northwestern Arkansas at river mile 609.0 on the White River approximately 14.5 km (9 mi) northwest of Eureka Springs, Carroll County, Arkansas. As finally constituted, the immediate Beaver Lake project area totals 16,375 ha (40,463 ac). The area includes 11,421 ha (28,220 ac) within the lake proper at normal conservation-water supply pool level and an extremely narrow perimeter of contiguous lands comprising 4,955 ha (12,243 ac). An area amounting to approximately 13,650 ha (34,000 ac) of the project land and water was licensed to the Arkansas Game and Fish Commission (AGFC) on December 2, 1965, for the conservation and management of fish and wildlife. Estimates of post-project wildlife resource use in a 557-km^2 (217-mi^2) project impact area contiguous to Beaver Lake were provided by the AGFC under terms of a subcontract negotiated by the Sport Fishing Institute in 1978. Hunting use during the post-project survey period, which extended from October 7, 1978, through January 20, 1979, was estimated at 36,919 man-days, or some 236 percent greater than the 11,000 without-the-project man-days use prediction made by the FWS. Only 1,000 hunting man-days were predicted with the project in place. Estimates of post-project hunter man-day use in the lower White River were not available. Documented average annual post-project angling pressure at Beaver Lake (172,000 man-days) was 23 percent higher than the 140,000 man-days predicted by the FWS. Annual post-project fishing pressure in the Beaver Lake tailwater and upper end of Table Rock Lake (37,337 man-days) was 1200 percent higher than the predicted 3,000 man-days per year, and recreational fishing in the lower White River was some 266 percent higher than predicted.

Sport Fishing Institute, "Evaluation of Planning for Fish and Wildlife at Corps of Engineers Reservoirs--Allegheny Reservoir Project, Pennsylvania", July 1982a, Washington, D.C.

Fish and wildlife resource related aspects of the Allegheny Lake project, which is located on the Allegheny River in northwestern Pennsylvania and southwestern New York, were initially described in an August, 1958 U.S. Fish and Wildlife Service (FWS) report. As delineated by the FWS, the immediate project impact area comprised a total of 8,750 ha (21,175 ac) located below maximum design flood pool elevation 416 (1,365 ft) and included approximately 7,905 ha (19,533 ac) of land and 665 ha (1,642 ac) of water. Fishery resources were expected to be affected over an additional 105 km (65 mi) of the Allegheny River below the dam as a result of alteration of stream flow and temperature regimes associated with project operation.

The FWS predicted that the total number of hunter man-days spent on the project impact area would decline severely with the project in place. The most severe reduction in hunting effort was predicted to occur in the Pennsylvania portion, particularly for white-tailed deer. However, the FWS predicted that the additional hunting effort expected from improved hunter access planned for contiguous areas of the Allegheny National Forest in Pennsylvania, would fully compensate the loss of hunter man-days incurred in the Pennsylvania sector. An analysis of available post-project hunter use data (both hunting license sale and deer harvest statistics) tended to verify the validity of this FWS prediction. The FWS prediction that hunting effort in the New York sector would decline with the project in place was not supported by post-project observations. Based on hunting license sales within the Allegheny Indian Reservation, estimated average annual hunter man-day use in post-project years was almost four times greater than predicted by the FWS.

FWS predictions of post-project angler man-day use of Allegheny Lake have proven to be overly optimistic. A lakewide creel survey conducted by Pennsylvania State University (PSU) from April 1, 1979 through March 31, 1980, indicated that Allegheny Lake supported less than 20,000 man-days of angling per year, or some 83 percent less than the minimum FWS report prediction. The authors of the FWS report should have considered a much smaller and more realistic area of project influence (40 to 120 km (25 to 75 mi) driving distance)--as representing the primary source of potential anglers that could be reasonably expected to frequent the project. The overly optimistic FWS angling man-day use prediction appeared to have been further exacerbated by the failure to properly assess the potential angler use of Allegheny Lake in relation to the amount, proximity and productivity of other nearby waters (Lake Erie, hundreds of kilometers (miles) of trout streams, etc.).

Sport Fishing Institute, "Evaluation of Planning for Fish and Wildlife at Corps of Engineers Reservoirs--Eufaula Reservoir Project, Oklahoma", Aug. 1982b, Washington, D.C.

Eufaula Dam is located in eastern Oklahoma at river mile 27 on the Canadian River approximately 19 km (12 mi) east of Eufaula and 50 km (31 mi) south of Muskogee, Oklahoma. Construction of the dam was initiated in December, 1956, and final closure was made for flood control in February, 1964. At full power pool, Eufaula Lake inundates 41,360 ha (102,200 ac) extending upstream from the dam on the Canadian River and three tributary streams (North Canadian River, Deep Fork River and Gaines Creek). A total of 74,167 ha (183,264 ac) were acquired for the project. Project lands acquired above the power pool total some 20,715 ha (51,187 ac) extending in a narrow band around the highly convoluted 965 km (600 mi) lake shoreline. Much of the land in the vicinity of the lake is devoted to agriculture, particularly livestock production, or has been subdivided for home sites for year-round or seasonal occupancy. The accuracy of U.S. Fish and Wildlife Service (FWS) planning report predictions of post-impoundment hunter and angler man-day use proved highly variable. Post-impoundment hunting man-day use for white-tailed deer was much greater than predicted. Hunting efforts for upland game and waterfowl were reasonably on target. The FWS considerably underestimated the actual extent of the post-impoundment recreational fishery, both in Eufaula Lake and the Canadian River tailwater below the dam. Post-impoundment recreational angling man-day use documented for the total project impact area was more than 2.5 times greater than predicted by the FWS. FWS predictions of the extent of the post-impoundment Eufaula Lake commercial fishery, on the other hand, were considerably exaggerated. Overall, the construction of the Eufaula Lake project greatly increased recreational angling opportunity within the area of project influence. Documented post-impoundment angling man-day use at Eufaula Lake and tailwater (316,508 man-days) was more than 2,100 percent greater than the FWS estimate of 14,000 angling man-days per year without the project.

Sport Fishing Institute, "Evaluation of Planning for Fish and Wildlife at Corps of Engineers Reservoirs--Deer Creek Lake Project, Ohio", Jan. 1983a, Washington, D.C.

Deer Creek Lake is located on Deer Creek, a tributary of the Scioto River, in Pickaway, Madison, and Fayette Counties, Ohio, approximately 56 km (35 mi) south of metropolitan Columbus and 11 km (7 mi) southeast of Mount Sterling, Ohio. Constructed as a unit of the comprehensive flood control plan for the Ohio River basin, the project was authorized under authority of the Flood Control Act of June, 1938, for the control of floods on Deer Creek and the Scioto River. Secondary purposes include fish and wildlife enhancement and general recreation. The Deer Creek Lake project purchase area totals approximately 2,923 ha (7,223 ac) and includes 2,406 ha (5,946 ac) of land circumscribing the 517 ha (1,277 ac) Deer Creek Lake at summer seasonal pool elevation 246.9 m (810 ft) msl. Some 700 ha (1,731 ac) of project land above summer seasonal pool elevation is maintained by the Ohio Department of Natural Resources (ODNR) as a state park (Deer Creek State Park) and 1,565 ha (3,867 ac) for wildlife management purposes (Deer Creek Wildlife Area).

Post-impoundment assessments of fishing and hunting man-day use were considerably below the levels predicted by the U.S. Fish and Wildlife Service (FWS). Documented post-impoundment hunting man-day use, 10,218 man-days, was 61 percent lower than predicted. Hunting effort for upland game species (pheasants, rabbits, and squirrels) was estimated at 8,311 man-days during the 1980-1981 survey, or some 66 percent lower than the 24,280 man-days predicted in the final 1964 FWS report. Hunting effort for waterfowl, estimated at only 345 man-days during the 1980-1981 survey, was 83 percent lower than the 2,000 man-days predicted by the FWS. Contrary to FWS predictions that the project impact area would not afford any post-project hunting opportunity for deer, the survey report estimated a total of 1,562 hunting man-days were spent exclusively in pursuit of deer during the 1980-1981 hunting season.

FWS report predictions of post-impoundment angling man-day use proved to be substantially overstated. Post-impoundment creel surveys conducted by the ODNR in 1979 indicated an annual angling effort of only 21,691 man-days (42/ha (17/ac)) in the 517 ha (1,277 ac) lake, which was less than one-eighth of the level of fishing pressure as predicted in the 1961 FWS reports (331 man-days/ha (134/ac)). Conversely, the 1961 FWS planning report prediction of post-project angling man-day use in the tailwater (5,175 man-days spread over some 12.5 km (7.75 mi)) was several-fold less than the estimate derived from the 1979 creel survey conducted by the ODNR (a total of 31,945 man-days from an abbreviated 0.6 ha (1.4 ac) area located immediately below the dam). Including both the lake and tailwater, the Deer Creek Lake project impact area supported an estimated total of 53,637 fishing man-days, or approximately 70 percent less than the post-impoundment prediction contained in the FWS planning report.

The estimated total post-impoundment hunting man-day use was twelve times greater and fishing man-day use was eleven times greater than predicted by the FWS without the project.

Sport Fishing Institute, "Evaluation of Planning for Fish and Wildlife at Corps of Engineers Reservoirs--Pine Flat Lake Reservoir Project, California", Jan. 1983b, Washington, D.C.

The Pine Flat project, located on the Kings River near Fresno, California, was authorized by the Flood Control Act of 1944 for flood control, irrigation, and other purposes. The lake has a storage capacity of 1.23 x $10^9 m^3$ (1,000,000 ac ft) and provides flood protection to about 32,376 ha (80,000 ac) along the Kings River. Total area within the project boundary is 5,376 ha (13,284 ac) of land and water. The reregulating function which the project serves has resulted in a normal recreation pool averaging 1,821 ha (4,500 ac). Construction began in 1950 and the project became operational in 1954. It appears reasonably assured that the wildlife values of project and near-project lands for small game are significantly greater than anticipated by project planners. Changes in agricultural use of riparian lands below Pine Flat were correctly predicted by the early planners. The value of these lands for wildlife following completion of the Pine Flat project was significantly underestimated, however. Fishing pressure is generally higher in all areas than was expected. However, the higher angling interest is supported by an expensive management strategy of stocking catchable trout.

Sport Fishing Institute, "Evaluation of Planning for Fish and Wildlife at Corps of Engineers Reservoirs--Pat Mayse Lake Project, Texas", July 1983c, Washington, D.C.

The Pat Mayse project lies wholly within Lamar County in northeastern Texas about 16 km (10 mi) north of Paris, Texas. The project was authorized by the Flood Control Act, approved October, 1962, for purposes that included flood control and municipal and industrial water supply. At conservation pool elevation, 137.5 m (451.0 ft), the lake covers 2,425 ha (5,993 ac). About 70 percent, or 5,819 ha (14,378 ac), of the total project purchase area (8,244 ha (20,371 ac)) lies above the lake periphery at normal conservation pool elevation.

Predictions of total post-project hunting man-day use contained in the 1965 U.S. Fish and Wildlife Service (FWS) report (10,400 man-days/yr) appeared to be reasonably accurate, as they fell within the range of estimated "minimum" (5,277 man-days/yr) and "maximum" (23,700 man-days/yr) post-project occurrences for every hunting category (white-tailed deer, upland game, and waterfowl). Contrary to FWS predictions, the abundance of sport fish species increased substantially over time, while nongame fish density declined. Much of the sport fish increase was attributable to the successful introduction of the striped bass hybrid in 1973. The FWS report predictions of 50,000 angling man-days per year in Pat Mayse Lake without implementation of FWS fishery-management-oriented recommendations, and 75,000 man-days per year if implemented by the CE, proved much higher than the 27,700 man-days actually documented by the 1980-1981 creel survey. The FWS correctly predicted (at least for drought years) that fishery resources in the tailwater would be severely damaged unless the CE provided for a minimum release from the reservoir (0.23 m^3/sec (8 cfs)) and eliminated plans to channelize some 3.2 km (2 mi) of the tailwater. However, the level of angler usage in both tailwater and lake may have been adversely affected by a typical weather conditions (drought and high summer temperature) plus the substantial increase in the price of gasoline which occurred over the 1980-1981 creel survey period. Overall, the post-impoundment recreational fishery provided by Pat Mayse Lake (27,000 man-days) was some 55 times higher than the FWS prediction of only 500 man-days per year without the project in place. The minimum estimate of post-impoundment hunting man-day use was approximately 3.3 times greater than predicted by the FWS without the project.

Sport Fishing Institute, "Evaluation of Planning for Fish and Wildlife at Corps of Engineers Reservoirs--J. Percy Priest Reservoir Project, Tennessee", Sept. 1983d, Washington, D.C.

J. Percy Priest dam is located on the Stones River approximately 10.9 km (6.8 mi) above the river's confluence with the Cumberland River. The project was authorized by the Flood Control Act of 1938 for flood control and hydroelectric power. Land acquisition began in February, 1963, and the project was completed in September, 1967. At summer pool, the impoundment covers 5,747 ha (14,200 ac) and extends 51.3 km (31.9 mi) up the Stones River. Total project land area includes 13,623 ha (33,662 ac), including 151 ha (373 ac) in flowage easement and 13,472 ha (33,289 ac) in fee lands. Studies in 1979-80 documented that wildlife resources at J. Percy Priest attracted approximately 5,000 man-days of recreational use, including approximately 3,750 man-days of hunting. This is significantly greater than the hunting use value predicted by the U.S. Fish and Wildlife Service (FWS) in 1961 of 2,100 man-days. J. Percy Priest lands possess much greater potential to replace the wildlife values which existed in the privately owned (no wildlife management) area prior to project construction. Major fishery problems currently exist, including water quality problems in the lake, and water quality and quantity deficiencies in the tailwater. Angler-use has averaged approximately 35 percent of the projected value.

Sposito, G., Page, A.L. and Frink, M.E., "Effects of Acid Precipitation on Soil Leachate Quality, Computer Calculations", EPA 600/3-80-015, Jan. 1980, U.S. Environmental Protection Agency, Washington, D.C.

The multipurpose computer program GEOCHEM was employed to calculate the equilibrium speciation in 23 examples of acid precipitation from New Hampshire, New York, and Maine, and in the same number of mixtures of acid precipitation with minerals characteristic of soils in the same three states. Between 100 and 200 soluble inorganic and organic complexes were taken into account in each speciation calculation. The calculations performed on the acid precipitation samples showed that the metals (including heavy metals) and the sulfate, chloride, and nitrate ligands would be almost entirely in their free ionic forms, while the phosphate, carbonate, ammonia, and organic ligands would be in their protonated forms. This result was independent of the geographic location of the acid precipitation and the month of the year in which the sample was collected. The speciation calculations on the precipitation-soil mineral mixtures showed that aluminum and iron levels in a soil solution affected by acid precipitation would be significantly higher than in one whose chemistry is dominated by carbonic acid. The higher levels found were caused by the lower pH value of acid precipitation as well as by complexes formed with inorganic and organic ligands. It was also shown that soil cation exchangers would absorb preferentially heavy metals, such as Cd and Pb, which are found in acid precipitation.

Vick, H.C. et al., "West Point Lake Impoundment Study", EPA 904/9-77-004, Nov. 1976, National Technical Information Service, U.S. Department of Commerce, Springfield, Virginia.

A preliminary postimpoundment study was conducted spring through fall, 1975, in West Point Lake, 100 mi below Atlanta, Georgia on the Chattahoochee River, to assess water quality and eutrophication potential. The lake was formed by closure of West Point dam in October 1974; concern for potential effects of upstream metropolitan areas prompted this study. The lake's waters are highly productive; Vollenweider, Dillon, and Larsen-Mercier phosphorus-based models predict the lake will become highly eutrophic, although a 95 percent reduction in phosphorus loadings from Atlanta will produce more mesotrophic conditions within the eutrophic classification. Acute dissolved iron and manganese problems have been encountered in downstream water supplies, partially alleviated by a portion of a coffer dam left in place 375 ft above the turbine penstocks. Downstream water temperatures are not significantly altered by the impoundment, probably due to partial blockage of colder hypolimnetic water by the submerged coffer dam (the top edge is 50 ft below the water surface). No problem is foreseen in meeting fecal coliform standards for recreational water quality; nor are there major problems with pesticides or toxic metals. The lake should start stratifying at the downstream end by April, with destratification to occur by late September.

Vlachos, E. and Hendricks, D.W., Technology Assessment for Water Supplies, 1977, Water Resource Publications, Fort Collins, Colorado.

This book provides a general discussion of the principles of technology assessment as applied to the development of water supplies.

Chapters are included on water resources and technology assessment, the water resources infrastructure, problems and planning options in the final development stage, evaluation principles, determination of effects, and forecasting consequences.

Watson, C.W., Barr, S. and Allenson, R.E., "Rainout Assessment: The ACRA System and Summaries of Simulation Results", LA-6763, Sept. 1977, Los Alamos Scientific Laboratory, Los Alamos, New Mexico.

A generalized, three-dimensional, integrated computer code system was developed to estimate collateral-damage threats from precipitation-scavenging (rainout) of airborne debris-clouds from defensive tactical nuclear engagements. This code system, called ACRA for Atmospheric-Contaminant Rainout Assessment, is based on Monte Carlo statistical simulation methods that allow realistic, unbiased simulations of probablistic storm, wind and precipitation fields that determine actual magnitudes and probabilities of rainout threats. Detailed models (or data bases) are included for synoptic-scale storm and wind fields; debris transport and dispersal (with the roles of complex flow fields, time-dependent diffusion, and multi-dimensional shear effects accounted for automatically); microscopic debris-precipitation interactions and scavenging probabilities; air-to-ground debris transport; and local demographic features for assessing actual threats to multishot scenarios. The authors simulated several hundred representative shots for West European scenarios and climates to study single-shot and multishot sensitivities of rainout effects to variations in pertinent physical variables.

Williams, D.F., "Postimpoundment Survey of Water Quality Characteristics of Raystown Lake, Huntington and Bedford Counties, Pennsylvania", Geological Survey Water-Resources Investigation 78-42, July 1978, National Technical Information Service, U.S. Department of Commerce, Springfield, Virginia.

Water quality data, collected from May 1974 to September 1976 at thirteen sites within Raystown Lake and in the inflow and outflow channels, define the water quality characteristics of the lake water and the effects of impoundment on the quality of lake outflow. Depth-profile measurements show Raystown Lake to be dimictic. Thermal stratification is well developed during the summer. Generally high concentrations of dissolved oxygen throughout the hypolimnion during thermal stratification, low phytoplankton concentrations, and small diel fluctuations of dissolved oxygen, pH, and specific conductance indicate that the lake is low in nutrients, or oligotrophic. Algal assays of surface samples indicate that orthophosphate was a growth-limiting nutrient. The diatoms (Chrysophyta) were the dominant phytoplankton group found throughout the study period. The lake waters contained very low populations of zooplankton. Fecal coliform and fecal streptococcus densities measured throughout the lake indicated no potentially dangerous areas of water-contact recreation. The most apparent effect that the impoundment had on water quality was the removal of nutrients, particularly orthophosphate, through phytoplankton uptake and sediment deposition.

INDEX